# 1 Progress in Clinical Biochemistry and Medicine

# Essential and Non-Essential Metals
# Metabolites with Antibiotic Activity
# Pharmacology of Benzodiazepines
# Interferon Gamma Research


With Contributions by
M. Costa, H. Kirchner, U. Klotz
J. Kraker, R. Patierno and G. Werner


With 42 Figures


Springer-Verlag
Berlin Heidelberg New York Tokyo 1984

As a rule, contributions to this series are specially commissioned. The editors and publishers will, however, always be pleased to receive suggestions and supplementary information. Papers are accepted for "Progress in Clinical Biochemistry and Medicine" in English.

**Library of Congress Cataloging in Publication Data.** Main entry under title: Essential and non-essential metals. (Progress in clinical biochemistry and medicine ; 1) Contents: Toxicity and carcinogenicity of essential and non-essential metals / M. Costa, A. J. Kraker, and S. R. Patierno – Secondary metabolites with antibiotic activity from the primary metabolism of aromatic amino acids / R. G. Werner – Clinical pharmacology of benzodiazepines / U. Klotz – [etc.]. 1. Biological chemistry–Addresses, essays, lectures. 2. Metals–Toxicology–Addresses, essays, lectures. 3. Antibiotics–Addresses, essays, lectures. 4. Amino acids–Metabolism–Addresses, essays, lectures. 5. Benzodiazepines–Addresses, essays, lectures. 6. Interferon–Addresses, essays, lectures. I. Forman, Donald T., 1932–. II. Costa, Max. III. Series. [DNLM: 1. Metals–pharmacodynamics. 2. Carcinogens. 3. Aromatic Amino Acid Decarboxylases–metabolism. 4. Benzodiazepines–pharmacodynamics. 5. Interferon Type II. W1 PR66BEM v. 1 / QU 130 E78]. QP509.E86. 1984. 612'.015. 84-13849

ISBN-13: 978-3-642-69874-3　　e-ISBN-13: 978-3-642-69872-9
DOI: 10.1007/978-3-642-69872-9



Softcover reprint of the hardcover 1st edition 1984



2152/3140-543210

# Editorial Bord

# Foreword

Scientific progress, more and more, makes it possible to relate disease to irregularities on a molecular basis. Both, diagnosis and cure can be targeted to a well defined biological structure. Modern medical research aims at the investigation of the interaction of molecules and bio-macro-molecules, formerly the area of chemistry, biochemistry and pharmacology.
This series was founded to provide a vehicle for the dissemination of results in a multidisciplinary area of contemporary research. It will publish extensive review articles at a high level, that will relate a particular subject to the scopes of medical chemistry. Thus, information and useful references will be spread among the scholars.

# Table of Contents

# Toxicity and Carcinogenicity of Essential and Non-essential Metals

**Max Costa, Alan J. Kraker and Steven R. Patierno**

Department of Pharmacology, University of Texas, Medical School at Houston, Houston, Texas 77025, USA

The major toxic effects of selected metals and their compounds have been considered, along with discussions of their essentiality, distribution and environmental exposure. The carcinogenic and mutagenic properties of these metals are discussed in detail with critical consideration of epidemiological and *in vitro* studies conducted with these metals and their compounds. Extensive individual consideration has been given Al, As, Be, Cd, Cr, Pb, Hg and Ni. Other metals such as Co, Cu, Mn, Ag and V have been discussed in less detail grouped together. It is hoped that the discussion of these metals will provide a useful general reference for understanding their toxicity and a more complete reference for consideration of their carcinogenic properties.



Progress in Clinical Biochemistry
and Medicine, Vol. 1

# Introduction

The major toxic effects of selected metals and their compounds have been considered, along with discussions of their essentiality, distribution and environmental exposure. The carcinogenic and mutagenic properties of these metals are discussed in detail with critical consideration of epidemiological and *in vitro* studies conducted with these metals and their compounds. Extensive individual consideration has been given to Al, As, Be, Cd, Cr, Pb, Hg, and Ni. Other metals such as Co, Cu, Fe, Mn, Ag and V have been grouped together. It is hoped that the discussion of these metals will provide a useful general reference for understanding their toxicity and a more complete reference for consideration of their carcinogenic properties.

# 1 Aluminum

Although aluminum metabolism has been the object of study for many years, the role of aluminum in the function of mammalian systems is not clear. The nonessential nature of Al is suggested by the net loss of administered Al in man[1]. An attempt to induce an Al-deficiency disease in rats via an Al-deficient prepared diet led to the conclusion that if Al was an essential element, the dietary requirement could be met by as little as 1 μg Al/day[2].

Parenteral and oral administration of large experimental doses of Al in animals as the chloride, sulfate, or hydroxide salts results in lethargy and death in rats[3]. Skin lesions, gastrointestinal disturbance, growth retardation, perihepatic granulomas and fibrous peritonitis are other toxic effects caused by Al[1]. Many of the effects of dietary Al intoxication are thought to be the result of perturbation of phosphate metabolism in the affected organism[1, 4–6].

At high levels of Al, a net loss of phosphorus occurs as a result of the formation of insoluble $AlPO_4$ in the gut which is then excreted. This loss of inorganic phosphate may shift the equilibrium of phosphorylated compounds in the organism toward inorganic phosphate causing a decrease in the amount of phosphorylated intermediates necessary in nucleic acid and energy metabolism.

The neurotoxicity of Al in man is evident from studies of occupational exposure to Al dust[7]. Inhalation of Al particles, less than 2 μm in diameter leads to encephalopathy as well as pulmonary fibrosis. Patients undergoing renal dialysis are subject to dementia as a result of elevated Al levels in the dialysate[8]. The whole tissue Al levels in dialysis encephalopathy patients are about 10 times the level of Al in control brain tissue[9].

Further implication of Al in neurotoxicity is offered by studies which show the presence of Al in neurofibrillary filament tangles in Alzheimer's disease[10]. Neurofibrillary tangles in the brain cells of Alzheimer's disease patients are one feature characteristic of the disease. The brain cells containing the degenerated neurofibrillary structures also contain Al in the nuclear region of the cell.

The two major sources of environmental exposure to Al are occupational contact and dietary ingestion[6,8]. As mentioned earlier, pulmonary fibrosis is one result of prolonged inhalation of Al dust. Aluminum-containing food and food additives, non-prescription drugs containing Al, and Al cookware have been cited as major dietary sources of Al contact[6].

The distribution of Al upon ingestion or inhalation is dependent on the solubility of its salts; most Al salts are relatively water insoluble. Most inhaled insoluble Al compounds remain in the lung over longer times[1]. The absorption of Al in the gastrointestinal tract is poor (vide supra) and is thus mainly excreted. That Al which is absorbed, presumably by phagocytosis carried out by macrophages, is distributed in liver, kidneys, testis, brain and skeleton. As with most metals, the turnover of Al in most soft tissues is greater than that of the metal in bone. However, the persistent high levels of Al in the brain in dialysis induced encephalopathy and in Alzheimer's disease remain unexplained.

The absence of published data regarding the carcinogenic potential of Al suggests that such activity has not been observed. The apparent effects of long-term Al exposure in man do not extend to carcinogenesis. The use of cell transformation assays as an indicator of the carcinogenicity of Al also lead to the conclusion that Al is not carcinogenic. An *in vitro* assay involving the induction of transformation in Syrian hamster embryo cells or the enhancement of viral transformation of SHE cells[12] demonstrates no transforming activity due to Al.

Aluminum is not a mutagenic metal when mutagenesis is taken as decreased fidelity of proper base insertion into DNA synthesized *in vitro* using a viral DNA polymerase which lacks a repair mechanism[13]. The use of a rec assay also failed to show Al possesses mutagenic activity in the bacterial system employed for the assay[14].

Since no evidence for the carcinogenicity or mutagenicity of Al is available, the lack of study of the effects of Al on DNA structure and function is not surprising. Although perturbations in phosphate metabolism due to the binding of Al to inorganic phosphate have been documented, the effect of phosphate depletion by Al on DNA metabolism has not been investigated. Similarly, no data is available on the interaction of Al with the structure of the DNA molecule; interaction of the positively charged Al with the phosphate may have an effect on the conformation of the nucleic acid.

Chromosomal aberration as a result of treatment of 3 mammalian species with Al has been reported[15] but doubts exist as to the interpretation of these findings as a result of the experimental method. Aluminum hydroxide gel was injected i.p. into animals after which non-characterized peritoneal cells were obtained by lavage. No attempt to differentiate peritoneal cells from cells involved in an immune response such as macrophages was reported, leaving the significance of the chromosomal aberration open to question.

## 2 Arsenic

There is substantial evidence that arsenic is an essential trace element for normal growth and development of experimental animals. For example, arsenic deficient goats had a 77% incidence of mortality whereas only 13% of the control goats died during the same time interval[16]. After weaning, 60% of As-deficient lambs died within 140 days while no mortality was evident in the control group[16]. A number of physiological parameters are effected by arsenic deficiency in goats including decreased conception rate, increased abortion rate, and decreased hematocrit[16].

The toxicity of arsenic is highly dependent upon the chemical form of the metal. The trivalent form is generally considered to be more toxic than the pentavalent form although this is probably due to its less rapid excretion from the body[17]. The trivalent form is thought to be approximately 4 fold more toxic than the pentavalent form (oral $LD_{50}$ values of trivalent form in rats are 10–293 mg/kg). Organic arsenicals are cleared more rapidly than the inorganic forms and are considered less toxic[17]. The most toxic compound containing arsenic is arsine gas ($AsH_3$), and it is relatively persistent in the body while most of the other arsenic compounds are completely eliminated *in vivo* within 48 hrs. The cellular mechanism of toxicity of the pentavalent arsenic form differs from that of the trivalent form. Arsenate ($As^{+5}$) uncouples mitochondrial oxidative phosphorylation by its ability to substitute for inorganic phosphate, forming unstable arsenate esters that spontaneously decompose[2] (arsenolysis). Trivalent arsenicals on the other hand react with sulfhydryl groups and inhibit the tricarboxylic acid cycle. The pyruvate dehydrogenase system is especially sensitive to the trivalent arsenicals since this form of arsenic interacts with the sulfhydryl groups present in lipoic acid. The ability of arsenic to inhibit the production of ATP suggests that it will effect virtually any organ in the body, dependent only upon its distribution. In both the chronic and acute exposure situation, the liver and kidney become degenerated since arsenic distributes well to these organs. Chronic exposure to arsenic also has effects upon the skin, producing hyperkeratosis and cancer at that site. CNS encephalopathy and peripheral nerve degeneration have also been reported following chronic exposure to arsenic. Arsenic is also highly teratogenic. Arsine gas produces severe hemolysis of red blood cells, leading to anoxic tissue damage. Environmental exposure to arsenic comes about primarily as a result of metal smelting operations, its use as a pesticide and herbicide, and the burning of fossil fuels such as coal. Since arsenic containing pesticides and herbicides have been used extensively in agriculture, arsenic taken up by plants is a significant source of As. Arsenic is also concentrated in marine organisms such as crustaceans; therefore water pollution is also a significant problem.

The carcinogenic activity of arsenic is an extremely controversial issue. Several authors have stated that there is a lack of good experimental evidence to establish arsenic as a carcinogen. They felt that the issue of arsenic being a carcinogen has been oversimplified and unjustifiably plagued with a hazardous symbol. For example, Frost[18] has indicated in a review of arsenic carcinogenesis that arsenic has been swept into a common pattern of our time as stated by a prediction made a century ago, "our time would be the age of the great simplifiers and that the essence of tyranny would be the denial of complexity". The point being made here is while extensive exposure to arsenic is certainly hazardous there is probably a safe and perhaps even beneficial level of exposure.

Numerous epidemiological studies support a role for arsenic in the induction of human lung and skin cancer. In a number of these studies exposure to arsenic was monitored and a dose-response relationship exists in the induction of malignancies as shown in the accompanying tables and figures.

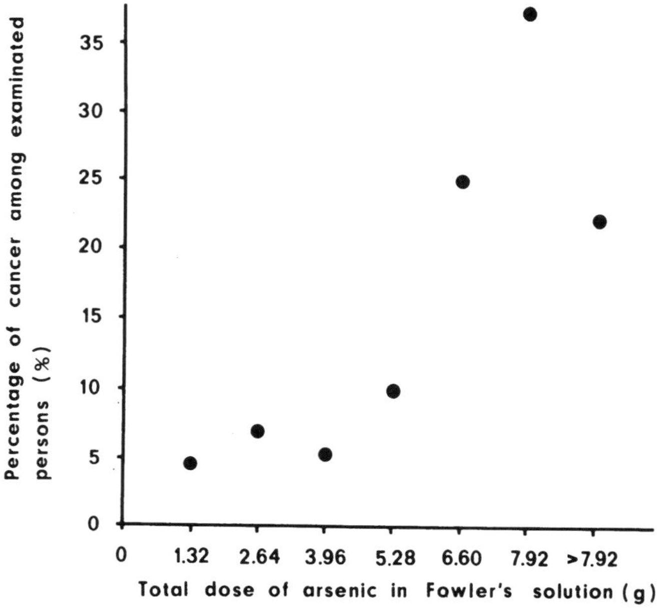


**Fig. 1.** The relative frequency of skin cancer with increasing doses of arsenic. (Recalculated from Fierz, 1965)[19]

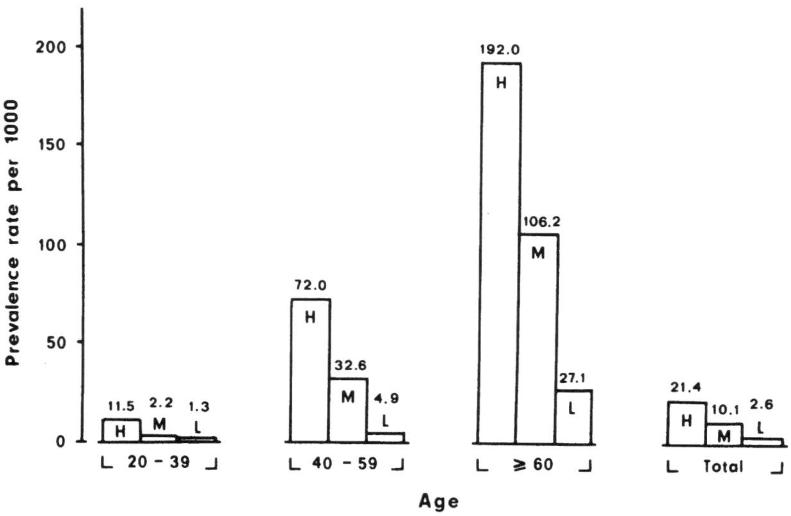


**Fig. 2.** Age-specific prevalence rate (1/1000) for skin cancer at different arsenic concentrations in well water in Taiwan. Arsenic concentration in well water: H – high, ≳0.60 mg/l; M – medium, 0.30–0.59 mg/l; L – low, 0.00–0.29 mg/l. (From Tseng et al., 1968)[20]

**Table 1.** Respiratory cancer deaths and SMRs by cumulative arsenic exposure lagged 0 and 10 years, Tacoma smelter workers

| Cumulative Exposure (μg/As/l) (urine-years) | Lag | | | |
|---|---|---|---|---|
| | 0 Lag | | 10 Year Lag | |
| | Observed Deaths | SMR | Observed Deaths | SMR |
| 600 (302) | 8 | 202.0 | 10 | 155.4 |
| 500–1500 (866) | 18 | 158.4 | 22 | 176.6[a] |
| 1300–3000 (2173) | 21 | 203.2[b] | 26 | 226.4[b] |
| 3000–5000 (4543) | 26 | 184.1[b] | 22 | 177.6[a] |
| 1000+ (13457) | 31 | 243.4[b] | 24 | 246.2[b] |

[a] $p < .05$
[b] $p < .01$
( ) Mean of class interval
Source: Enterline and Marsh (1982)[21]

The reported skin cancers with arsenic exposure appear to be of two histological types: squamous carcinomas in the keratotic areas, and basal cell carcinomas. The amount of data on the histological classification of lung tumors associated with exposure to arsenic is limited. Since arsenic was once used in a medication (Fowler's solution) to treat syphilis, accurate human exposure data is available. Some of the epidemiological studies of arsenic carcinogenesis have also been corrected for smoking histories[22]. Therefore the epidemiological evidence for arsenic being a carcinogen to humans is excellent[22]. However, numerous attempts to induce cancer in experimental animals with arsenic have been unsucessful[22]. In eighteen separate studies, rats and mice have received salts of $As_2O_3$ by oral, i.v., skin painting, intramuscular, and intratrachael administration but only two of these reported any statistically significant excess of tumors over controls[22]. However, most of these studies were conducted in the 1950's and 1960's and there is clearly a need to further investigate the carcinogenic role of arsenic in experimental animals[22].

Arsenic has clearly been shown to produce chromosomal aberrations *in vivo*[23] and *in vitro*[24]. It also increases the frequency of sister chromatid exchanges in humans that had taken Fowler's solution for psoriasis[25]. These results suggest that arsenic is genotoxic. One study has shown that arsenic produces point mutations[26] while two other studies have demonstrated no such induction of mutations in bacteria by arsenic[27, 28]. Arsenic has been shown to induce transformation in Syrian hamster embryo cells in a dose dependent fashion[29]. This effect is more closely linked to a carcinogenic response than the genotoxic-mutagenic effects described above. Thus at the present time, it seems reasonable to consider arsenic as a potential carcinogen since many good epidemiological studies support this claim. The fact that arsenic is not highly mutagenic despite its having carcinogenic activity is not inconsistent with the results obtained with other carcinogenic metals[30]. For example, nickel compounds are clearly implicated as carcinogenic yet they are not mutagenic in many systems[30].

## 3 Beryllium

Beryllium has been industrially manipulated in quantity only since the 1930's so concern about its interaction with organisms has developed only relatively recently. Beryllium is not an essential element and is found in non-exposed tissue at very low (< 20 µg/kg dry pulmonary tissue) to undetectable levels[31)].

Exposure of humans to Be dust in a chronic fashion results in pulmonary granulomatosis, pneumonitis, cardiac enlargement, and granulomatous lesions of skin, liver and lymph nodes[31–33)]. This exposure is a consequence of occupational contact as well as the combustion of Be-containing fossil fuel[31–37)].

Beryllium is extensively used in a copper alloy for electrical switching devices, in nuclear weapons production and nuclear reactor structures, and was formerly used as a fluorescent tube phosphor[31)]. Workers employed in industries utilizing Be have in the past been exposed to Be resulting in the dermal and pulmonary responses noted above[38)]. The general population has also been exposed to low levels of Be since coal contains Be and the combustion products of coal are widely dispersed in the environment.

The distribution and excretion of Be once absorbed has been the subject of a number of studies[31, 39)]. The inhalation of $BeSO_4$ in rats results in a concentration dependent retention of Be in the lungs under conditions of continuous exposure. After exposure has ceased, the level of pulmonary Be falls with a half life of about 2 weeks. Subsequent migration of Be from lungs is much less rapid with smaller amounts of Be remaining for several months[31)]. Injection of carrier-free $BeCl_2$ i.p., i.v. or i.m. in rats resulted in deposition of about 50% of Be in the skeleton[39)]. The absorption of ingested Be is limited by the poor solubility of $Be_3(PO_4)_2$ formed in the bulk of the gastrointestinal tract at alkaline pH. Of $BeSO_4$ ingested by rats, only 20% of metal was absorbed, probably in the stomach where the low pH increases the solubility of the metal[40)]. Beryllium circulates as a colloidal phosphate adsorbed on plasma α-globulin. The half life of carrier free $BeCl_2$ in blood is 3 hours[41)]. Since Be is tightly bound to plasma protein it is not filtered by the glomerulas but appears in the urine by active secretion into the tubules.

The carcinogenicity of Be compounds has been the subject of study in a number of systems. Experimental studies done on rats, rabbits, guinea pigs, hamsters and monkeys and epidemiological analysis of data from industrially exposed workers lead to differing conclusions regarding Be carcinogenicity. More workers exposed to Be than statistically expected died as a result of lung cancer (47/3055 observed, 34/3055 expected) based on records maintained over a 20 year period[38)]. In a study comparing lung cancer mortality between Be-exposed workers and employees in another manufacturing industry, the incidence of lung cancer was also greater (80 observed, 57 expected) in the Be-exposed individuals[43)]. In the face of the data indicating a greater occurrence of cancer in Be exposed humans, one IARC group has concluded Be is a potential but not a confirmed carcinogen[39)]. This classification was based primarily on questions regarding the methodology of the epidemiological studies. This conclusion stands in contrast to that of another committee that concluded Be is the cause of increased cancer mortality in Be exposed groups[35)].

Studies in animals demonstrate an apparent species specificity in the response to Be exposure. Intravenous zinc beryllium silicate induced osteosarcoma in the long bones of rabbits[35)]. These experiments were followed by hosts of others in which Be was studied in a number of species, utilizing different salts, administered at varied concentrations by different routes. Guinea pigs and hamsters of the species studied failed to develop tumors under conditions that induced tumors in rabbits, pigs, rats, and monkeys[34, 36)].

Cell transformation assays have confirmed the carcinogenic properties of Be in mammalian systems. The morphological transformation of Syrian hamster embryo (SHE) cells was increased in a dose dependent manner by $BeSO_4$[47, 48)]. Further, the viral transformation of SHE cells was also increased by the presence of $BeSO_4$[49)].

In contrast to the demonstrated ability of Be to transform mammalian cells in cell culture assay systems stands the effects of Be in some mutagenicity assays. A number of bacterial systems have been utilized to determine the mutagenic potential of Be. Beryllium did not cause the development of mutant Salmonella colonies[50)] nor did Be show an effect on an *E. coli* induct test[51)]. The failure of the induct test may be the result of defects in the design of the assay. Another *B. subtilis* test system which measured the growth differences of recombination repair-deficient strains treated with metals did demonstrate a mutagenic effect of Be[52)].

A Chinese hamster cell line was used to determine the mutagenicity of Be in a mammalian systems[53)]. Mutants resistant to 8-azaguanine were induced from a cell line sensitive to 8-azaguanine by the action of $BeCl_2$ but the extent of induction was less than that of the control agent N-methyl-N′-nitro-N-nitrosoguanidine under equivalent levels of exposure.

Since Be has been shown to be carcinogenic and mutagenic in a number of systems, its effects on DNA structure and metabolism have been studied. Deoxythymidine kinase has been shown to be inhibited by millimolar concentrations of Be *in vitro*[54)]. However, the physiological significance of this inhibition is open to question because of the relatively high levels of Be necessary to produce this effect.

The fidelity of DNA synthesis has been used to measure the effect of Be on the enzymes of DNA metabolism[55, 56)]. In an assay employing a purified viral DNA polymerase and a synthetic homopolymer DNA template, Be was found to increase the number of incorrect bases inserted into the newly synthesized strand of DNA. The decreased fidelity has been attributed to effects of Be on the DNA polymerase at a site removed from the catalytic site of the enzyme since preincubation of only the polymerase with Be caused the misincorporation of bases[57–59)].

In addition to decreasing the fidelity of DNA replication, Be also has been shown to cause sister chromatid exchange (SCE) to a small extent in a P 388 cell line but did not cause SCE in human lymphocytes[31)].

In view of the demonstrated effects of Be in the mutagenesis of cells, in carcinogenesis in a number of species, and on the nucleic acid metabolism of cells, Be is correctly considered a metal with potent carcinogenic and toxicological activity.

# 4 Cadmium

Cadmium has been associated with metal toxicity since at least 1858 when a report was recorded of inhalation of $CdCO_3$ dust and subsequent pulmonary effects[60]. In the intervening period, Cd has been the object of numerous assessments of its effects on organisms which are exposed to it.

Major toxic effects in man follow two primary routes of exposure pulmonary and gastrointestinal absorbtion. Inhalation of large quantities of Cd as CdO fumes results in bronchial and pulmonary irritation followed by dizziness, weakness, chills, fever, chest pain and dyspena. Pulmonary edema is the primary cause of death resulting from acute exposure. Acute renal failure and cardiopulmonary depression brought about by Cd exposure may also result in death[60, 61]. Chronic inhalation of Cd causes emphysema, liver damage, anemia, proteinuria, and renal tubular damage[63]. Cadmium causes atrophy of proximal tubules resulting in increased excretion of low molecular weight proteins. Glomerular damage may also follow Cd intoxication leading to aminoaciduria, glucosuria and phosphaturia[60].

Ingestion or systemic adsorption of large amounts of Cd ($>3$ mg in an adult) causes nausea, vomiting, abdominal cramps and headache which may be followed by acute renal failure and cardiopulmonary collapse. Chronic ingestion, particularly well studied in Japan as a result of widespread environmental contamination, leads to renal tubular defects and a painful degenerative bone disease called "itai-itai" (literally, "ouch-ouch"). Cadmium-induced renal tubular damage perturbs calcium metabolism, resulting in osteoporosis and osteomalacia. Pressure on the bones cause pain, hence the name of the disease[61, 63].

The testis is perhaps the organ most sensitive to damage produced by Cd. Injection of $Cd^{2+}$ caused rapid atrophy of the testes with resulting sterility in experimental animals[68].

Environmental exposure to Cd occurs via the diet and by atmospheric contamination. Various dietary components contain Cd in the range of 0.01–0.2 µg/kg wet weight in areas not contaminated by this metal. Rice grown in some of the many Cd-contaminated areas of Japan contains an average of about 1 mg Cd/kg[60]. Soil contamination can result from airborne particulate matter arising from metal refining processes and fossil fuel combustion or from the application of Cd containing fertilizer or sewage sludge[61, 63, 64]. Cigarette smoke also contains Cd and is a source of human exposure[65].

Occupational exposure is another avenue of Cd intoxication. Cadmium is used as a coating for iron and steel since Cd-plated items are highly rust-resistant. Alloys containing Cd are employed in solders, welding rods and automobile radiators. Some pigments and plastics also contain Cd. Cadmium is a by-product of Zn smelting as well. Workers employed in industries utilizing Cd are at risk of Cd exposure if adequate measures are not taken to reduce contact with the metal[61, 62, 66, 67].

Differences exist in the absorption of Cd depending upon the route of exposure. Approximately 35% of inhaled Cd is absorbed based on calculations from exposure studies in animals[63, 69]. The size of the particulate matter and the solubility of the particle influence the absorption with smaller, more soluble particles resulting in greater absorption. The uptake of Cd from the gut extends to only about 6% of the ingested amount[70]. After absorption, approximately one third of the total body burden of Cd is found in the kidneys with half that amount in the liver after long times.

Once absorbed, Cd remains in the body for extended periods and accumulates in tissues over time with continued exposure. The half life in humans is greater than 10 years[69, 71]. The finding that most of the Cd present in liver and kidney was bound to a protein called metallothionein led to the focus of work on the role of this protein as a natural protective chelator against toxic exposure to this metal[72, 76]. Cadmium has been found to induce the synthesis of metallothionein mRNA[77, 81]. Once synthesized, the biodegradation of Cd-metallothionein has been found to occur with a half time of between 3 and 4 days; the Cd released from the biodegraded protein then causes the induction of new metallothionein synthesis which results in Cd continually bound to metallothionein in the steady state[82]. This mechanism may explain the long term retention of Cd in tissue since the association constant of Cd bound to the protein is relatively high[83–85].

Since Zn and smaller amounts of Cu are bound to metallothionein in unperturbed systems and since metallothionein is present even in the absence of induction, a physiological role of metallothionein may be to serve as a donor of Zn to Zn-requiring sites in cells[85–88]. The perturbation of this function of metallothionein in Zn homeostasis may be in part responsible for the expression of the toxic effects of Cd. The effects of Cd on the proximal tubules and its relationship to metallothionein may serve as an illustrative example of the expression of Cd effects in the kidney. The proteinuria resulting from Cd exposure has been correlated histologically to degeneration of the tubular epithelium[89]. Metallothionein is found in the proximal convoluted tubular cells of rats after Cd treatment but more interestingly, also prior to Cd administration[90]. If the physiological role of metallothionein is disrupted by the binding of Cd to metallothionein causing the proximal tubular cells to degenerate, the subsequent proteinuria is not surprising. The competition of Cd and Zn for the same metabolic pathways may also explain the amelioration of the toxic effects of Cd by administration of Zn[60].

In addition to renal toxicity as a result of Cd exposure, the carcinogenic potential of Cd is a factor in its interaction with organisms. A number of epidemiological studies have been carried out following cancer morbidity and mortality resulting from Cd exposure. An early tabulation of cancer cases in workers exposed to CdO dust showed four observed instances of prostatic cancer in a cohort of workers in which 0.58 cases were statistically expected[91, 92]. Another study of workers in a Cd smelting plant found 27 deaths from cancer in a population where only 17.5 were expected. Furthermore, 4 of the deaths resulted from prostatic cancer when 1.2 were expected[93]. Still another study found no significant difference in cancer mortality between Cd-exposed and non-exposed workers[94]. An interpretation of these studies suggests that overall, Cd tends to increase the incidence of prostatic cancer among workers exposed to Cd[95] but that this association is tenuous[96].

The carcinogenicity of Cd has also been studied in rodents but evaluation of the result must be tempered by noticing the differences in the prostate gland between rodents and humans[95]. When rats were given subcutaneous injections of $CdSO_4$ for 2 years, 4 animals (of 25 original) receiving the highest dose of Cd developed sarcomas at the injection site. No signs of prostate tumors were evident[97]. Administration of Cd by stomach tube in rats and mice over a 2 year period also failed to produce postatic tumors or tumors in other organs in excess of controls[98, 99].

Cadmium has been studied in mammalian cell transformation assays which have been used as one measure of the carcinogenicity of these compounds[100]. Syrian hamster embryo (SHE) cells were transformed by cadmium acetate[101]; both $CdCl_2$ and $Cd(CH_3COO)_2$ increased the viral transformation of SHE cells as well[102].

In view of the equivocal carcinogenicity of Cd based on epidemiological and experimental studies, the evaluation of the mutagenic properties of the metal may contribute to understanding the interaction of Cd with organisms[103]. Cadmium caused a difference in the growth patterns of recombinant repair-deficient strains of *B. subtilis,* indicating mutagenic potential[104]. However, in two other bacterial assay systems, Cd was not shown to be mutagenic[105, 106]. Cadmium also gave mixed results in two different mammalian cell lines used to detect mutagenicity. Mutants were induced by Cd in L 5178 Y/$TK^{+/-}$ cells[48] but not in a C3H mouse cell HGPRT assay system[108]. It should be borne in mind, however, that metal compounds as a class are not reliably detected in many mutagenesis assays[103].

The effects of Cd on nucleic acid structure and function are varied and have been the object of a recent study[109]. The enzymes of nucleic acid metabolism responsible for the catalysis of the reactions necessary for nucleic acid replication are adversely affected by treatment with Cd *in vivo* and *in vitro.* The incorporation of $^3H$ thymidine into the DNA of Erlich ascites tumor cells maintained in mice treated with $CdCl_2$ was decreased over the level of incorporation of non-treated cells[84]. The misincorporation of CMP into a poly(rA-rU) product directed by poly(aD-dT) using *E. coli* RNA polymerase was increased in a dose dependent manner by the presence of $CdCl_2$[51]. The incidence of incorrect bases incorporated into DNA synthesized from poly(dA-dT) using a viral DNA polymerase also was greater when $CdCl_2$ or $Cd(CH_3COO)_2$ was present[111].

In addition to affecting the enzymes of nucleic acid metabolism, Cd also has effects on DNA structure itself. Single strand breaks of DNA as a result of Cd exposure in intact rat hepatocytes have been reported[112] and cadmium disrupts the structure of isolated helical double-stranded DNA as shown by a decrease in the melting temperature of the DNA[113].

Since Cd affects the structure of DNA, effects of Cd on chromosomal structure might also be expected. Sister chromatid exchange in a P388D carcinoma cell line was increased by Cd but the results obtained using human lymphocytes were equivocal. The conclusion was drawn that Cd has weak chromosome-breaking ability[114].

In spite of the lack of clear evidence of the absolute carcinogenicity of Cd, the sum of studies thus far completed shows Cd interacts with organisms in a deleterious fashion and as a result its toxic potential alone is sufficient reason for caution in dealing with Cd in the environment.

# 5 Chromium

Chromium possesses both essential and toxic characteristics in its interactions with organisms; furthermore, the toxic effects occur at levels of the metal which may resonably be encountered in the environment. Deficiency of Cr leads to impaired growth and disturbances in glucose, lipid, and protein metabolism[115, 116]. The carbohydrate metabolism disturbances in rats fed Cr-deficient diets were eliminated by administration of $Cr^{3+}$ [117]. Subsequent studies identified a $Cr^{3+}$ complex of nicotinic acid, glycine, glutamic acid, and cysteine as the so-called glucose tolerance factor (GTF) responsible for effective metabolism of glucose due to the postulated role of GTF as a cofactor to insulin necessary for initiation of peripheral insulin effects[118, 119]. Administration of $Cr^{6+}$ or complexes of $Cr^{6+}$ showed no activity in alleviating these symptoms of Cr deficiency.

As is suggested above, the oxidation state of Cr determines the effects of Cr on the organism with which it interacts. Chromium exists in a number of oxidation states of which $Cr^{6+}$ and $Cr^{3+}$ are most commonly encountered in biological systems. The most stable state of Cr is $Cr^{3+}$, with $Cr^{6+}$ becoming reduced to $Cr^{3+}$ by NADH or NAHPH upon interaction with cells[120, 121]. Cell membranes are impermeable to $Cr^{3+}$ but are permeable to $Cr^{6+}$ as $CrO_4^{2-}$ which is transported by the sulfate transport system[121, 122].

The major toxic effects of Cr compounds are related to the presence of $Cr^{6+}$; trivalent Cr is not toxic in humans[122]. Since $Cr^{6+}$ is the toxicologically active agent, consideration of the toxic effects of Cr will deal primarily with this ion. Chronic ulcers of the skin may result from contact with $Cr^{6+}$ persisting for months after exposure[119]. Acute irritative dermatitis and allergic eczematous dermatitis also may result from contact with $Cr^{6+}$ containing compounds[119, 120]. Corrosion of the nasal septum has been reported in individuals occupationally exposed to Cr. Chromate becomes deposited on the septum, perforating the mucosa and causing necrosis of the underlying cartilage[119]. Pneumoconiosis has been observed in occupationally exposed individuals as well[119]. Bronchial asthma resulting from the inhalation of chromate dust is another of the toxic effects of Cr[120]. Ingestion of large amounts of $Cr^{6+}$ ($>1.5$ g) leads to hemorrage in the gastrointestinal tract and death[120, 123].

Exposure to Cr compounds results from contact in the general environment for the majority of the populace and, to a more significant extent, from contact in the workplace for a far smaller number of individuals. Chromium is used in the metallurgical industry, in paint pigments and dyes, in leather tanning, in fungicides, and in wood perservatives[124]. Workers in occupations utilizing Cr are at higher risk for suffering the toxic effects of Cr[125].

Particulate Cr in the air results from combustion of fossil fuels[126, 127], metal refineries and cement plants[120], and from particles generated by automobile brake linings which also contain asbestos[128]. Waterways may become contaminated by the discharge of Cr-containing waste from metal plating plants with subsequent effects on the organisms which populate the waterways[124]. The mean daily uptake of Cr from food (primarily $Cr^{3+}$) has been estimated as approximately 100–200 μg Cr/day with that ingested from water as 4 μg/day and that Cr inhaled as 0.28 μg/day[119, 124]. The oxidation state of Cr must be considered when assessing the relative amount of metal taken up since the primary species in air is $Cr^{6+}$ [120], and not the relatively innocous $Cr^{3+}$ present in most foodstuffs.

The distribution of Cr once absorbed follows some organ specificity. In organs from individuals known to have been exposed to Cr, lungs, spleen, liver and adrenal glands contained significant levels of Cr. The concentration of Cr in organs from workers in the chromate industry was as much as 4 orders of magnitude greater than that of non-exposed individuals[120].

The time course of Cr uptake, distribution, and excretion has been reviewed by a number of investigators[119, 120, 122]. In animals, the marginally soluble $Cr^{3+}$ ion is absorbed to the extent of only 1% of ingested dose as opposed to 3–6% of the dose for $Cr^{6+}$. Similarly, $Cr^{6+}$ is absorbed relatively rapidly from the lungs while $Cr^{3+}$ is not. Once taken up into the bloodstream, $Cr^{6+}$ can enter cells where it becomes reduced to $Cr^{3+}$ (*vide supra*). Once Cr has been reduced intracellularly, excretion occurs only slowly. In the instance of red blood cells, this behavior is responsible for the relatively high levels of Cr found in the spleen since the degradation products of erythrocytes are localized to some extent in that organ[120]. The whole body kinetics of Cr excretion in rats were measured by utilizing $^{51}CrCl_3$[129]. The elimination of $^{51}Cr$ was found to occur in 3 phases with half lives of 0.5 days, 5.9 days and 83.4 days corresponding to 3 different compartments containing Cr.

Evaluation of the carcinogenic properties of Cr compounds has fruitfully occupied many investigators over the course of quite a number of years. Epidemiological assessment of the incidence of cancer primarily in occupationally exposed individuals has clearly shown the effects of Cr exposure on the cancer mortality rates of these workers. Beginning with early studies[130–132] and extending to recent reports[133], the incidence of lung cancer in $Cr^{6+}$-exposed workers was found to be 5 to 40 times that statistically expected.

Experimental studies of the incidence of cancer in animals as a result of Cr exposure using different salts and different routes of administration in a variety of species have been summarized[120]. Inhalation studies in mice have given positive results for the induction of lung adenomas by $CaCrO_4$[134]. However, with the exception of $CaCrO_4$, few other Cr compounds were found to induce pulmonary tumors in experimental animals[121, 135].

Use of mammalian cell transformation assays have shown that Cr compounds possess carcinogenic potential. Calcium chromate was found to cause morphological changes in a line of baby hamster kidney cells[136]. Sodium chromate induced a dose dependent morphological transformation in SHE cells as well[137]. The viral transformation of SHE cells by a simian adenovirus was enhanced to a significant extent by the presence of either $CaCrO_4$ or $K_2CrO_4$ in the assay medium[138]. The transformation of SHE cells by benzo(a)pyrene was also increased by $Cr^{6+}$ [139]. The *in vitro* transformation of mammalian cells by Cr has been summarized with the conclusion that $Cr^{6+}$ is a far more potent transforming agent *in vitro* than $Cr^{3+}$ [140].

Because the mutagenicity of a compound may be used as an indication of possible carcinogenesis, Cr-containing compounds have been screened in a number of assays[141]. The *B. subtilis* rec-assay has been used to determine the mutagenicity of many metal compounds[142, 143]. A DNA repair deficient strain of bacteria is treated with the potenial mutagen and its growth is compared to that of a strain possessing an intact repair mechanism. Slower growth in the repair deficient strain may be considered a result of unrepaired DNA damage. Three different $Cr^{6+}$ compounds gave a significant rec effect. The same studies also demonstrated the mutagenicity of $Cr^{6+}$ in a reverse mutation assay in one *E. coli* strain; however, no reversions were induced in another strain.

The *Salmonella* mutagenesis system, which follows the induction of histidine prototrophy in histidine auxotrophic mutants, has been successfully employed in the detection of mutagenic properties of Cr. A number of $Cr^{6+}$ compounds induced mutations in several strains of *Salmonella*, while trivalent Cr compounds exerted no effect on the cultures tested[144, 145].

The reduction of $Cr^{6+}$ by $Na_2SO_3$, ascorbic acid, or the biological reduction systems of the S-9 rat liver microsomal metabolic activation system decreased the mutagenicity of this ion[146]; conversely, oxidation of $Cr^{3+}$ by $KMnO_4$ increased the mutations caused by the otherwise inactive trivalent chromium[147]. These results reinforce the explanation of the activity of Cr in mutagenesis and carcinogenesis as being dependent upon the oxidation state of the metal[121, 127, 148, 149]. Hexavalent chromium (as $CrO_4^{2-}$) is the primary form of Cr transported through the cellular membrane. Once inside the cell, the ion becomes reduced to $Cr^{3+}$ and essentially trapped in the cell; the $Cr^{3+}$ is the active mutagenic species.

Mammalin cell mutagenesis has also been used to evaluate effects of Cr compounds. Chromate and dichromate caused the induction of 8-azaguanine resistence in a Chinese hamster cell line[150]. A mouse lymphoma cell line contain heterozygous for thymidine kinase activity underwent forward mutation when treated with $CrO_4^{2-}$ and $Cr_2O_7^{2-}$ [151].

Since the carcinogenic and mutagenic effects of Cr are without question, the interaction of Cr with DNA and its impact on DNA metabolism ought to illuminate to some degree the chemical and molecular mechanism of carcinogenesis. The synthesis of DNA measured by thymidine incorporation is reduced in baby hamster kidney cells treated with $K_2Cr_2O_7$ even though the uptake of thymidine is stimulated by $Cr^{6+}$ [152]. Another effect of this metal on the enzymes of DNA metabolism involves the misincorporation of bases into DNA. The fidelity of base incorporation in DNA synthesized by a viral DNA polymerase using a synthetic [d(A-T)] template was decreased by both $Cr^{3+}$ and $Cr^{6+}$ [153].

Since the carcinogenic effects of Cr in organisms may be the result of Cr-DNA interactions, lesions in DNA as a function of Cr treatment have been studied. Treatment of baby hamster kidney cell DNA *in vitro* with $Cr^{3+}$ resulted in changes in the thermal stability of DNA as well as the UV spectral characteristics. However, no change in the absorption spectrum of the DNA was detected after *in vitro* treatment with $Cr^{6+}$ [154]. Breaks in DNA strands and DNA-protein crosslinks are two lesions found in DNA exposed to Cr[155]. The molecular weight of DNA from CHO cells treated with $CaCrO_4$ was decreased over controls following 24 hr treatment of the cells[156]. The DNA from liver and kidney of $Na_2Cr_2O_7$ treated rats was found to be crosslinked to protein in addition to suffering strand breaks[157].

Another measure of Cr effects on DNA structure is determination of chromosomal aberrations and sister chromatid exchange in treated cells, with DNA damage serving as another potential indicator of gentoxicity[155]. Hexavalent Cr has been shown to cause gaps, breaks, and exchanges in chromosomes from a variety of species both *in vitro* and *in vivo*[158–160].

Induction of DNA repair is still another means used to identify effects of potential carcinogens or mutagens upon cells. Unscheduled DNA synthesis or incorporation of nucleotides into DNA at portions of the cell cycle not corresponding to DNA synthesis which normally occurs has been detected in human fibroblasts treated with $Cr^{6+}$[161].

In view of the wealth of data which indicates Cr compounds are carcinogenic and mutagenic, Cr has been correctly characterized as a carcinogen[120, 127, 162].

# 6 Lead

Although some studies have suggested a possible nutritional role for lead, the evidence that lead is essential for life is not convincing. Lead toxicity must be considered in terms of two chemical forms of lead, the organic and inorganic forms, and in terms of two exposure populations, children and adults. These differences are illustrated in Table 2 below. In the case of inorganic lead the major toxicologically significant route of exposure is ingestion, and adults only absorb 7% of the ingested lead while children may absorb as much as 40% of the ingested lead from the GI tract. Organic lead compounds are very lipid soluble and absorption may occur by a variety of routes as shown in Table 2. Lead effects primarily three organ systems, 1) the bone marrow, 2) the kidney and 3) the nervous system. Since adults have a well developed blood brain barrier, inorganic lead does not penetrate into the CNS as readily as in children. Therefore children tend to have more CNS problems when exposed to inorganic lead. With lead or with other metal compounds, distribution of the metal to a particular organ appears to be the key feature that will ultimately determine its toxicological effects. Lead blocks the synthesis of heme primarily by inhibiting the enzyme aminolevulinic acid (ALA) dehydratase but coprooxidase and ferrochelatase are also inhibited. This depression of heme biosynthesis results in lead affecting the maturation of red blood cells (rbc), although heme is synthesized in many other cells as well. In some cases of severe lead poisoning, stippling of erythrocytes is seen, indicating the presence of poorly differentiated red blood cells. Lead also effects the kidney as shown in Table 2. The inhibition of heme synthesis results in the accumulation of by-products of the heme pathway in the blood and urine. Thus measurement of ALA concentration in urine is a good indicator of Pb exposure. Although lead has toxic effects on maturation of rbc and on the kidney, in children lead has a predominant effect on the CNS as indicated in Table 2. Although lead does not penetrate into the CNS as readily in adults as it does in children, the peripheral nervous system of adults is affected by lead exposure in a manner which decreases the conductance velocities of motor nerves.

A number of studies have suggested that lead bears a striking similarity to calcium in the way it is handled by biological systems. In fact, the reason children absorb more lead from the GI tract may be because they are absorbing more Ca than adults and lead interacts with a Ca-binding protein, facilitating its intestinal absorption. Additionally, lead is distributed and stored in the bones and is there regulated in its desposition and mobilization in a manner analogous to that of calcium.

**Table 2.**

| | Population | Toxic Category | Tissue Levels | Toxic Manifestation | Diagnostic Sign | Treatment |
|---|---|---|---|---|---|---|
| Inorganic Lead | Adults<br>(7% GI) | Chronic | 1. Bone<br>2. Liver<br>3. Kidney | Microcytic Anemia<br>Colic<br>Kidney<br>Peripheral Neuropathy | ALA (+)<br>Constip, or Diarr.<br>Fanconi (+)<br>Wrist & Ankle Drop | $CaNa_2EDTA$ |
| | | Acute | 1. Liver<br>2. Kidney | Severe Anemia<br>Kidney | ALA (5+)<br>Fanconi (++) | $CaNa_2EDTA_+$<br>Bal, Pen |
| | Children<br>(1–5)<br>(40% GI) | Chronic | 1. Bone<br>2. Liver<br>3. Kidney | Pb Encephalopathy (+) | Behavior Intelligence | $CaNa_2EDTA$<br>Bal |
| | | Acute | 1. CNS<br>2. Liver<br>3. Kidney | Pb Encephalopathy (++) | Convulsion Coma | Diazepam<br>$CaNa_2EDTA$<br>Bal, Pen |
| Organic Pb | Inhal. Skin<br>Ingestion | Acute and<br>Chronic | Bone (Chronic)<br>Liver<br>CNS | CNS | Sleep<br>Delerium<br>Coma | No Treatment |

Abbreviations used: Pen. – Penacillamine, ALA = delta aminolevulinic acid in urine or blood

Human exposure to lead in the non-occupational setting arises primarily as the result of combustion of leaded gasoline which in recent years has been reduced by substitution of other antiknock reagents in gasoline. The burning of fossil fuels such as coal also contributes to lead pollution. The use of lead in paint has caused numerous cases of toxicity in children who have ingested paint chips. Eating utensils painted with lead containing paint pigments has also contributed to lead toxicity.

## *6.1 Human Epidemiological Studies of Lead Carcinogenesis*

Epidemiological studies of industrial workers whose potential for lead exposure could have been greater than that of a normal population have been carried out in an attempt to understand the role of lead in the induction of human neoplasia[163–171)]. In general, these studies made no attempt to consider the type of lead compounds to which workers were exposed or to determine the probable route of exposure. Some information on the specific lead compounds encountered in the various occupational settings, along with probable exposure routes, would have made the studies more interpretable and useful. If exposure occurred by ingestion, the ability of water-insoluble lead salts such as lead oxide and lead sulfide to dissolve in the gastrointestinal tract may greatly facilitate an understanding of their ultimate systemic effects in comparison to their local actions in the gastrointestinal tract. Factors such as particle size are also important in the dissolution of relatively water insoluble compounds in the gastrointestinal system[172)]. When considering other routes of exposure, such as inhalation, the water solubility of the lead compound in question as well as the particle size are extremely important, both in terms of systemic absorption and contained injury in the immediate locus of the retained particle.

The studies of Cooper[163, 164)] and Cooper and Gaffey[165)] examined the incidence of cancer in a large population of indutrial workers exposed to lead. Two groups of individuals were identified as the lead-exposed population under consideration: smelter workers from six lead production facilities and battery plant workers[165)]. The investigators reported (see Table 3) that total mortality from cancer was higher in lead smelter workers than in a control population in two ways:

1. the difference between observed and expected values for the types of malignancies reported, and
2. the standardized mortality ratios, which indicates a greater than "normal" response if it is in excess of 100%.

These studies report not only an excess of all forms of cancer in smelter workers but also a greater level of cancer in the respiratory and digestive systems in both battery and smelter workers. The incidence of cancer of the urinary organs was also elevated in the smelter workers (but not in the battery plant workers), although the number of individuals who died from these neoplasms was very small. As the table indicates, death from neoplasia at other sites was also elevated compared with a normal population, but these results were not discussed in the report. Kang et al.[173)] examined the report of Cooper and Gaffey[165)] and noted an error in the statistical equation they had used to assess the significance of excess cancer mortality. Table 4 is taken from Kang et al., who used a corrected form of the statistical equation

**Table 3.** Expected and observed deaths for malignant neoplasms Jan. 1, 1947–Dec. 31, 1979 for lead smelter and battery plant workers

| Causes of Death | | | Smelters | | Battery plant | |
|---|---|---|---|---|---|---|
| (ICD[a] Code) | Obs | Exp | SMR[b] | Obs | Exp | SMR[c] |
| All malignant neoplasms (140–205) | 69 | 54.95 | 133 | 186 | 180.34 | 111 |
| Buccal cavity & pharynx (140–248) | 0 | 1.89 | – | 6 | 6.02 | 107 |
| Digestive organs peritoneum (150–159) | 25 | 17.63 | 150 | 70 | 61.48 | 123 |
| Respiratory system (160–164) | 22 | 15.76 | 148 | 61 | 49.51 | 132 |
| Genital organs (170–179) | 4 | 4.15 | 101 | 8 | 18.57 | 46 |
| Urinary organs (180–181) | 5 | 2.95 | 179 | 5 | 10.33 | 52 |
| Leukemia (204) | 2 | 2.40 | 88 | 6 | 7.30 | 88 |
| Lymphosarcoma, lymphatic and hematopoietic (200–203, 205) | 3 | 3.46 | 92 | 7 | 9.74 | 77 |
| Other sites | 8 | 6.71 | 126 | 23 | 17.39 | 142 |

[a] International Classification of Diseases
[b] Correction of + 5.55% applied for 18 missing death certificates
[c] Correction of + 7.52% applied for 71 missing death certificates
Source: Cooper and Gaffey (1975)[165]

employed by Cooper and Gaffey and also utilized another statistical test claimed to be more appropriate. Statistical significance was observed in every category listed in the table with the exception of battery plant workers, whose deaths from all forms of neoplasia were not different from those of a control population.

Although Cooper and Gaffey[165] did not discuss the types of lead compounds to which these workers may have been exposed in smelting operations, workers thus employed may have ingested or inhaled oxides and sulfides of lead. Since these and other lead compounds produced in the industrial setting are not readily soluble in water, the cancers arising in respiratory or gastrointestinal systems may have been caused by exposure to water-insoluble particulates. Although Cooper and Gaffey's[165] study had a large sample (7032), only 2275 of the workers (32.4%) were employed when plants monitored urinary lead. Urinary lead values were only available for 9.7% of the 1356 deceased employees on whom the cancer mortality data were based. Only 23 (2%) of the 1356 decedents had blood lead levels measured. Cooper and Gaffey[165] did report some average urinary and blood lead levels, where 10 or more urine or at least three blood samples were taken (viz., battery plant workers: urine lead = 129 μg/l, blood lead = 67 μg/dl; smelter workers; urine lead = 73 μg/l, blood lead = 79.7 μg/dl). Cooper noted that these workers were potentially exposed to other materials, including arsenic, cadmium, and sulfur dioxide, although no data on exposure to these agents were reported. In these and other epidemiological studies in which selection of subjects for monitoring exposure to an agent such as lead is left to company discretion, it is possible that individual subjects are selected primarily on the basis of frank indications of lead exposure, while other individuals who show no symptoms of such intoxication would not be monitored[165]. It is also not clear from these studies when the lead levels were measured, although the timing of the lead measurement would make little difference since no attempt was made to match an individual's lead exposure to any disease process.

**Table 4.** Expected and observed deaths resulting from specified malignant neoplasms for lead smelter and battery plant workers and levels of significance by type of statistical analysis according to one-tailed tests

| Causes of death (ICD[a] code) | Number of deaths | | SMR[b] | Probability | | |
|---|---|---|---|---|---|---|
| | Observed | Expected | | Poisson[c] | This analysis[d] | Cooper and Gaffey[e] |
| | Lead smelter workers | | | | | |
| All malignant neoplasms (140–205) | 69 | 54.95 | 133 | <0.02 | <0.01 | <0.02 |
| Cancer of the digestive organs peritoneum (250–159) | 25 | 17.63 | 150 | <0.03 | <0.02 | <0.05 |
| Cancer of the respiratory system (160–164) | 22 | 15.76 | 148 | <0.05 | <0.03 | >0.05 |
| | Battery plant workers | | | | | |
| All malignant neoplasms (140–205) | 186 | 180.34 | 111 | >0.05 | >0.05 | >0.05 |
| Cancer of the digestive organs, peritoneum (150–159) | 70 | 61.48 | 123 | <0.05 | <0.04 | >0.05 |
| Cancer of the respiratory system (160–164) | 61 | 49.51 | 132 | <0.03 | <0.02 | <0.03 |

[a] International Classification of Diseases
[b] SMR values were corrected by Cooper and Gaffey for missing death certificates under the assumption that distribution of causes of death was the same in missing certificates as in those that were obtained
[c] Observed deaths were recalculated as follows: adjusted observed deaths – (given SMR/100) × expected deaths
[d] Given $z = (\text{SMR} - 100)\sqrt{\text{expected}/100}$
[e] Given $z = (\text{SMR} - 100/\sqrt{100 \times \text{SMR}/\text{expected}}$
Source: Kang et al. (1980)

In a follow-up study of the same population of lead smelter and battery plant workers, Cooper[164)] claimed that lead had no significant role in the induction of neoplasia. However, an examination of several of his tables reveals standardized mortality ratios (SMRs) of 149% and 125% for all types of malignant neoplasms in lead battery plant workers with less than and greater than 10 years of employment, respectively. SMR is a percentage value that is based upon a comparison of an experimental population relative to a control population. If the value exceeds 100%, the incidence of death is greater than normal. In battery workers employed for 10 years or more there was an unusually high incidence of cancer listed as "other site" tumors (SMR = 229%; expected = 4.85, observed = 16). Respiratory cancers were elevated in the battery plant workers employed for less than 10 years (SMR = 172%). Similarly, in workers involved with lead production facilities for more than 10 years the SMR was 151%. Again, in the absence of good lead exposure documentation, it is difficult to assess the role of lead in the induction of the observed cancer.

Cooper[164] claimed that the excess of respiratory cancers may have been due to a lack of correction for smoking histories, which unfortunately were not compiled in the present study.

A recent study[169] examined the historical incidence of cancers in a population of smelter workers diagnosed as having lead poisoning. The incidence of cancer in a relatively small group of 241 workers was compared with 695 deceased employees from the same company. The control group had been employed during approximately the same period and was asserted to be free from lead exposure, although there were no data to indicate lead levels in either the control or the experimental group. Based upon diagnoses of lead poisoning made in the 1920s and 1930s for a majority of the deaths, the authors concluded that there was a considerably lower incidence of cancer in lead-poisoned workers. However, there is no indication of how lead poisoning was diagnosed. It is difficult to draw any conclusion from this study with regard to the role of lead in human neoplasia.

Evaluation of the ability of lead to induce human neoplasia must await further epidemiological studies in which factors that may contribute to the observed effects are well controlled and the disease process is assessed in individuals with well controlled documented exposure to lead. There is little that can be reliably concluded from the epidemiological studies available at the present time.

## *6.2 Induction of Tumors in Experimental Animals*

Experiments testing the ability of lead to cause cancer in experimental animals are an essential aspect of understanding its oncogenicity in humans[174]. However, a proper lifetime animal feeding study to assess the carcinogenic potential of lead following National Cancer Institute guidelines[175] has not been conducted. The costs of such studies exceed $ 1 million and consequently they are limited only to those agents in which sufficient evidence based upon *in vitro* or epidemiological studies warrants such an undertaking. The literature on lead carcinogenesis contains a number of smaller studies where only one or two doses were employed and where toxicological monitoring of experimental animals exposed to lead was generally absent. Some of these studies are summarized in Table 5. Most of them serve merely to illustrate that a number of different laboratories can induce renal tumors in rats by feeding them diets containing 0.1% or 1.0% lead acetate. In some cases other lead formulations were tested, but the dosage selection was not based upon lethal dose values and in most cases only one dosage was used. An additional problem with many of these studies was that the actual concentration of lead administered to the animal was not measured. These studies will be discussed very briefly and some have even been omitted from consideration because they add little knowledge to our understanding of lead carcinogenesis.

Other lead compounds have also been tested in experimental animals, but in these studies only one or two dosages, generally quite high were employed, making it difficult, again, to assess the important question of the carcinogenic activity of lead compounds at relatively nontoxic concentrations. Moreover, it is difficult to assess

**Table 5.** Examples of the incidence of cancer in experimental animals exposed to lead compounds

| Species | Pb compound | Dose and mode | Incidence (and type) of neoplasms | Reference |
|---|---|---|---|---|
| Rat | Pb phosphate | 120–680 mg (total dose s.c.) | 19/29 (renal tumors) | Zollinger[176)] |
| Rat | Pb acetate | 1% (in diet) | 15/16 (kidney tumors)<br>14/16 (renal carcinomas) | Boyland et al.[177)] |
| Rat | Pb subacetate | 0.1% and 1.0% (in diet) | 11/32 (renal tumors)<br>13/24 (renal tumors) | Van Esch et al.[178)] |
| Mouse | Pb naphthenate | 20% in benzene (dermal 1–2 times weekly) | 5/59 (renal neoplasms) (no control with benzene) | Baldwin et al.[179)] |
| Rat | Pb phosphate | 1.3 g (total dosage s.c.) | 29/80 (renal tumors) | Balo et al.[180)] |
| Rat | Pb subacetate | 0.5–1% (in diet) | 14/24 (renal tumors) | Hass et al.[181)] |
| Rat | Pb subacetate | 1% (in diet) | 31/40 (renal tumors) | Mao and Molnar[182)] |
| Mouse | Tetraethyl lead in tricaprylin | 0.6 mg (s.c.) 4 doses between birth and 21 days | 5/41 (lymphomas) in females, 1/26 in males, and 1/39 in controls | Epstein and Mantel[183)] |
| Rat | Pb acetate | 3 mg/day for 2 months;<br>4 mg/day for 16 months (p.o.) | 72/126 (renal tumors)<br>23/94 males (testicular [Leydig cell] tumors) | Zawirska and Medraś[184)] |
| Hamster | Pb subacetate | 1.0% (in 0.5% diet) | No significant incidence renal neoplasms | Van Esch and Kroes[185)] |
| Mouse | Pb subacetate | 0.1% and 1.0% (in diet) | 7/25 (renal carcinomas) at 0.1%<br>Substantial death at 1.0% | Van Esch et al.[186)] |
| Rat | Pb nitrate | 25 g/l in drinking water | No significant incidence of tumors | Schroeder et al.[187)] |
| Rat | Pb acetate | 3 mg/day (p.o.) | 89/94 (renal, pituitary, cerebral gliomas, adrenal, thyroid, prostatic, mammary tumors) | Zawirska and Medraś[188)] |
| Rat | Pb acetate | 0, 10, 50, 100, 1000, 2000 ppm (in diet) for 2 yr | No tumors 0–100 ppm; 5/50 (renal tumors) at 500 ppm; 10/20 at 1000 ppm; 16/20 males, 7/20 females at 2000 ppm | Azar et al.[189)] |
| Hamster | Pb oxide | 10 intratracheal administrations (1 mg) | 0/30 without benzopyrene, 12/30 with benzopyrene (lung cancers) | Kobayashi and Okamoto[190)] |
| Rat | Pb powder | 10 mg orally 2 times each month | 5/47 (1 lymphoma, 4 leukemias) | Furst et al.[191)] |
| | | 10 mg/monthly for 9 months; then 3 monthly injections of 5 mg | 1/50 (fibrosarcoma) | |

the true toxicity caused by these agents, since in many cases properly designed toxicity studies were not performed in parallel with these cancer studies.

As shown in Table 5, lead nitrate produced no tumors in rats when given at very low concentrations, but lead phosphate administered subcutaneously at relatively high levels induced a high incidence of renal tumors in two studies. Lead powder administered orally resulted in lymphomas and leukemia; when given intramuscularly only one fibrosarcoma was produced in 50 animals. Lead naphthenate applied as a 20% solution in benzene two times each week for 12 months resulted in the development of four adenomas and one renal carcinoma in a group of 50 mice[179)]. However, in this study control mice were not painted with benzene. Tetraethyl lead at 0.6 mg given in four divided doses between birth and 21 days to female mice resulted in 5/36 surviving animals developing lymphomas while 1/26 males treated similarly and 1/39 controls developed lymphomas[183)].

Lead subacetate has also been tested in the mouse lung adenoma bioassay[192)]. This assay measures the incidence of nodules forming in the lung of strain A/Strong mice following parenteral administration of various test agents. Nodule formation in the lung does not actually represent the induction of lung cancer but merely serves as a general measure of carcinogenic potency independent of lung tissue[192)]. Lead subacetate was administered to mice at 150, 74 and 30 mg (total dose), which represented the maximum tolerated dose (MTD), 1/2 MTD, and 1/5 MTD, respectively, over a 30-week period using 15 separate i.p. injections[192)]. Survivals at the three doses were 15/20 (MTD), 12/20 (1/2 MTD), and 17/20 (1/5 MTD), respectively, with 11/15, 5/12, and 6/17 survivors having lung nodules. Only at the highest doses was the incidence of nodules greater than in the untreated 1 or 2 highest groups. However, these authors concluded that on a molar-dose basis lead subacetate was the most potent of all the metallic compounds examined. Injection of 0.13 mmol/kg lead subacetate was required to produce one lung tumor per mouse, indicating that this compound was about three times more potent than urethane (at 0.5 mmol/kg) and approximately 10 times more potent than nickelous acetate (at 1.15 mmol/kg). The mouse lung adenoma bioassay has been one of the most utilized systems for examining carcinogenic activity in experimental animals and is well recognized as a highly accurate test system for assessing potential carcinogenic hazard[192)]. Lead oxide combined with benzopyrene administered intratracheally resulted in 11 adenomas and 1 adenocarcinoma in a group of 15 hamsters, while no lung neoplasias were observed in groups receiving benzopyrene or lead oxide alone[190)].

Administration of lead acetate to rats has been reported to produce other types of tumors, e.g. testicular, adrenal, thyroid, pituitary, prostate, lung[188)] and cerebral gliomas[188)]. However, in other animal species, such as dogs[189)] and hamsters[185)], lead acetate induced either no tumors or only kidney tumors.

The above studies seem to implicate some lead compounds as carcinogens in experimental animals, but these studies were not designed to address the question of lead carcinogenesis in a definitive manner. In contrast, the study of Azar et al.[189)] examined the oncogenic potential of lead acetate at a number of doses and in addition monitored a number of toxicological parameters in the experimental animals. Azar et al.[189)] gave 0, 10, 50, 100, 1000 and 2000 ppm dose levels of lead (as lead acetate) to rats during a two-year feeding study. Fifty rats of each sex were utilized doses up to 500 ppm, while 100 animals of each sex were used as controls.

During this study, the clinical appearance and behavior of the animals was observed. Food consumption, growth, and mortality were recorded. Blood, urine, fecal, and tissue lead analyses were done periodically using atomic absorption spectrophotometry. A complete analysis was done periodically on the blood specimens. This analysis included blood count, hemoglobin, hematocrit, stippled cell count, prothrombin time, alkaline phosphatase, urea nitrogen, glutamic-pyruvate transaminase, and albumin-to-globulin ration. The activity of the enzyme alpha-aminolevulinic acid dehydrase (ALA-D) in the blood and the excretion of its substrate, delta-aminolevulinic acid (δ-ALA) in the urine were also determined. A thorough necropsy, including both gross and histologic examination, was performed on all animals. Reproduction was also assessed.

At 550 ppm (0.05%) and above, male rats developed a significant number of renal tumors. Female rats did not develop tumors except when fed 2000 ppm lead acetate. The number of stippled red blood cells increased at the 10 ppm dose in the rats utilized in the study, and ALA-D was decreased at 50 ppm. Hemoglobin and hematocrit, however, were not depressed in the rats until they received a dose of 1000 ppm lead. These results illustrate that the induction of kidney tumors coincides with moderate to severe toxicological doses of lead acetate, for it was at 500–1000 ppm lead in the diet that a significant increase in mortality occurred (see Table 6). At 1000 and 2000 ppm lead, 21-day-old weanling rats showed no

**Table 6.** Mortality and kidney tumors in rats fed lead acetate for two years

| Nominal (actual)[a] concentration in ppm of Pb in diet | No. of rats of each sex | % Mortality[b] | | % Kidney tumors | |
|---|---|---|---|---|---|
| | | Male | Female | Male | Female |
| 0 (5) | 100 | 37 | 34 | 0 | 0 |
| 10 (18) | 50 | 36 | 30 | 0 | 0 |
| 50 (62) | 50 | 36 | 28 | 0 | 0 |
| 100 (141) | 50 | 36 | 28 | 0 | 0 |
| 500 (548) | 50 | 52 | 36 | 10 | 0 |
| 0 (3) | 20 | 50 | 35 | 0 | 0 |
| 1000 (1130) | 20 | 50 | 50 | 50 | 0 |
| 2000 (2102) | 20 | 80 | 35 | 80 | 35 |

[a] Measured concentration of lead in diet
[b] Includes rats that either died or were sacrificed *in extremis*
Source: Azar et al.[189)]

tumors but did show histological changes in the kidney comparable to those seen in adults receiving 500 ppm or more lead in their diet. This study showed that the induction of renal tumors by lead acetate was linearly proportional to the dietary levels of lead fed to male rats. It may be concluded therefore, that chronic exposure of rats to lead at levels where clinical signs of toxicity would be evident in human results in a significant elevation in the incidence of kidney tumors.

## 6.3 Cell Transformation

Although strictly speaking cell transformation is an *in vitro* experimental system, its end point is a neoplastic change. There are two types of cell transformation assays: (A) those employing continuous cell lines, and (b) those employing cell cultures prepared from embryonic tissue. Use of continuous cell lines has the advantage of ease in preparation of the cell cultures, but these cells generally have some properties of cancer cells. The absence of a few characteristics of a cancer cell in these continuous cell lines allows for an assay of cell transforming activity. End points include morphological transformation (ordered cell growth to disorder cell growth), ability to form colonies in soft agar-containing medium (a property characteristic of cancer cells), and ability of cells to form tumors when inoculated into experimental animals. Assays that utilize freshly isolated embryonic cells are generally perferred to those that use cell lines, because embryonic cells have not yet acquired any of the characteristics of a transformed cell. The cell transformation assay system has been utilized to examine the potential carcinogenic activity of a number of chemical agents; the results seem to agree generally with the results of carcinogenesis tests using experimental animals. Cell transformation assays can be made quantitative by assessing the percentage of surviving colonies exhibiting morphological transformation. Verification of a neoplastic change can be accomplished by cloning these cells and testing their ability to form tumors in animals.

Lead acetate has been shown to induce morphological transformation in Syrian hamster embryo cells following a continuous exposure to 1 or 2.5 μg/ml of culture medium for nine days[194)]. The incidence of transformation increased from 0 percent in untreated cells to 2.0 and 6.0% of the surviving cells following lead treatment, these transformants being capable of forming fibrosarcomas when cloned and administered to "nude" mice and Syrian hamsters; no tumors growth resulted from similar inoculation of untreated cells[194)]. In another study lead acetate was shown to enhance the incidence of simian adenovirus (SA-7) induction of Syrian hamster embryo cell transformation. Lead acetate also caused significant enhancement of viral transformation (2–3 fold) at 100 and 200 μg/ml following three hours of exposure[195)]. Lead oxide also enhanced SA-7 transformation of Syrian hamster embryo cells almost 4 fold at 50 μM following three hours of exposure[195)]. The significance of enhanced virally-induced carcinogenesis in relationship to the carcinogenic potential of an agent is not known.

Morphological transformation induced by lead acetate was correlated with the ability of the transformed cells to form tumors in appropriate hosts (see above), indicating that a truly neoplastic change occurred in cell culture. The induction of neoplastic transformation by lead acetate suggests that this agent is potentially carcinogenic at the cellular level. However, with *in vitro* systems such as the cell transformation assay it is essential to compare the effects of other, similar types of carcinogenic agents in order to evaluate the response and determine the reliability of the assay. The incidence of transformation obtained with lead acetate was greater than the incidence following similar exposure to $NiCl_2$, but less than that produced by $CaCrO_4$[196)]. Both nickel and chromium have been implicated in the etiology of human cancer[174)]. Results from this assay thus suggest that lead acetate has effects that are similar to those caused by other metal carcinogens. In particular, the ability of lead acetate to induce neoplastic transformation in cells in a concentration-dependent manner is highly suggestive of potential carcinogenic activity. It should also be noted that lead acetate induced these transformations at concentrations that decreased cell survival by only 27%[195]. Further studies from other laboratories utilizing the cell transformation assay and other lead compounds are needed.

## 6.4 Chromosomal Aberrations

Two approaches may be taken in the analysis of the effect of an agent such as lead on chromosomal structure. The first approach involves culturing lymphocytes either from humans exposed to lead or from experimental animals given a certain dosage of lead. The second approach involves exposing cultured lymphocytes directly to lead. Both approaches have been employed in assessing whether lead is capable of inducing chromosomal aberrations. For present purposes emphasis will not be placed on the type of chromosomal aberration induced, since most of the available studies do not appear to associate any specific type of chromosomal aberration with lead exposure. It should be noted, however, that moderate aberrations include gaps and fragments, whereas severe aberrations include dicentric rings, translocations, and exchanges.

Contradictory reports exist on the effect of lead in inducing chromosomal aberrations (Tables 7 and 8). These studies have been grouped in two separate tables based upon their conclusions. Those studies reporting a positive effect of lead on chromosomal aberrations are indexed in Table 7, whereas studies reporting no association between lead exposure and chromosomal aberrations are indexed in Table 8. Unfortunately, these studies are difficult to fully evaluate because of a number of unknown variables (e.g., absence of sufficient evidence of lead intoxication, no dose-response relationship, and absence of information regarding lymphocyte culture time). To illustrate, in a number of the studies where lead exposure correlated with an increased incidence of chromosomal aberrations (Table 7), lymphocytes were cultured for 72 hours. Most cytogenetic studies have been conducted with a maximum culture time of 48 hours to avoid high background levels of chromosomal aberrations due to multiple cell divisions during culture. Therefore, it is possible that the positive effects of lead in inducing chromosomal aberrations may have been due to the longer culture period. Nonetheless, it is evident that in the negative studies the blood lead concentration was generally lower than in the studies reporting a positive effect of lead on chromosomal aberrations, although in many of the latter instances, blood lead levels indicated severe exposure. In some of these positive studies there was a correlation in the incidence of gaps, fragments, chromatid exchanges, and other chromosomal aberrations with blood lead levels[208]. However, as indicated in Table 6 in other studies there were no direct correlations between indices of lead exposure (i.e., δ-ALA excretion) and numbers of chromosomal aberrations. Nutritional factors such as $Ca^{2+}$ levels *in vivo* or *in vitro* are also important since it is possible that the effects of lead on cells may be antagonized by $Ca^{2+}$[213]. As is usually the case in studies of human populations

**Table 7.** Cytogenetic investigations of cells from individuals exposed to lead: 9 positive studies

| Number of exposed subjects | Number of controls | Cell culture time (hrs.) | Blood (μg/dl) or urine (μg/l) level | Exposed subjects | Type of damage | Remarks | Ref. |
|---|---|---|---|---|---|---|---|
| 8 | 14 | ? | 62.–89. (blood) | Workers in a lead oxide factory | Chromatid and chromosome | Increase in chromosomal damage correlated with increased δ-ALA excretion | Schwanitz et al.[197] |
| 10 | 10 | 72 | 60.–100. (blood) | Workers in a chemical factory | Chromatid gaps, breaks | No correlation with blood lead levels | Gath & Thiess[198] |
| 14 | 5 | 48 | 155–720 (urine) | Workers in a zinc plant, exposed to fumes & dust of cadmium, zinc & lead | Gaps, fragments, exchanges, dicentrics, rings | Thought to be caused by lead, not cadmium or zinc | Deknudt et al.[199] |
| 105 | – | 72 | 11.6–97.4 mean, 37.7 (blood) | Blast-furnace workers, metal grinders, scrap metal processers | "Structural abnormalities", gaps, breaks, hyperploidy | No correlation with δ-ALA excretion or blood lead levels | Schwanitz et al.[200] |
| 11 (before and after exposure) | – | 68–70 | 34.–64. (blood) | Workers in a lead-acid battery plant and a lead foundry | Gaps, breaks, fragments | No correlation with ALA-D activity in red cells | Forni et al.[201] |
| 44 | 15 | 72 | 30.–75. (blood) | Individuals in a lead oxide factory | Chromatid and chromosome aberrations | Positive correlation with length of exposure | Garza-Chapa et al.[202] |
| 23 | 20 | 48 | 44.–95. (not given) | Lead-acid battery melters, tin workers | Dicentrics, rings, fragments | Factors other than lead exposure may be required for severe aberrations | Deknudt et al.[203] |
| 20 | 20 | 46–48 | 53.–100. (blood) | Ceramic, lead & battery workers | Breaks, fragments | Positive correlation with blood lead levels | Sarto et al.[204] |
| 26 (4 low, 16 medium, 6 high exposure) | not given | 72 | 22.5–65. (blood) | Smelter workers | Gaps, chromatid and chromosome aberrations | Positive correlation with blood lead levels | Nordenson et al.[205] |
| 12 | 18 | 48–72 | 24–49 (blood) | Electrical storage battery workers | Chromatid and chromosome aberrations | | Forni et al.[206] |

Source: International Agency for Research on Cancer (1980), with modifications

**Table 8.** Cytogenetic investigations of cells from individuals exposed to lead: 6 negative studies

| Number of exposed subjects | Number of controls | Cell culture time (hrs.) | Blood lead level (μg/dl) | Exposed subjects | References |
|---|---|---|---|---|---|
| 29 | 20 | 46–48 | Not given, stated to be 20–30% higher than controls | Policemen "permanently in contact with high levels of automotive exhaust" | Bauchinger et al.[207] |
| 32 | 20 | 46–48 | Range not given; highest level was 590 mg/l [sic] | Workers in lead manufacturing industry; 3 had acute lead intoxication | Schmid et al.[208] |
| 35 | 35 | 45–48 | Control, < 4.; exposed, 4.–> 12. | Shipyard workers employed as "burners" cutting metal structures on ships | O'Riordan and Evans[209] |
| 24 | 15 | 48 | 19.3 (lead) 0.4 (cadmium) | Mixed exposure to zinc, lead, and cadmium in a zinc-smelting plant; significant increase in chromatid breaks and exchanges. Authors suggest that cadmium was the major cause of this damage | Bauchinger et al.[210] |
| 9 | 9 | 72 | 40.0 ± 5.0, 7 weeks | Volunteers ingested capsules containing lead acetate | Bulsma & De France[211] |
| 30 | 20 | 48 | Control, 11.8–13.2; exposed, 29–33 | Children living near a lead smelter | Bauchinger et al.[212] |

Source: International Agency for Research on Cancer (1980)

exposed to lead, exposure to other metals (zinc, cadmium, and copper) that may produce chromosomal aberrations was also prevalent. No study has been done to determine the specific lead compound to which the individuals were exposed.

In a more recent study by Forni et al.[206], 18 healthy females with occupational exposure to lead were evaluated for chromosomal aberrations in their lymphocytes cultured for 48 or 72 hours. More aberrations at the 72-hour culture time were present when compared with 48-hour culture period in both control and lead-exposed groups; however, this difference was not statistically significant. Statistically significant differences from the 72-hour controls were noted in the 72-hour culture obtained from the lead exposed group. These results demonstrate the extended 72-hour culture time results in increased chromosomal aberrations in the control lymphocytes and that the longer culture time was apparently necessary to detect the effects of lead on chromosomal structure. However, the blood lead levels in the exposed females ranged from 24 to 59 μg/dl, while control females had blood lead levels ranging from 22 to 37 μg/dl. Thus there was a marginal effect of lead on chromosomal aberration but the two groups may not have been sufficiently different in their exposure to lead to show clear differences in frequency of chromosomal aberrations.

Some studies have also been conducted on the direct effect of soluble lead salts on cultured human lymphocytes. In a study by Beek[214], a longer (72-hr) culture time was used with the result that lead acetate was found to induce chromosomal aberrations at 100 μM. Lead acetate had no effect on chromatid aberrations induced with X-rays or alkylating agents[217)]. In another study[215)], lead acetate at 1 and 0.1 mM caused minimal chromosomal aberrations.

Chromosomal aberrations have been demonstrated in lymphocytes from cynomolgus monkeys treated chronically with lead acetate (6 mg/day, 6 days/week for 16 months), particularly when they were kept on a low calcium diet[216)]. These aberrations accompanying a low $Ca^{2+}$ diet were characterized by the authors as severe. The effect of low calcium on chromosomal aberrations induced by lead is most likely due to interaction of $Ca^{2+}$ and $Pb^{2+}$ at the level of the chromosome.

Sister chromatid exchange represents the normal movement of DNA in the genome. The sister chromatid exchange assay offers a very sensitive probe for the effects of genotoxic compounds on DNA rearrangement as a number of chemicals with carcinogenic activity are capable of increasing these exchanges. The effect of lead on such movement has been examined in cultured lymphocytes[217)] with no increase in exchanges observed at a lead acetate concentration of 0.01 mM. However, one study with lead at one dose on one system is probably not sufficient to rule our whether lead increases the incidence of these exchanges.

## *6.5 Effect of Lead on Bacterial and Mammalian Mutagenesis Systems*

Bacterial and mammalian mutagenesis test systems examine the ability of chemical agents to induce changes in DNA sequences of a specific gene product that is monitored by selection procedures. They measure the potential of a chemical agent to produce a change in the DNA, although this change is not likely to be the same alteration in gene expression that occurs during oncogenesis. However, if an agent is capable of affecting the expression of a particular gene product that is being monitored, it could potentially affect other sequences which may result in cancer. Since many carcinogens are also mutagens, it is appropriate to use such systems in evaluating the genotoxic effects of lead. Use of bacterial systems for assaying metal genotoxicity must await further development of bacterial strains that are appropriately responsive to known mutagenic metals[218)].

## *6.6 Effect of Lead on Parameters of DNA Structure and Function*

There are a number of very sensitive techniques for examining the effect of metals on DNA structure and function in intact cells. Although these techniques have not been extensively utilized with respect to metal compounds, future research will probably be devoted to this area. Considerable work has been done to understand the effects of metals on enzymes involved in DNA transcription.

Sirover and Loeb[219)] examined the effect of lead and other metal compounds upon the fidelity of transcription of DNA by a viral DNA polymerase. Relatively high concentrations of metal ions (in some cases in the millimolar range) were required to decrease the fidelity of transcription, but there was a good correlation between metal ions that are carcinogenic or mutagenic and their activity in decreasing the fidelity of transcription. This assay system measures the ability of a metal ion to cause the incorporation of incorrect (non-homologous) bases by a polymerase using a defined polynucleotide template. In an intact cell, this would result in the induction of a mutation if the insertion of an incorrect base is phenotypically expressed. Since the interaction of metal ions with cellular macromolecules is relatively unstable, misincorporation of a base during synthesis could alter the base sequence of DNA in an intact cell. Lead at 4 mM was among the metals listed as mutagenic or carcinogenic that caused a decrease in the fidelity of transcription[219]. Other metals active in decreasing fidelity in addition to Pb included $Ag^{+}$, $Be^{2+}$, $Cd^{2+}$, $Co^{2+}$, $Cr^{3+}$, $Cu^{2+}$, $Mn^{2+}$, and $Ni^{2+}$. No change in fidelity was produced by $Al^{3+}$, $Ba^{2+}$, $Ca^{2+}$, $Fe^{3+}$, $K^{+}$, $Rb^{+}$, $Mg^{2+}$, $Mg^{+}$, $Se^{2+}$, $Sr^{2+}$, and $Zn^{2+}$. Metals that decreased fidelity were, generally, also those metals implicated as carcinogenic or mutagenic[219].

In a similar study, Hoffman and Niyogi[220] demonstrated that lead chloride was the most potent of 10 metals tested in inhibiting RNA synthesis (i.e., $Pb^{2+} > Cd^{2+} > Co^{2+} > Mn^{2+} > Li^{+} > Na^{+} > K^{+}$) for both types of templates tested, i.e. calf thymus DNA and $T_4$ phage DNA. These results were explained in terms of the binding of these metal ions more to the bases than to the phosphate groups of the DNA (i.e., $Pb^{2+} > Cd^{2+} > Zn^{2+} > Mn^{2+} > Mg^{2+} > Li^{+} = Na^{+} = K^{+}$). Additionally, metal compounds such as lead chloride with carcinogenic or mutagenic activity were found to stimulate mRNA chain initiation at 0.1 mM concentrations.

These well-conducted mechanistic studies provide evidence that lead can affect a molecular process associated with the normal regulation of gene expression. Although far removed from the intact cell situation, these effects of lead suggest that it can alter the genetic system and thus must be considered potentially genotoxic.

## 7 Mercury

There is no evidence that Hg is essential for living organisms and in fact of the toxicologically important metals only Cd and Hg have even no suggestive essentiality for life. As with most metals, the toxicity of Hg must be considered in terms of its chemical form. Table 9 is a summary of the toxicity of inorganic, elemental and organic mercury. Inorganic mercury is absorbed only in the GI tract and since it is so polar its effects are primarily upon the kidney. In contrast, elemental mercury is very lipid soluble and is readily volatile under normal atmospheric conditions; therefore, absorption by inhalation is the primary avenue of toxicological exposure. Metallic mercury is not well absorbed from the GI tract since it forms water insoluble globules there; however, some absorption through the skin may occur due to its high lipid solubility. As shown in Table 9 metallic mercury owing to its high lipid solubility can penetrate into the CNS and may become trapped there due to intracellular oxidation by the catalase/peroxidase system[221]. $Hg^{2+}$ is an extremely soft metal

**Table 9.**

| | Route of Absorption | Type of Toxicity | Organs Affected | Diagnostic Sign |
|---|---|---|---|---|
| Inorganic $Hg^{2+}$, $Hg^{+}$ | GI (10%) | Acute | Kidney<br>Oral & GI | Oliguria<br>Foul Breath<br>Salivation (↑)<br>Diarrhea |
| | | Chronic | Kidney | Proteinuria |
| Elemental Hg | GI (not Absorbed) | Acute | Lung<br><br>CNS<br>Kidney | Pneumonitis, Emphysema<br>Tremors, Lethargy<br>Tubular Epithelium |
| | Inhalation (well Absorbed) | Chronic | CNS | Neurological & Psychiatric |
| Organic Hg | GI (90% Absorbed) | Acute | CNS (Sensory) | Lethargy Excitement |
| | Inhalation | Chronic | CNS (Sensory) | Ataxia<br>Paresthesia<br>Deafness |

ion having a high degree of chemical reactivity toward sulfhydryl groups found in proteins[222]. Its reactions with these molecules are probably responsible for its retention in the CNS[221]. CNS toxicity is also a feature of organic mercury toxicity. Organic mercury (i.e. methyl mercury) is extremely lipid soluble and is readily absorbed at any site of exposure. The CNS toxicity of organic Hg appears to involve sensory inputs into the CNS (i.e. constriction of visual fields) whereas the toxicity of elemental Hg appears to have psychiatric components.

The predominant form of mercury in the earth's crust is as the sulfide, which is found in relatively low concentrations. Environmental exposure to Hg comes about primarily as a result of combustion of fossil fuels. Organic-mercurials have been widely used as fungicides in the wood pulp and paper industries and exposure to these agents is extremely hazardous. Mercury can be methylated by microbial action, resulting in the generation of organic mercury. This process is extremely important in the aquatic environment where fish accumulate the methylated mercury.

No evidence exists based upon epidemiological studies or experimental animal studies that mercury is carcinogenic. In one study, metallic mercury injected intraperitioneally in 39 rats caused 5 of 12 survivors to develop peritoneal sarcomas[223]. However, this effect was ascribed to smooth surface or "solid state" carcinogenesis. Mercuric chloride was shown to enhance viral transformation of Syrian hamster embryo cells[224] at 50 μM and was judged to be of similar potency in such enhancement as cobalt, lead and manganese. Mercury has been shown to induce chromosomal aberrations both *in vitro* and *in vivo*[225]. Humans ingesting methyl mercury-contaminated fish had a dose related increase in chromosomal aberrations in their lymphocytes. At blood methyl-mercury concentrations $> 100$ μg/L, a significant increase in chromosomal aberrations was observed[226]. Mercuric ion has been found to possess weak mutagenic activity in bacterial and mammalian systems[227, 228].

The DNA lesions induced by $HgCl_2$ have been studied in considerable detail. Mercuric chloride produces a concentration dependent induction of DNA single strand breaks and, with time, DNA-DNA crosslinks[229–231]. Mercuric chloride also causes the production of oxygen radicals in cells; some similarity is apparent in the way X-rays and $HgCl_2$ cause these DNA lesions[229]. However, the X-ray induced DNA damage is rapidly repaired by the cell in contrast to the damage induced by $HgCl_2$ which is not readily repaired since $Hg^{2+}$ inhibits repair systems[231].

# 8 Nickel

Among the trace elements currently being considered for essentiality, nickel is one of the most ubiquitous in the environment and is readily available for human exposure. With the exception of nickel carbonyl, nickel is considered relatively non-toxic; however, certain compounds of nickel carry the distinction of possessing extremely potent carcinogenic activity. Thus, this apparent anomaly concerning the essentiality and carcinogenicity of nickel merits further consideration.

Naturally occurring nickel, as a major component of the lithosphere, is for the most part stationary, i.e. complexed with other metallic ores in igneous rock. Human exposure to nickel occurs primarily as a result of industrialization via processing of nickel ores into nickel alloys used in numerous products. Mobilization into the atmospheric, aquatic, and terrestrial ecosystems occurs via the processes of mining, smelting and refining with minor contributions from melting, casting of alloys, grinding, electroplating and combustion of petroleum products such as diesel fuel (for reviews see[232–237]). The latter is of particular importance since respiratory cancers are the major toxicity associated with nickel exposure and nickel accounts for a large percentage of the total trace metal particulate content of urban air (for review see[237]). Approximately one-half of the atmospheric nickel is associated with fine particles (ca. 1 μm diameter) of fly ash as nickel sulfides and oxides and it is estimated that urban residents may inhale up to 2–14 μg of nickel per day depending on conditions[232]. Cigarette smokers significantly enhance nickel intake by inhalation since this metal is present in mainstream smoke; the average individual who smokes two packs a day would inhale 3–15 μg of nickel daily[238, 239]. The majority of human intake of nickel occurs via food consumption though, since total dietary intact averages 300 to 500 μg daily in the United States[240]. Dermal contact with nickel occurs via handling of hard goods such as stainless steel kitchen knives and nickel-plated jewelry[232].

The essentiality of nickel to human metabolism has been the object of speculation for some time and has been the subject of numerous reviews[232, 241–244]. Nickel is known to be necessary for the activity of plant and microbial urease, microbial carbon monoxide dehydrogenase, and coenzyme F 430 (a microbial analog to vitamin B 12). Nutritional essentially has been demonstrated in rats, swine, goats and chicks. Changes in nickel distribution in humans have been observed to be associated with several pathological states. It appears that nickel partially fulfills Mentz's requirements for essentiality[245)].

Pharmacokinetic parameters of nickel absorption and distribution are determined by the whole absorption, quantities inhaled or ingested and by the chemical and physical forms of the compound. Nickel carbonyl, for example, is a volatile, highly toxic compound that is absorbed readily by the pulmonary route[246)]; it is known that humans and experimental animals exposed to nickel carbonyl vapor have highest levels of nickel in lung, kidney, liver and brain[232, 247)]. Parenteral administration of nickel results in highest accumulation in kidney, endocrine glands, lung and liver[248, 249)]. Low doses of nickel salts given orally have little effect on tissue levels and distribution; however, high doses apparently overwhelm homeostatic regulatory mechanisms resulting in accumulation of nickel in kidneys, liver, heart, testes, pancreas and bone[250, 251)]. Gastrointestinal absorption is about 1–10% of the ingested dose[252)]. Inhaled particulate nickel deposition is primarily determined by particle size, i.e. deepest penetration is achieved by particles 1–2 μm mass median diameter[253, 254)]. Deposited nickel oxide and sulfide particles are long-lived (> 45 days) and little if any systemic absorption occurs[255)]; however, nickel absorbed after treatment with a nickel chloride aerosol was 75% cleared by 4 days[256)]. Nickel can be absorbed percutaneously to a small extent and absorption can be enhanced by hydration[257, 258)].

Absorbed nickel is carried by the blood stream, primarily bound to serum albumin and $\alpha_1$-macroglobulin[259)]. Injected nickel chloride undergoes 2-compartment metabolism, has a body half-life of several days and is primarily excreted through the urine[260)]. Nickel is also excreted in sweat and can be deposited to a significant extent in the hair[261, 262)].

As previously alluded to, nickel chloride is relatively non-toxic when ingested orally; in some studies, rats, cats and dogs were fed up to 12 mg/kg daily for up to 200 days without adverse effects[263)]. However, large acute doses of nickel chloride may be nephrotoxic[263)] and human exposure has resulted in renal edema and parenchymatous degeneration[264, 265)]. Also, while overt toxicity is often not apparent, nickel adversely affects spermatogenesis[266)] and decreases litter size of females. Nickel also crosses the placental barrier causing both teratogenic and embryotoxic effects in experimental animals[267, 268)]. Parenteral administration seems to affect endocrine physiology, especially pancreatic function and insulin action[269, 270)]. It also inhibits release of prolactin from the pituitary and decreases iodine uptake by the thyroid[271, 272)]. Inhalation of the soluble nickel chloride salt and insoluble nickel oxide results in bronchial epithelial hyperplasia, focal proliferative pleuritis and adenomatosis[273)]. The phagocytic ability of a alveolar marcophages is inhibited by exposure to nickel[274, 275)].

An exception to the generally low toxicity of nickel is the compound nickel carbonyl which is a byproduct of the nickel purification process (known as the Mond process). Nickel carbonyl has been causally associated with a severe respiratory affliction resembling viral pneumonia. Pathological damage ranges from pulmonary hemorrhage to fibrosis. Interestingly, the lung is the target organ for nickel carbonyl toxicity regardless of the route of administration[276)].

The spectrum of nickel toxicity also includes a unique allergic response, apparently the result of complexation of nickel with other biomacromolecules in hapten-like fashion[232]. The response takes the form of atopic dermatitis, leaving recurrent, itchy, pustular vesicles[277]. This toxicity has been a special problem with stainless steel prosthetic devices which contain up to 35% nickel[232].

Little doubt presently exists that certain forms of nickel, especially the subsulfide and sulfide particles, are human carcinogens. The status of nickel in occupational and experimental carcinogenesis has been reviewed[232–236, 278–280]. In these reviews, the epidemiological data from studies of nickel refinery workers is summarized and conclusively demonstrates that these workers are at high risk for cancer of lungs, nasal cavity and stomach. Compounds implicated are particulate nickel subsulfide and nickel oxides, nickel carbonyl, and aerosols of nickel sulfate, nitrate and chloride.

In experimental animal carcinogenesis studies, nickel has been frequently implicated as a carcinogen. However, qualitative and quantitative aspects vary depending upon chemical and physical form, dose, route of exposure and animal species and strain. This literature has been extensively reviewed by others[232–236, 278–280].

The insoluble forms of nickel, such as the crystalline sulfide, oxide and subsulfide, induce malignant sarcomas at the site of injection; however, adenocarcinomas are also commonly observed[281, 282]. Amorphous nickel sulfide, however, appears to lack carcinogenicity when administered in a similar manner[282, 283]. More soluble forms of nickel yield less well defined results. Nickel acetate given intraperitoneally causes pulmonary carcinoma; inhalation or intravenous exposure to nickel carbonyl causes pulmonary carcinoma or carcinomas and sarcomas of liver and kidney. While no strong statistical correlation exists, there is a trend towards the most insoluble compounds being the most potent carcinogens[283]. Since the activity of soluble compounds implicates nickel ion as the ultimate carcinogen, it is likely that factors which govern longevity of the insoluble nickel particles, such as size, solubility, charge and other factors, which may influence the particles' ability to be internalized by target cells, affect the carcinogenic potential of the compound.

*In vitro* mammalian cell transformation experimentation has provided fairly conclusive support for the experimental animal carcinogenesis and human epidemiological studies of nickel. Several investigators using various cell lines have shown that nickel compounds can induce transformation in a dose-dependent manner with nickel subsulfide having potency equivalent to benzo(a)pyrene[284, 285]. Nickel also enhanced viral transformation[286] and potentiated benzo(a)pyrene induced transformation[287]. Interestingly, DiPaolo and Casto[285] and Costa and Mollenhauer[288] have shown that amorphous nickel sulfide, a compound that is chemically similar to nickel subsulfide but which possesses different physical structure, does not induce transformation. Subsequent studies attributed this difference to the inability of the amorphous compound to be internalized by nonphagocytic cells, a property which seems to be influenced by a combination of surface charge and solubility[289–293].

In contrast to nickel being a documented human carcinogen and a potent transforming agent in mammalian transformation assays, this compound is relatively inactive in mutagenesis assays. Nickel compounds were not mutagenic in four different bacterial mutagenesis assays: T4 bacteriaphage assay, *B. subtilis* rec assay, *E. coli* WP2 fluctuation test and the Ames *Salmonella* test[294–303]. Mammalian mutagenesis assays i.e. $C_3H$ HGPRT, L5178YTK$^{+/-}$ or V79 cell HGPRT assay, have at best identified nickel as a weak mutagen; however, there is little consistency in the mutagenic response of nickel[299–302].

Substantial evidence exists to suggest that alterations in the DNA are a necessary step in transforming a normal cell to a tumorigenic cell[303]. Thus nickel, as a potent inducer of several types of DNA lesions, may be carcinogenic via this mechanism. Treatment of embryonic muscle cells with nickel subsulfide decreased $^3$H-thymidine incorporation, indicating inhibited DNA synthesis[304]. Chinese hamster ovary cells were blocked in S-phase after treatment with nickel chloride and crystalline nickel sulfide[305]. Nickel seems to have a weak affect in several purified systems such as DNA polymerase catalyzed synthesis of poly[d(A-T)] and poly[d(C)], used to determine whether metal ions can interfere with the fidelity of nucleic acid synthesis[306]. Misincorporation of CMP during synthesis of poly[(rA-rU)] by RNA polymerase (a measure of fidelity of transcription) was increased only 1.5 times over control by nickel whereas cadmium produced 50-fold increases in misincorporation[307].

It has been shown that nickel ions can exert profound effects on RNA and DNA. Nickel ions bind to the phosphate groups and the heterocyclic bases of DNA and RNA[308]. Heating a nickel-bound RNA results in depolymerization of the RNA molecule[309]. Interestingly, *in vitro* studies demonstrated that the presence of protein decreased the equilibrium binding of nickel to DNA but increased the affinity of nickel for DNA, suggesting that a ternary nickel-protein-DNA complex may be formed[310].

DNA damage in the form of DNA-protein crosslinks and single-strand breaks was detected using alkaline sucrose gradients and alkaline elution in CHO cells treated with nickel chloride and in kidney tissue of rats treated with nickel carbonate[311, 312]. DNA repair synthesis has also been shown to be stimulated in CHO and SHE cells following treatment with $NiCl_2$ or crystalline NiS[313]. Finally, chromosomal aberrations were induced in $C_3H$ mouse mammary cells by various nickel compounds including nickel chloride and nickel sulfide[314, 315]. Weak sister chromatid exchange was noted in human lymphocytes treated with nickel; SCE was not detected in lymphocytes from nickel refinery workers[316, 317].

# 9 Other Metals

## *9.1 Cobalt*

Cobalt is clearly established as an essential nutrient to humans since it is a component of vitamin $B_{12}$. Cobalt is a relatively rare element but its usage is increasing, particularly as a component of steel alloys. Cobalt has effects on a number of organs including the pancreas, kidney, thyroid and erythropoietic system. One of its most serious toxic effects is on the heart. Cobalt was at one time used as a defoaming additive in beer, and a number of cardiac failures were reported among beer drinkers. Similar effects on the myocardium have been described in experimental animals following dietary cobalt administration.

There is no evidence from epidemiological studies thus far to suggest that cobalt is a human carcinogen, although a number of experimental studies have indicated that this metal does possess oncogenic potential. Subcutaneous administration of cobalt powder produced sarcomas in rats at the site of injection[318)], as did intramuscular administration of CoO to rats[319)]. Cobalt acetate and crystalline CoS produced a weak to moderate incidence of transformation in Syrian hamster embryo cells[320, 321)]. Cobalt was found to be a weak or ineffective inducer of DNA damage in the bacterial rec assay and did not induce mutations in E. coli, Salmonella or $T_4$ phage assays[322, 324, 325)]. It did produce a small incidence of 8-azaguanine resistant mutants in Chinese hamster V 79 cells but was far less potent than was Be[325)]. Cobalt also produced stable thymidine kinase mutants in L 5178 Y/$TK^{+/-}$ cells[326)]. $CoCl_2$ was effective in decreasing the fidelity of DNA replication[327)] and treatment of cultured CHO cells with crystalline CoS resulted in the induction of DNA single strand breaks[328)]. Cobalt did not induce chromosomal aberrations in cultured human lymphocytes[329)] but its ability to induce chromosomal aberrations *in vivo* has not been tested.

These results indicate some concern about the carcinogenic activity of cobalt, although clearly it is not well established as a carcinogen.

## *9.2 Copper*

Copper is clearly established as an essential nutrient since it is an obligatory component of several enzymes (i.e. tyrosinase, cytochrome oxidase, uricase, superoxide dismutase, and amine oxidase). Damage to the liver, kidney and the hematopoietic systems are the major toxic effects of water soluble copper salts. Inhalation of copper has also been associated with metal fume fever.

Epidemiological studies of copper miners and smelter workers have demonstrated a higher incidence of lung cancer than in control populations[330)]. However, copper metal does not induce tumors in experimental animals[331)]. In contrast, copper acetate and crystalline copper sulfide induced a moderate amount of morphological transformation[320, 321)]. Copper acetate also caused a decrease in fidelity of DNA synthesis[327)]. Crystalline CuS also caused an induction of DNA single strand breaks in cultured cells based upon analysis with alkaline sucrose gradient sedimentation[328)].

## *9.3 Manganese*

The essential nature of Mn is well established; Mn functions as a cofactor in a number of enzyme reactions. Mn is used in a number of metallurgical processes and its major toxic effects in humans is in the CNS where a Parkisonian like state develops. Inhalation of dust containing Mn is severely irritating to the respiratory system.

There is no evidence that excessive manganese exposure causes cancer in humans; however, in rats intramuscular administration of an organic manganese compound produced a statistically significant increase in fibrosarcomas at the site of injection[332]. Manganese powder or manganese oxide did not produce injection-site sarcomas[332]. In the mouse lung adenoma bioassay, manganese sulfate produced a significant increase in pulmonary adenomas compared to untreated animals[333]. Metallic manganese did not induce transformation of Syrian hamster embryo cells while $MnCl_2$ produced a weak-moderate response in this bioassay system[320, 334]. Mixed results were obtained when determining the mutagenicity of Mn in mammalian and bacterial systems[235, 323, 326, 335]. However, $MnCl_2$ did decrease the fidelity of DNA synthesis *in vitro*[327]. $MnCl_2$ at 10 and 100 μM concentration did not produce single strand breaks in the DNA of cultured mammalian cells[328].

## *9.4 Silver*

Silver is not considered an essential element for life. The major toxic effect of silver is an agyriosis or an ashen-grey discoloration of the skin. Silver foils implanted into rats induced sarcomas at the implantation site[339] however administration of silver powder did not induce tumors in rats[340]. The ability of the foil to induce tumors has been described as a smooth surface carcinogenesis phenomena since implantation of plastics and ivory produce similar results. Silver nitrate has been shown to induce a weak to moderate incidence of transformation in Syrian hamster embryo cells[320]. Silver nitrate has been shown to decrease the fidelity of DNA synthesis *in vitro*[327].

## *9.5 Vanadium*

Vanadium is well established as an essential nutrient in chickens and rats but its requirement by humans has not been tested. Deficiency of vanadium results in disturbances of lipid metabolism. Environmental exposure to vanadium comes about as a result of its use in the production of steel and non-ferrous alloys. Exposure and accumulation of vanadium in humans result primarily from inhalation of the metal which has irritant effects on the respiratory system; systemic effects on the liver, kidney, nervous system, blood and cardiovascular system has also been noted. These varied effects may be related to the ability of vanadium to interfere with lipid metabolism in these target cells.

No evidence is apparent which demonstrates vanadium compounds are carcinogenic to humans and experimental animals. Vanadium has not been tested in mutagenesis and cell transformation systems.

## 10 References

*Aluminum: 1–15*

1. Venugopal, B., Luckey, T. D.: Metal Toxicity in Mammals, Vol. 2, p. 104, New York, Plenum 1978
2. Hove, E., Elvehejm, C. A., Hart, E. B.: Am. J. Physiol. *123,* 160 (1938)
3. Berlgene, G. M., Yagil, R., Ben-Air, J., Weinberger, G., Knopf, E., Danovitch, G. M.: Lancet *1,* 564 (1972)
4. Underwood, E. J.: Trace Element in Human and Animal Nutrition, 4th ed., p. 430, New York, Academic Press 1977
5. Norseth, T.: in Handbook on the Toxicology of Metals, Friberg, L., Nordberg, G. F., Vouk, V. B. (eds.), p. 275, Amsterdam, Elsevier 1979
6. Lione, A.: Food Chem. Toxicol., *21,* 103 (1983)
7. McLaughlin, A. I., Nagantzie, G., King, E., Tease, D., Porter, R. J., Owen, R.: Br. J. Inds. Med. *19,* 253 (1962)
8. Berman, E.: Toxic Metals and Their Analysis, p. 677, London, Heyden 1980
9. Crapper, D. R., Quittkat, S., Krishman, S. S., Dalton, A. J., DeBoni, H.: Acta Neuropathologica *50,* 19 (1980)
10. Crapper, D. R., Krishnan, S. S., Dalton, A. J.: Science *108,* 511 (1973)
11. Perl, D. P., Brody, A. R.: Science *208,* 297 (1980)
12. Casto, B. C., Meyers, J., DiPaolo, J. A.: Cancer Res. *39,* 193 (1979)
13. Sirover, M. A., Loeb, L. A.: Science *194,* 1434 (1976)
14. Kanematsu, N., Hara, M., Kada, T.: Mutat. Res. *77,* 109 (1980)
15. Nashed, N.: Mutat. Res. *30,* 407 (1976)

*Arsenic: 16–30*

16. Anke, M., Groppel, B., Grun, M., Hennig, A., Meissner, D.: Arsenic 3 Spurene Lement-Symposium, p. 25, 1980
17. Fowler, B. A.: Toxicology of Environmental Arsenic in Toxicology Trace Elements (Goyer, R. A., Mehlman, M. A. ed.) Hemisphere Publishing Corp. Washington, D. C. pp. 79–122 1977
18. Frost, D. V.: The Arsenic Problems, in: Inorganic and Nutritional Aspects of Cancer (ed. Schrauzer, G. M.) p. 254, Plenum Press, N.Y. 1978
19. Fierz, U.: Dermatologica *131,* 41 (1965)
20. Tseng, W. P., Chu, H. M., How, S. W., Fong, J. M., Lin, C. S., Yeh, S. J.: J. Natl. Cancer Inst. *40,* 453 (1968)
21. Enterline, P. E., Marsh, G. M.: Am. J. Ep. *116,* 895 (1982)
22. EPA Health Assessment Document for Inorganic Arsenic
23. Petres, J., Baron, D., Hagedorn, M.: Env. Health Perspective *19,* 223 (1977)
24. Nakamuro, M., Sajato, Y.: Mut. Res. *88,* 7380 (1981)
25. Burgddorf, W. F., Kurvink, K., Cerrenka, J.: Hum. Genet. *36,* 69 (1977)
26. Mishioka, H.: Mut. Res. *31,* 185 (1975)
27. Fiscor, G., Piccolo, G.: EMS Newsl. *6,* 6 (1972)
28. Rossman, J. G., Stone, J. D., Molina, M., Troll, W.: Env. Mut. *2,* 371 (1980)
29. DiPaolo, J. A., Casto, B. C.: Cancer Res. *39,* 1008 (1979)
30. Heck, J. D., Costa, M.: Biol. Trace Element Res. *4,* 319 (1982)

*Beryllium: 31–59*

31. Reeves, A. L.: in: Handbook on the Toxicology of Metals, Friberg, L., Nordberg, G. F., Vouk, V. B. (eds.); Elsevier, p. 329, Amsterdam 1979
32. Berman, E.: Toxic Metals and Their Analysis, p. 48–53, Heyden, London 1980
33. Venugopal, B., Luckey, T. D.: Metal Toxicity in Mammals, Vol. 2, p. 43–50, Plenum Press, New York 1978
34. Andersen, O.: Environ. Health Perspectives, *47,* 239 (1983)

35. Kuschner, M.: Environ. Health Perspectives, *40,* 101 (1981)
36. Fishbein, L.: Environ. Health Perspectives, *40,* 43 (1981)
37. Vouk, V. B., Piver, W. T.: Environ. Health Perspectives, *47,* 201 (1983)
38. Wagoner, J. K., Infante, P. F., Bayliss, D. L.: Environmental Res. *21,* 15 (1980)
39. IARC Monographs on the Evaluation of the Carcinogenic Risk of Chemicals to Humans, Vol. 23, p. 143, International Agency for Research on Cancer, Lyon 1980
40. Reeves, A. L.: Arch. Environ. Health *11,* 209 (1965)
41. Vacher, J., Stoner, H. B.: Brit. J. Exp. Pathol. *49,* 315 (1968)
42. Van Cleave, C. D., Kaylor, C. T.: Arch. Ind. Health *11,* 375 (1955)
43. Mancuso, T. F.: Environ. Res. *21,* 48 (1980)
44. Doll, R., 10 others: Environ. Health Perspectives *40,* 11 (1981)
45. Gardner, L. U., Heslington, H. F.: Fed. Proceedings *5,* 221 (1946)
46. Reeves, A. L.: in: Inorganic and Nutritional Aspects of Cancer, Schrauzer, G. N. (ed.), p. 13, Plenum Press, New York 1978
47. DiPaolo, J. A., Casto, B. C.: Cancer Res. *39,* 1008 (1979)
48. Pienta, R. J., Poiley, J. A., Lebherz, W. B.: Int. J. Cancer *19,* 642 (1977)
49. Casto, B. C., Meyers, J., DiPaolo, J. A.: Cancer Res. *39,* 193 (1979)
50. Simmon, V. F.: J. Natl. Cancer Inst. *62,* 893 (1979)
51. Speck, W. T., Santella, R. M., Rosenkranz, H. S.: Mut. Res. *77,* 109 (1980)
52. Kanematsu, N., Hara, M., Kada, T.: Mut. Res. *77,* 109 (1980)
53. Miyaki, M., Akamatsu, M., Ono, J., Koyama, H.: Mut. Res. *68,* 259 (1979)
54. Mainigi, K. D., Bresnick, E.: Biochem. Pharmacol. *18,* 2003 (1969)
55. Sirover, M. A., Loeb, L. A.: Proc. Natl. Acad. Sci. USA *73,* 2331 (1976)
56. Sirover, M. A., Loeb, L. A.: Science *194,* 1434 (1976)
57. Zakour, R. A., Tkeshelashvili, L. K., Shearman, C. W., Koplitz, R. M., Loeb, L. A.: J. Cancer Res. Clin. Oncol. *99,* 187 (1981)
58. Loeb, L. A., Zakour, R. A.: in: Nucleic Acid-Metal Ion Interactions, Spiro, T. (ed.), Wiley, p. 115, New York 1980
59. Loeb, L. A., Mildvan, A. S.: in: Metal Ions in Genetic Information Transfer, Eichhorn, G. L., Marzilli, L. G. (eds.), p. 126, Elsevier/North Holland, Amsterdam 1981

*Cadmium: 60–114*

60. Singhal, R. L., Merali, F.: in: Cadmium Toxicity, Mennear, J. H. (ed.), Dekker, p. 61, New York 1979
61. Winter, H.: J. Appl. Toxicol. *2,* 61 (1982)
62. Berman, E.: Toxic Metals and Their Analysis, Heyden, p. 65, London 1980
63. Friberg, L., Kjellstrom, T., Nordberg, G., Piscator, M.: Handbook on the Toxicology of Metals, Friberg, L., Nordberg, G. F., Vouk, V. B. (eds.), p. 355, Elsevier/North Holland, Amsterdam 1979
64. Vouk, V. B., Piver, W. T.: Environ. Health Perspectives *47,* 201 (1983)
65. Lewis, G. P., Coughlin, L., Jusko, W., Hartz, S.: Lancet *1,* 291 (1972)
66. Chang, C. C., Lauwerys, R. L., Bernard, A., Roels, H., Buchet, J. P., Garvey, J. S.: Environ. Res. *23,* 422 (1980)
67. Suzuki, Y., Toda, K., Koike, S., Yoshikawa, H.: Indust. Health *19,* 223 (1981)
68. Gunn, S. A., Gould, T. C.: in: The Testis, Vol. 3, Johnson, A. D., Games, W. R., van Denmark, N. L., Academic Press, New York 1970
69. Friberg, L., Piscator, M., Nordberg, G. F., Kjellstrom, T.: Cadmium in the Environment, 2nd ed. CRC Press, Cleveland 1974
70. Rahola, T., Aaron, R.-K., Miettinen, J. K.: in: Assessment of Radioactive Contaminants in Man, AEM-SM-150/13, Proceedings Series, International Atomic Energy Agency, Unipublishers, p. 563, New York 1972
71. Kjellstrom, T.: Nord. Hyg. Tidskr. *53,* 111 (1971)
72. Margoshes, M., Vallee, B.: J. Am. Chem. Soc. *79,* 4813 (1957)
73. Kagi, J. H. R., Vallee, B. L.: J. Biol. Chem. *235,* 3460 (1960)
74. Kagi, J. H. R., Vallee, B. L.: J. Biol. Chem. *236,* 2435 (1961)
75. Piscator, M.: Nord. Hyg. Tidskr. *45,* 76 (1964)
76. Webb, M.: Biochem. Pharm. *21,* 2751 (1972)

77. Shaikh, Z. A., Smith, J. C.: Chem. Biol. Interact. *19,* 161 (1977)
78. Swerdel, M. R., Cousins, R. J.: J. Nutr. *112,* 89 (1982)
79. Gick, G. C., McCarty, K. S.: J. Biol. Chem. *257,* 9049 (1982)
80. Hildebrand, C. E., Enger, M. D.: Biochemistry *19,* 5850 (1980)
81. Onasaka, S., Cherian, G.: Toxicol. *23,* 11 (1982)
82. Chen, R. W., Whanger, P. D., Weswig, P. H.: Biochem. Med. *12,* 95 (1975)
83. Li, T. Y., Kraker, A. J., Shaw, C. F., Petering, D. H.: Proc. Natl. Acad. Sci. USA *77,* 6334
84. Koch, J., Wieglus, S., Shankara, B., Saryan, L. A., Shaw, C. F., Petering, D. H.: Biochem. J. *189,* 95 (1980)
85. Li, T. Y., Minkel, D. T., Shaw, C. F., Petering, D. H.: Biochem. J. *193,* 441 (1981)
86. Kraker, A. J., Petering, D. H.: Biol. Trace Elem. Res. *5,* 363 (1983)
87. Udom, A. O., Brady, F. O.: Biochem. J. *187,* 329 (1980)
88. Webb, M., Cain, K.: Biochem. Pharm. *31,* 137 (1982)
89. Castano, P.: Pathol. Microbiol. (Basel) *37,* 280 (1971)
90. Danielson, K. G., Ohi, S., Huang, P. C.: J. Histochem. Cytochem. *30,* 1033 (1982)
91. Potts, C. L.: Ann. Occup. Hyg. *8,* 55 (1965)
92. Kipling, M. D., Waterhouse, J. A. H.: Lancet *1,* 730 (1967)
93. Lemen, R. A., Lee, J. S., Wagnoer, J. K., Blejer, H. P.: Ann. N.Y. Acad. Sci. *271,* 273 (1976)
94. Kjellstrom, T., Friberg, L., Rahnster, B.: Environ. Health Perspec. *28,* 199 (1972)
95. Piscator, M.: Environ. Health Perspec. *40,* 107 (1981)
96. Malcom, D.: in: Cadmium Toxicity, John Mennear (eds.), Dekker, p. 173, New York 1979
97. Levy, L. S., Roe, F. J. C., Malcom, D., Kazantzis, G., Clark, J., Platt, H. S.: Ann. Occup. Hyg. *16,* 111 (1973)
98. Levy, L. S., Clack, J.: Ann. Occup. Hyg. *17,* 205 (1975)
99. Levy, L. S., Clack, J., Roe, F. J. C.: Ann. Occup. Hyg. *17,* 213 (1975)
100. Heck, J. D., Costa, M.: Biol. Trace Elem. Res. *4,* 71 (1982)
101. DiPaolo, J. A., Casto, B. C.: Cancer Res. *39,* 1008 (1979)
102. Casto, B. C., Meyers, J., DiPaolo, J. A.: Cancer Res. *39,* 193 (1979)
103. Heck, J. D., Costa, M.: Biol. Trace Elem. Res. *4,* 319 (1982)
104. Kanematsu, N., Hara, M., Kada, T.: Mut. Res. *77,* 109 (1980)
105. Tso, W.-W., Fund, W.-P.: Toxicol. Lett. *8,* 195 (1981)
106. Nishioka, H.: Mut. Res. *31,* 185 (1975)
107. Amacher, D. A., Paillet, S. C.: Mut. Res. *78,* 279 (1980)
108. Nishimura, M., Umeda, M.: Mut. Res. *54,* 246 (1978)
109. Christie, N. T., Costa, M.: Biol. Trace Elem. Res. *5,* 55 (1983)
110. Niyogi, S. K., Feldman, R. P.: Nucl. Acids Res. *9,* 2615 (1981)
111. Sirover, M. A., Loeb, L. A.: Science *194,* 1434 (1976)
112. Sina, J. F., Bean, C. L., Dysart, G. R., Taylor, V. I., Bradley, M. O.: Mut. Res. *113,* 357 (1983)
113. Eichhorn, G. L., Shin, Y. A.: J. Am. Chem. Soc. *90,* 7323 (1968)
114. Andersen, O.: Environ. Health Perspect. *47,* 239 (1983)

*Chromium: 115–162*

115. Underwood, E. J.: Trace Elements in Human and Animal Nutrition, Academic, p. 258, New York 1977
116. Prasad, A. S.: Trace Elements and Iron in Human Metabolism, Plenum, p. 3, New York 1978
117. Schwartz, K., Mertz, W.: Arch. Biochem. Biophys. *72,* 515 (1957)
118. Martz, W., Toepfer, E. W., Roginski, E. E., Polansky, M. M.: Fed. Proc. Fed. Am. Soc. Exp. Biol. *33,* 2275 (1974)
119. Langard, S., Norseth, T.: in: Handbook on the Toxicology of Metals, Friberg, L., Nordberg, G. F., Vouk, V. B. (eds.), Elsevier/North Holland, p. 383, Amsterdam 1979
120. IARC (International Agency for Research on Cancer) Evaluation of the Carcinogenic Risk of Chemicals to Humans: Some Metals and Metallic Compounds, Vol. 23, IARC, p. 205, Lyon 1980
121. Jennette, K. W.: Environ. Health Perspect. *40,* 233 (1981)

122. National Academy of Sciences: Medical and Biological Effects of Environmental Pollutants: Chromium, National Academy Press, Washington, D.C. 1974
123. Venugopal, B., Luckey, T. D.: Metal Toxicity in Mammals, Vol. 2, Plenum, p. 248, New York 1978
124. Fishbein, L.: Environ. Health Perspect. *49,* 43 (1981)
125. Sunderman, F. W. Jr.: in: Toxicology of Trace Elements, Goyer, R. A., Mehlman, M. A. (eds.), Hemisphere Publishing Corp., p. 262, Washington, D. C. 1977
126. Vouk, V. B., Piver, W. T.: Environ. Health Perspect. *47,* 201 (1983)
127. Andersen, O.: Environ. Health Perspect. *47,* 239 (1983)
128. Fishbein, L.: J. Toxicol. Environ. Health *2,* 77 (1976)
129. Mertz, W., Roginski, E. E., Reba, R. C.: Am. J. Physiol. *209,* 489 (1965)
130. Machel, W., Gregorius, F.: Public Health Rep. *63,* 1112 (1948)
131. Baetjer, A. M.: Arch. Ind. Hyg. Occup. Med. *2,* 487 (1950)
132. Baetjer, A. M.: Arch. Ind. Hyg. Occup. Med. *2,* 505 (1950)
133. Fentzel-Beyne, R. J.: Cancer Res. Clin. Oncol. *105,* 183 (1983)
134. Nettesheim, P., Hanana, Jr., M. G., Doherty, D. G., Newell, R. F., Hellman, A.: J. Natl. Cancer Inst. *47,* 1129 (1971)
135. Kazantzis, G., Lilly, L. J.: in: Handbook on the Toxicology of Metals, Friberg, L., Nordberg, G. F., Vouk, V. B. (eds.), Elsevier/North Holland, p. 237, Amsterdam 1979
136. Fradkin, A., Janoff, A., Lane, B. P., Kuscher, M.: Cancer Res. *35,* 1058 (1975)
137. DiPaolo, J. A., Casto, B. C.: Cancer Res. *39,* 1008 (1979)
138. Casto, B. C., Meyers, J., DiPaolo, J. A.: Cancer Res. *39,* 193 (1979)
139. Rivedal, E., Sanner, T.: Cancer Res. *41,* 2950 (1981)
140. Heck, J. D., Costa, M.: Biol. Trace Elem. Res. *4,* 71 (1982)
141. Heck, J. D., Costa, M.: Biol. Trace Elem. Res. *4,* 319 (1982)
142. Nishioka, H.: Mut. Res. *31,* 185 (1975)
143. Kanematsu, N., Hara, M., Kada, T.: Mut. Res. *77,* 109 (1980)
144. Petrilli, F. L., DeFlora, S.: Appl. Environ. Microbiol. *33,* 805 (1977)
145. Tso, W.-W., Fung, W.-P.: Toxicol. Lett. *8,* 195 (1981)
146. Petrilli, F. L., DeFlora, S.: Mut. Res. *54,* 139 (1978)
147. DeFlora, S.: Nature *271,* 455 (1978)
148. Nordberg, G. F., Andersen, O.: Environ. Health Perspect. *40,* 65 (1981)
149. Norseth, T.: Environ. Health Perspect. *40,* 121 (1981)
150. Newbold, R. F., Amos, J., Connell, J. R.: Mut. Res. *67,* 55 (1979)
151. Oberly, T. J., Piper, C. E., McDonald, D. S.: J. Toxicol. Environ. Health *9,* 367 (1982)
152. Levis, A. G., Bianchi, V., Tamino, G., Pegararo, B.: Brit. J. Cancer *38,* 110 (1978)
153. Sirover, M. A., Loeb, L. A.: Science *194,* 1434 (1976)
154. Tamino, G., Peretta, L., Levis, A. G.: Chem. Biol. Interact. *37,* 309 (1981)
155. Christie, N. T., Costa, M.: Biol. Trace Elem. Res. *5,* 55 (1983)
156. Robison, S. H., Cantoni, O., Costa, M.: Carcinogenesis *3,* 657 (1982)
157. Tsapakos, M. J., Hampton, T. H., Jennette, K. W.: J. Biol. Chem. *256,* 3623 (1981)
158. Umeda, M., Nishimura, M.: Mut. Res. *67,* 221 (1979)
159. Majone, F., Levis, A. G.: Mut. Res. *67,* 231 (1979)
160. Venier, P., Montaldi, A., Majone, F., Bianchi, V., Levis, A. G.: Carcinogenesis *3,* 1331 (1982)
161. Whiting, R. F., Stich, H. F., Koropatrick, D. J.: Chem. Biol. Interact. *26,* 267 (1979)
162. Nordberg, G. F., Andersen, O.: Environ. Health Perspect. *40,* 65 (1981)

*Lead: 163–220*

163. Cooper, W. C.: Ann. N.Y. Acad. Sci. *271,* 250 (1976)
164. Cooper, W. C.: XIX International Congress on Occupational Health. Dubrovnik, Yugoslavia, N.Y. 14 p. 1978
165. Cooper, W. C., Gaffey, W. R.: J. Occup. Med. *17,* 100 (1975)
166. Chrusciel, H.: Czas. Stomatol. *28,* 103 (1975)
167. Dingwall-Fordyce, I., Lane, R. E.: Br. J. Ind. Med. *20,* 313 (1963)
168. Lane, R. E.: Arch. Environ. Health *8,* 243 (1964)
169. McMichael, A. J., Johnson, H. M.: J. Occup. Med. *24,* 375 (1982)

170. Neal, P. A., Dreessen, W.-C., Edwards, T. I., Reinhart, W. H., Webster, S. H., Castberg, H. T., Fairhall, L. T.: Washington, D. C.: Government Printing Office; U.S. Public Health Bulletin No. 267, 1941
171. Nelson, D. J., Kiremidgian-Schumacher, L., Stotzky, G.: Environ. Res. *28,* 154 (1982)
172. Mahaffey, K. R., Michaelson, I. A.: The interaction between lead and nutrition. In: Needleman, H. L. (ed.), Low level lead exposure: the clinical implications of current research. New York NY: Raven Press, pp. 159–200, 1980
173. Kang, H. K., Infante, P. F., Carra, J. S.: Science *207,* 935 (1980)
174. Costa, M.: Metal Carcinogenesis Testing Principles and In Vitro Methods, Humana Press, Clifton, N.J. 1980
175. Sontag, J. M., Page, N. P., Saffiotti, U.: U.S. Department of Health, Education and Welfare, National Cancer Institute; report No. NCI-CG-Tr-1. DHEW publication no. (NIH) 76, 1976
176. Zollinger, H. U.: Virchows Arch. Pathol. Anat. Physiol. *323,* 694 (1953)
177. Boylund, E., Dukes, C. E., Grover, P. L., Mitchlev, B. C. V.: Br. J. Cancer *16,* 283 (1962)
178. Van Esch, G. J., Kroes, R.: Br. J. Cancer *23,* 765 (1969)
179. Baldwin, R. W., Cunningham, G. J., Pratt, D.: Br. J. Cancer *18,* 503 (1964)
180. Balo, J., Batjai, A., Szende, B.: Magyar Onkol. *9,* 144 (1965)
181. Haas, T., Wieck, A. G., Schaller, K. H., Mache, K., Valentin, H.: Zentralbl. Bakteriol. Parasitenkd. Infektionskr. Hyg. Abt. 1: Orig. Reihe B *155,* 341 (1972)
182. Mao, P., Molnar, J. J.: Am. J. Pathol. *50,* 571 (1967)
183. Epstein, S. S., Mantel, N.: Experientia *24,* 580 (1968)
184. Zawirska, B., Medras, J.: Zentralbl. Allg. Pathol. Pathol. Anat. *3,* 1 (1968)
185. Van Esch, G. J., Kroes, R.: Br. J. Cancer *23,* 765 (1969)
186. Van Esch, G. J., Van Gendeven, H., Vink, H. H.: Br. J. of Cancer *16,* 289 (1962)
187. Schroeder, H. A., Mitchener, M., Nason, A. P.: J. Nutr. *199,* 59 (1970)
188. Zawirska, B., Medras, K.: Arch. Immunol. Ther. Exp. *20,* 257 (1972)
189. Azar, A., Trichiomowicz, H. J., Maxfield, M. E.: Env. Health Aspects of Lead, 199 (1973)
190. Kobayashi, N., Okamoto, T.: J. Natl. Cancer Inst. *53,* 1605 (1974)
191. Furst, A., Schlauder, M., Sasmore, D. P.: Cancer Res. *36,* 1779 (1976)
192. Stoner, G. D., Shimkin, M. B., Troxell, M. C., Thompson, T. L., Terry, L. S.: Cancer Res. *36,* 1744 (1976)
193. Oyasu, R., Battifora, H. A., Clasen, R. A., McDonald, J. H., Hass, C. M.: Cancer Res. *30,* 1248 (1970)
194. DiPaolo, J. A., Nelson, R. L., Casto, B. C.: Br. J. Cancer *38,* 452 (1978)
195. Casto, B. C., Meyers, J., DiPaolo, J. A.: Cancer Res. *39,* 193 (1979)
196. Heck, J. D., Costa, M.: Biol. Trace Elem. Res. *4,* 71 (1982 a)
197. Schwanitz, G., Lehnert, G., Gebhart, E.: Dtsch. Med. Wochenschr. *95,* 1636 (1970)
198. Gath, J., Theiss, A. M.: Zentralbl. Arbeitsmed. *22,* 357 (1972)
199. Deknudt, G., Leonard, A., Ivanov, B.: Environ. Physiol. Biochem. *3,* 132 (1973)
200. Schwanitz, G., Gebhart, E., Rott, H.-D., Schaller, K.-H., Essing, H.-G., Lauer, O.: Dtsch. Med. Wochenschr. *100,* 1007 (1975)
201. Forni, A., Cambiaghi, G., Secchi, G. C.: Arch. Environ. Health *31,* 73 (1976)
202. Garza-Chapa, R., Lead, L., Garza, C. H., Molina-Ballesteros: Arch. Invest. Med. *8,* 11 (1977)
203. Deknudt, G., Manuel, T., Gerber, G. B.: J. Toxicol. Environ. Health *3,* 885 (1977 b)
204. Sarto, F., Stella, M., Acqua, A.: Med. Lav. *69,* 172 (1978)
205. Nordenson, I., Beckman, G., Beckman, L., Nordstrom, S.: Hereditas *88,* 263 (1978)
206. Forni, A., Sciame, A., Bertazzi, P. A., Alessio, L.: Arch. Environ. Health *35,* 139 (1980)
207. Bauchinger, M., Schmid, E., Schmidt, D.: Mut. Res. *16,* 407 (1972)
208. Schmid, E., Bauchinger, M., Pietruck, S., Hall, G.: Mut. Res. *16,* 401 (1972)
209. O'Riordan, M. L., Evans, H. J.: Nature (London) *247,* 50 (1974)
210. Bauchinger, M., Schmid, E., Einbrodt, H. J., Dresp, J.: Mutat. Res. *40,* 57 (1976)
211. Bulsma, J. B., De France, H. F.: Int. Arch. Occup. Env. Health *38,* 145 (1976)
212. Bauchinger, M., Dresp, J., Schmid, E., Englert, M.: Mut. Res. *56,* 75 (1977)
213. Mahaffey, K. R.: Fed. Proc. Fed. Am. Soc. Exp. Biol. *42,* 1730 (1983)
214. Beek, B., Obe, G.: Experientia *30,* 1006 (1974)
215. Deknudt, G., Deminatti, M.: Toxicology *10,* 67 (1978)

216. Deknudt, G., Colle, A., Gerber, G. B.: Mutat. Res. *45,* 77 (1977b)
217. Beek, B., Obe, G.: Humangenetik *29,* 127 (1975)
218. Heck, J. D. Costa, M.: Biol. Trace Elem. Res. *4,* 319 (1982b)
219. Sirover, M. A., Loeb, L. A.: Science *194,* 1434 (1976)
220. Hoffman, D. J., Niyogi, S. K.: Science *198,* 513 (1977)

*Mercury: 221–231*

221. Clarkson, T. W.: Handbook of the Toxicology of Metals, Friberg, L., Nordberg, D., Vouk, V. B. (eds.), Elsevier/North Holland, 99, 1979
222. Williams, M. W., Hoeschele, J. D., Turner, J. E., Jacobson, K. B., Christie, N. T., Paton, C. L., Smith, L. H., Witschi, H. R., Lee, E. H.: Toxicol. Appl. Pharmacol. *63,* 461 (1982)
223. Furst, A.: Biol. Trace Element Res. *1,* 169 (1979)
224. Casto, B. C., Meyers, J., DiPaolo, J. A.: Cancer Res. *39,* 193 (1979)
225. Kato, R., Nakamura, A., Suwai, T.: Japan J. Human Genet. *29,* 256 (1976)
226. Skervfing, S., Hanson, K., Lundsten, J.: Ar. Env. Health *21,* 133 (1970)
227. Kanematsu, N., Hara, M., Kada, T.: Mutat. Res. *77,* 109 (1980)
228. Oberly, J. J., Piper, C. E., McDonald, D. S.: J. Toxicol. Environ. Health *9,* 367 (1982)
229. Cantoni, O., Robison, S. H., Evans, R. M., Christie, N. T., Drath, D. B., Costa, M.: Fed. Proc. *42,* 1135 (1983)
230. Cantoni, O., Evans, R. M., Costa, M.: Biochem. Biophys. Res. Commun. *108,* 614 (1982)
231. Cantoni, O., Costa, M.: Molecul. Pharm. *24,* 84 (1983)

*Nickel: 232–317*

232. NAS. Nickel, National Academy of Sciences, Washington, D.C. 1975
233. NIOSH. Criteria for a Recommended Standard. Occupational Exposure to Inorganic Nickel, National Institute for Occupational Safety and Health, U.S. Govt. Printing Office, Washington, D.C. 1977
234. IARC. Some Inorganic and Organometallic Compounds, (Monographs on the Evaluation of Carcinogenic Risk of Chemicals to Man, Vol. 2), International Agency for Research on Cancer, Lyon, pp. 126, 1973
235. EPA. Scientific and Technical Assessment Report on Nickel. U.S. Environmental Protection Agency, Washington, D.C. 1975
236. Norseth, T., Piscator, M.: Nickel In: Handbook on the Toxicology of Metals, Friberg, L., Nordberg, G. F., Vouk, V. B. (eds.), Elsevier/North Holland Biomedical Press, Amsterdam 1979
237. Fishbein, L.: Sources, transport and alterations of metal compounds: An Overview. I. Arsenic, Beryllium, Cadmium, Chromium and Nickel. Environ. Health Perspect. *40,* 43 (1981)
238. Sunderman, F. W., Sunderman, Jr., F. W.: Am. Clin. Lab. Sci. *35,* 203 (1961)
239. Sladkowski, D., Schultze, H., Schaller, K. H., Lehnert, G.: Arch. Hyg. *153,* 1 (1969)
240. Schroeder, H. A., Balassa, J. J., Tipton, I. H.: J. Chron. Dis. *15,* 15 (1962)
241. Spears, J. W., Hatfield, E. E.: Feedstuffs *49,* 24 (1977)
242. Nielsen, F. H., Ollerich, D. A.: Fed. Proc. *33,* 1767 (1974)
243. Nielsen, F. H.: Trace Elem. Human Health and Disease, *2,* 379 (1976)
244. Nielsen, F. H., Sandstead, H. H.: Am. J. Clin. Nutr. *27,* 515 (1974)
245. Mertz, W.: Fed. Proc. *29,* 1482 (1970)
246. Sunderman, F. W., Jr., Selin, C. E.: Toxicol. Appl. Pharmacol. *12,* 207 (1968)
247. Armit, H. W.: J. Hygiene *8,* 565 (1908)
248. Wase, A. W., Goss, D. M., Boyd, M. J.: Arch. Biochem. Biophys. *51,* 1 (1954)
249. Smith, J. C., Hackley, B.: J. Nutr. *95,* 541 (1968)
250. Schroeder, H. A., Mitchener, M., Nason, A. P.: J. Nutr. *104,* 239 (1974)
251. Whanzer, P. D.: Toxicol. Appl. Pharmacol. *25,* 323 (1973)
252. Horak, E., Sunderman, Jr., F. W.: Toxicol. Appl. Pharmacol. *33,* 388 (1973)
253. International Commissions on Radiological Protection. Recommendation of the International Commission of Radiological Protection. Pub. No. 26 adopted Jan. 17, 1977. Pergammon Press, Oxford, England 1977

254. Natusch, D. F., Wallace, J. R., Evans, Jr., C. A.: Science *183,* 202 (1974)
255. Leslie, A. C. D., Winchester, J. W., Leysieffer, F. W., Ahlberg, M. S.: In: Trace Substances in Environmental Health-X. Hemphill, D.D. (ed.) U. of Missouri, Columbia, MO., 1976
256. Graham, J. A., Miller, F. J., Daniels, M. J., Payne, E. A., Gardner, D. E.: EnViron. Res. *16,* 77–87 (1978)
257. Spruitt, D., Mali, J. W. H., de Groot, N.: J. Invest. Dermatol. *44,* 103 (1965)
258. Samitz, M. H., Pomerantz, H.: Arch. Ind. Health *18,* 473 (1958)
259. Sunderman, F. W., Jr.: In: Clinical Chemistry and Chemical Toxicology of Metals. Brown, S. S. (ed.) Elsevier/Amsterdam, p. 231, 1977
260. Onkelin, G., Becker, J., Sunderman, Jr., F. W.: Res. Comm. Chem. Pathol. Pharmacol. *6,* 663 (1973)
261. Hohnadel, D. C., Sunderman, Jr., F. W., Nechaj, M. W., McNeely, M. D.: Clin. Chem. *19,* 1288 (1973)
262. Schroeder, H. A., Nason, A. P.: J. Invest. Dermatol. *53,* 71 (1969)
263. Stokinger, H. E.: In: Industrial Hygiene and Toxicology. Vol. II. 2nd Rev. Fassett, D. W., Irish, D. D. (eds.) Interscience, New York, p. 987, 1963
264. Gitlitz, P. H., Sunderman, Jr., F. W., Goldblatt, P. J.: Toxicol. and Appl. Pharmacol. *34,* 430 (1975)
265. Carmichael, J. L.: Arch. Ind. Hyg. Occup. Med. *8,* 143 (1953)
266. Von Weltschewer, W., Slatewa, M., Michailow, I.: Exp. Pathol. *6,* 116 (1972)
267. Sunderman, F. W., Jr., Shen, S. K., Reid, M. C., Allpass, P. R.: Teratogen., Carcinogen., Mutagen. *1,* 223 (1980)
268. Lu, C. C., Matsumoto, M., Iijima, S.: Teratology *19(2),* 137 (1979)
269. Horak, E., Sunderman, Jr., F. W.: Toxicol. Appl. Pharmacol. *32,* 316 (1975)
270. Dormer, R. L., Kerbey, A. L., McPherson, M., Manley, S., Ashcroft, S. J. H., Schofield, J. G., Randle, P. J.: Biochem. J. *140,* 135 (1973)
271. LaBella, F., Dular, R., Lemon, P., Vivian, S., Queen, G.: Nature *245,* 331 (1973)
272. Lestrovoi, A. P., Itskova, A. I., Eliseev, I. N.: Gig. Sanit. *10,* 105 (1974)
273. Port, C. D., Fenters, J. D., Ehrilich, R., Coffin, D. L., Gardner, D.: Environ. Health Perspect. *10,* 268 (1975)
274. Waters, M. D., Gardner, D. E., Coffin, D. L.: Environ. Health Perspect. *10,* 267 (1975)
275. Graham, J. A., Gardner, D. E., Waters, M. D., Coffin, D. L.: Infect. Immun. *11,* 1278 (1975)
276. Sunderman, F. W., Jr.: In: Laboratory Diagnosis of Diseases caused by Toxic Agents. Sunderman, F. W., Sunderman, F. W., Jr. (eds.), Warren H. Green, St. Louis, MO., p. 387, 1970
277. Cahan, C. D.: Brit. J. Dermatol. *76,* 384 (1956)
278. Sunderman, F. W., Jr.: Ann. Clin. Lab. Sci. *3,* 156 (1973)
279. Sunderman, F. W., Jr.: Perventive Med. *5,* 279 (1976)
280. Sunderman, F. W., Jr.: Fed. Proc. *37,* 40 (1978)
281. Sunderman, F. W., Jr., Maeza, R. M., Allpass, P. R., Mitchell, J. M., Danjanov, I., Goldblatt, P. J.: In: Inorganic and Nutritional Aspects of Cancer. Schruzer, G. N. (ed.), Plenum Publishing Corp., New York 1978
282. Gilman, G. P. W.: Cancer Res. *22,* 158 (1962)
283. Sunderman, F. W., Jr., Maenza, R. M.: Res. Commun. Chem. Pathol. Pharmacol. *14,* 319 (1976)
284. Pienta, R. J., Poiley, J. A., Lebhertz, W. N.: Int. J. Cancer *19,* 642 (1977)
285. DiPaolo, J. A., Casto, B. C.: Cancer Res. *39,* 1008 (1979)
286. Casto, B. C., Meyers, J., DiPaolo, J. A.: Cancer Res. *39,* 193 (1979)
287. Rivedal, E., Sanner, T.: Cancer Lett. *8,* 203 (1980)
288. Costa, M., Mollenhauer, H. H.: Science *209,* 515 (1980)
289. Costa, M., Nye, J. S., Sunderman, Jr., F. W., Allpass, P. R., Gondos, B.: Cancer Res. *39,* 3591 (1979)
290. Costa, M., Abbracchio, M. P., Simmons-Hansen, J.: Toxicol. Appl. Pharm. *60,* 313 (1981)
291. Heck, J. D., Costa, M.: Cancer Lett. *15,* 19 (1982)
292. Abbracchio, M. P., Heck, J. D., Costa, M.: Carcinogenesis *3,* 2 (1982)
293. Heck, J. D., Costa, M.: Toxicol. Lett. *12,* 243 (1982)
294. Corbett, T. H., Heidelberger, C., Dove, W. F.: Molec. Pharmacol. *6,* 667 (1970)
295. Nishioka, H.: Mut. Res. *53,* 207 (1972)

296. Kanematsu, N., Hara, M., Kada, T.: Mut. Res. *77,* 109 (1980)
297. Green, M., Muriel, W., Bridges, B.: Mut. Res. *38,* 33 (1976)
298. Tso, W. W., Fung, W. P.: Toxicol. Lett. *8,* 195 (1981)
299. Amacher, D. A., Paillet, S. C.: Mut. Res. *78,* 279 (1980)
300. Miyaki, M., Akamatsu, M., Ono, J., Koyama, H.: Mut. Res. *68,* 259 (1979)
301. Nishimura, M., Umeda, M.: Mut. Res. *54,* 246 (1978)
302. Costa, M., Jones, M. K., Lindgerg, O.: Amer. Chem. Soc. Symp. *140,* 45 (1980)
303. Krontiris, J. G., Cooper, G. N.: Proc. Natl. Acad. Sci. *78,* 1181 (1981)
304. Pasrur, P. K., Gilman, J. P. W.: Cancer Res. *27,* 1168 (1967)
305. Harnett, P. B., Robison, S. R., Swartzendruber, D. E., Costa, M.: Toxicol. Appl. Pharmacol. *64,* 20 (1982)
306. Sirover, M. A., Loeb, L. A.: Science *194,* 1434 (1976)
307. Niyogi, S. K., Feldman, R. P.: Nucleic Acids Res. *9,* 2515 (1981)
308. Eichhorn, G. L., Shin, T. A.: J. Amer. Chem. Soc. *90,* 7323 (1968)
309. Butzow, J. J., Eichhorn, G. L.: Biopolymers *3,* 95 (1965)
310. Lee, J. E., Ciccarelli, R. B., Jennette, K. W.: Biochem. *21,* 771 (1982)
311. Robison, S. H., Costa, M.: Cancer Lett. *15,* 35 (1982)
312. Ciccarelli, R. B., Hampton, T. H., Jennette, K. W.: Cancer Lett. *12,* 349 (1981)
313. Robison, S. H., Cantoni, O., Heck, J. D., Costa, M.: Cancer Lett. *17,* 273 (1983)
314. Nishimura, M., Umeda, M.: Mut. Res. *68,* 337 (1979)
315. Majone, F., Levis, A. G.: Mut. Res. *67,* 55 (1979)
316. Newman, S. M., Summit, R. L., Nunez, A. H.: Mut. Res. *101,* 151 (1982)
317. Waksvik, H., Boysen, M.: Mut. Res. *103,* 185 (1982)

*Other Metals*

318. Heath, J. L.: Br. J. Cancer *10,* 668 (1956)
319. Gilman, J. P. A., Ruckerbauer, G. M.: Cancer Res. *22,* 152 (1962)
320. Casto, B. C., Meyers, J., DiPaolo, J. A.: Cancer Res. *39,* 193 (1979)
321. Costa, M., Heck, J. D., Robison, S. H.: Cancer Res. *42,* 2757 (1983)
322. Kanematsu, N., Hara, M., Kada, T.: Mut. Res. *77,* 109 (1980)
323. Tso, W. W., Fung, W. P.: Toxicol. Lett. *8,* 195 (1981)
324. Corbett, T. H., Heidelberger, C., Dove, W. F.: Molec. Pharmacol. *6,* 667 (1970)
325. Miyaki, M., Akamatsu, M., Ono, J., Koyama, H.: Mut. Res. *68,* 259 (1979)
326. Amacher, D. A., Paillet, S. C.: Mut. Res. *78,* 279 (1980)
327. Sirover, M. A., Loeb, L. A.: Science *194,* 1434 (1976)
328. Robison, S. H., Cantoni, O., Costa, M.: Carcinogenesis *3,* 657 (1982)
329. Paton, G. R., Allison, A. C.: Mut. Res. *16,* 332 (1972)
330. Newman, J. A., Archer, V. E., Saciomanner, G., Kuschner, M., Auerbach, O.: Ann. N.Y. Acad. Sci. *271,* 260 (1976)
331. Sunderman, F. W., Jr., Lau, T. J., Cralley, J.: Cancer Res. *34,* 92 (1974)
332. Furst, A.: J. Natl. Cancer Inst. *60,* 1171 (1978)
333. Stoner, G. D., Shinker, M. B., Troxell, M. C., Thompson, T. L., Terry, L. S.: Cancer Res. *36,* 1744 (1976)
334. Costa, M., Simmons-Hansen, J., Bedrossian, C. W. M., Bonura, J., Caprioli, R. M.: Cancer Res. *41,* 2868 (1981)
335. Oberly, T. J., Piper, C. E., McDonald, D. S.: J. Toxicol. Environ. Health *9,* 367 (1982)
336. Nishimura, M., Umeda, M.: Mut. Res. *54,* 246 (1978)
337. Demerec, M., Hanson, J.: Cold Spring Harbor Symp. Quant. Biol. *16,* 215 (1977)
338. Orgel, A., Orgel, L. E.: J. Mol. Biol. *14,* 453 (1965)
339. Oppenheimer, B. S., Oppenheimer, F. T., Danishefsky, F., Stout, A. P.: Cancer Res. *16,* 439 (1956)
340. Furst, A., Schlauder, M. C.: J. Environ. Pathol. Toxicol. *1,* 51 (1977)

# Secondary Metabolites with Antibiotic Activity From the Primary Metabolism of Aromatic Amino Acids

**Rolf G. Werner**

Dr. Karl Thomae GmbH, Department of Biological Research, D-7950 Biberach an der Riss, West-Germany

Secondary metabolites of microorganisms play an important commercial role as antibiotics and immunostimulating or immunosuppressiv agents in human and veterinary medicine, as growth promotors in animal farms and as insecticides and herbicides in plant protection. Although to the present knowledge their importance for the producing organisms is not well understood.

For the antibiotics as an example of secondary metabolites the branching points from the precisely regulated primary metabolism of aromatic aminoacids to the secondary metabolism are demonstrated. Depending on the state of experiments, antibiotic synthesis from biosynthesis of *shikimic acid* – 2,5-dihidrophenylalanine, bacilysin-, *chorismic acid*-chloramphenicol, candicidin-, *L-phenylalanine*-neoantimycin-, *L-tyrosine*-edeine, novobiocin-, *L-tryptophan*-indolmycin, pyrrolnitrin, streptonigrin, tryptanthrin- and from the *metabolism of L-tyrosine-lincomycin-, and kynurenine*-anthramycin, 11-demethyl-tomaymycin, antimycin A, actinomycin D-, and of antimicrobial active peptides – tyrocidin, gramicidin – are reviewed. Various regulatory mechanisms of secondary metabolites are outlined. The knowledge of such biosynthetic pathways gives advices for increase of antibiotic production by feeding specific precursors, creation of deregulated or auxothrophic mutants with higher antibiotic yields and for a directed biosynthesis of new antibiotic derivatives.

As necessary for the understanding of the antibiotic biosynthesis the intermediary metabolism of aromatic amino acids and the regulation of their specific pathways in different microorganisms is summarized.



Progress in Clinical Biochemistry
and Medicine, Vol. 1

# Introduction

Antibiotics besides unusual types of linkage and rare chemical fundamental units, are composed largely from primary metabolites. For this reason, as secondary metabolites, they necessarily originate from the primary metabolism of the corresponding producer organisms.

Primary metabolism can be defined as an economic, well regulated metabolism with high specificity which is absolutely necessary for the maintenance of the essential functions of the corresponding organism. This restricts secondary metabolism so that, in our present state of knowledge, the products arising from it have no immediate function for the producer. Therefore, an error in the synthesis of secondary metabolites not necessarily has consequence for the corresponding microorganism. On the other hand, because of the highly specialized synthesis of the antibiotic producers one should look for selection advantages.

The characteristics of antibiotic producers are: spore-forming saprophytes, occurring mainly in soil – and so in a variable milieu – in contrast to Enterobacteriaceae they are not subject to such strong regulation in primary metabolism and display the ability to utilize unusual substrates. In summary, because of their abundant enzyme system they possess high metabolic flexibility.

If evolution is seen as an alternation between increasing variation by mutation and decreasing variation by selection, primary metabolism can hardly be viewed as possible creativity. On the other hand, secondary metabolism plays a particularly clear part in evolution[577)].

If the biological system is divided into five important levels for the development of the organism – intermediary metabolism, regulation, substrate transport, differentiation, morphogenesis – diversification and lack of knowledge increase in the same sequence[575, 576)]. Metabolic products which enter into one of the five levels and are absolutely necessary for metabolism, can be assigned to primary metabolism.

Compounds which are not obligatory for the minimum existence of the organism, can be assigned to secondary metabolism and permit evolutionary progress. This means that if a potent enzyme or metabolic product is formed among the nonessential metabolites, under a given selection pressure, it may be of advantage to the producer.

Studies of individual secondary metabolites for their biogenesis, their origin in intermediary metabolism, provide a rational starting point for the increase in yield of the product. Knowledge of the biosynthesis of secondary metabolites offers the possibility of directed biosynthesis through feeding the corresponding modified precursors.

Besides these more economical criteria, which justify such studies, examinations of both, biogenesis and biosynthesis give knowledge on biogenetic origin, metabolic pathways, regulation, amino acid sequence and genetic coding of enzymes involved, which provide information about evolution and taxonomy of the microorganisms investigated, as well as starting points for genetic engineering.

Because of the numerous antibiotics, the abundance of coupling steps between primary and secondary metabolism can hardly be comprehended. Consequently, in this review, the points of intersection for the resulting secondary metabolites with antibiotic activity are shown using the aromatic amino acids as an example. The regulation of secondary metabolites is described with examples.

The antibiotics are selected according to adequate experimental data, which are essential for such a study.

The intermediary metabolism of the aromatic amino acids is summarized below to the extent that is necessary for the understanding of antibiotic biosynthesis. The biosynthesis of the individual antibiotics is given according to their origin in intermediary metabolism.

# 1 Biosynthesis of Aromatic Amino Acids

In order to elucidate the biogenesis of secondary metabolites from aromatic amino acids, the more general primary metabolism leading to aromatic amino acids is dealt with first. The three aromatic amino acids, L-phenylalanine, L-tyrosine and L-tryptophan are all essential amino acids in animals and in man and cannot be synthesized de novo. In plants and microorganisms they are synthesized via the shikimic acid pathway.

## *1.1 Common Biogenesis of Aromatic Amino Acids*

The biosynthesis of the aromatic amino acids has a common origin in the condensation of phosphoenolpyruvate and D-erythrose-4-phosphate to 3-desoxy-D-arabinoheptulonic acid 7-phosphate[111, 477, 482, 483)] (Fig. 1).

This reaction is catalyzed by 3-desoxy-D-arabinoheptulonic acid-7-phosphate synthetase ①[74, 530)]. Phosphoenolpyruvate is first bound to a nucleophilic group of the enzyme, which then reacts with the second substrate[72)]. 3-Dehydroquinic acid, the precursor of 3-dehydroshikimic acid[71, 229, 537)], is produced by cyclisation, catalyzed by 3-dehydroquinic acid synthetase ②. This enzyme is present in all organisms synthesizing aromatic amino acids[371, 372)] and requires cobalt ions and NAD[485)] for activity. 3-Dehydroquinic acid is rearranged to shikimic acid by 3-dehydroquinic acid dehydratase ③ and 3-dehydroshikimic acid reductase ④[229, 482, 486, 487, 568)]. Shikimic acid-5-phosphate is produced by phosphorylation from adenosine triphosphate by means of shikimic acid kinase, and is turned into 3-enolpyruvylshikimic acid-5-phosphate by condensation with phosphoenolpyruvate through 3-enolpyruvylshikimic acid-5-phosphate-synthetase ⑤. The formation of 3-enolpyruvylshikimic acid-5-phosphate is a relatively uncommon reaction in which the enolpyruvyl moiety of phosphoenolpyruvate is transferred unchanged to the shikimic acid-5-phosphate molecule. Presumably the 3-enolpyruvylshikimate-5-phosphate synthetase reaction proceeds by a reversible addition-elimination mechanism[299, 300)]. Protonation of the C-3 of phosphoenolpyruvate facilitates electron release of the enol-ester oxygen; it is associated with a synchronous nucleophilic attack on C-2 of the substrate by the 3-hydroxyl group of shikimic acid-5-phosphate. In the second

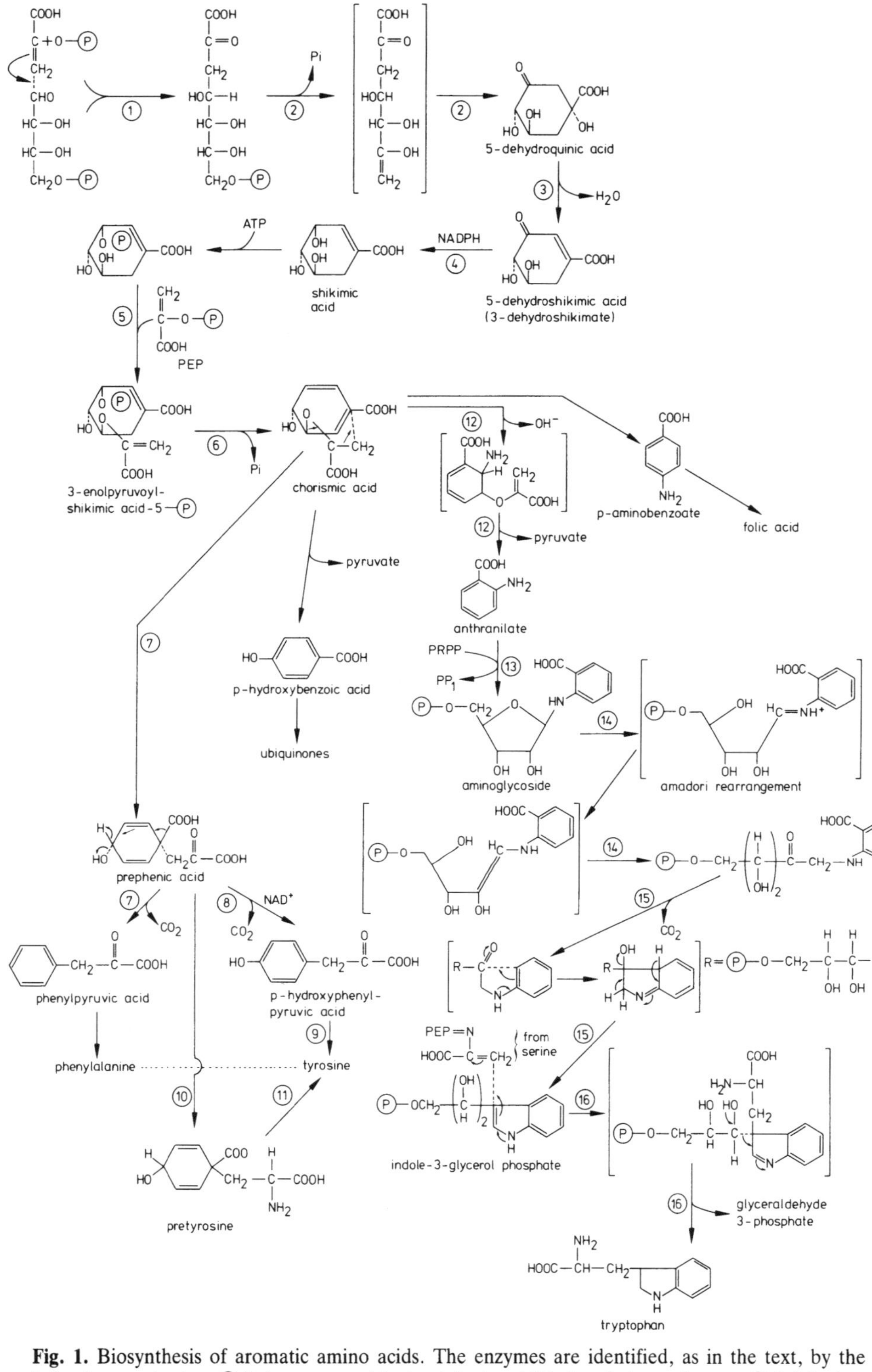


**Fig. 1.** Biosynthesis of aromatic amino acids. The enzymes are identified, as in the text, by the following numbering: ① 3-desoxy-D-arabinoheptulonic acid-7-phosphate synthetase, ② 3-dehydroquinic acid synthetase, ③ 3-dehydroquinic acid reductase, ④ 3-dehydroshikimic acid reductase, ⑤ 3-enolpyruvyl-shikimic acid 5-phosphate-synthetase, ⑥ chorismate mutase, ⑦ prephenate dehydratase, ⑧ prephenatedehydrogenase, ⑨ tyrosine transaminase, ⑩ prephenic acid transaminase, ⑪ pretyrosine dehydrogenase, ⑫ anthranilate synthetase, ⑬ 5-phosphoribosyl-pyrophosphate transferase, ⑭ N-5′-phosphoriboxylanthranilate isomerase, ⑮ indole-3-glycerophosphate-synthetase, ⑯ tryptophan synthetase

stage of the reaction, elimination of orthophosphate from the intermediate leads to 3-enolpyruvylshikimic acid-5-phosphate[33]. 3-Enolpyruvylshikimic acid-5-phosphate is converted by chorismatemutase ⑥ to chorismic acid by 1,4-conjugated elimination of phosphoric acid[111, 176, 399] (Fig. 1). Chorismic acid is the starting point for several essential metabolites, e.g. p-aminobenzoic acid and the folic acid moiety of coenzymes, the isoprenoid quinones and the three aromatic amino acids.

## *1.2 Biosynthesis of Aromatic Amino Acids from Chorismic Acid*

Two biosynthetic routes branch out from chorismic acid: the synthesis of L-phenylalanine and L-tyrosine via a common intermediate, and the synthesis of L-tryptophan from anthranilic acid as the initial metabolite.

### *1.2.1 Route of Synthesis of L-Phenylalanine and L-Tyrosine*

In *Escherichia coli, Salmonella typhimurium, Bacillus subtilis* and *Aerobacter aerogenes,* two multifunctional enzyme complexes are involved in the conversion of chorismic acid to L-phenylalanine and L-tyrosine[60–62, 70, 72, 314, 447]. The partially purified enzyme complex contains chorismate mutase ⑥ – and prephenate dehydratase ⑦ – activity and catalyzes the conversion of chorismic acid to phenylpyruvate. The two enzyme activities cannot be separated by chromatography on DEAE-cellulose. Kinetic studies on the isolated enzymes show that no direct conversion of chorismic acid to phenylpyruvate occurs. The prephenic acid formed first dissociates from the enzyme complex, accumulates in the medium and is then converted into phenylpyruvate after a lag phase. The basis for this lies in two functionally distinct active centres in the enzyme complex for chorismate mutase and prephenate dehydrogenase[449]. The enzyme complex can be dissociated into two subunits which are inactive when separated. It is assumed that association of the subunits leads to a steric configuration forming the active centres of the enzymes[268]. In contrast to *Escherichia coli, Salmonella typhimurium* and *Aerobacter aerogenes,* an independent chorismate-mutase ⑥ is present in *Neurospora crassa* which is not complexed with a prephenate dehydratase ⑦[16, 17].

Two chorismate mutases ⑥ have been demonstrated in yeasts which are both separate from prephenate-dehydratase ⑦[304]. In the further biosynthetic routes, L-phenylalanine is produced from phenylpyruvate by transamination. p-Hydroxyphenylpyruvate is formed by dehydration and simultaneous decarboxylation of prephenic acid by prephenate dehydrogenase ⑧, and L-tyrosine is produced from it by transamination with tyrosine transaminase ⑨.

As an alternative to this biosynthetic pathway, L-tyrosine is formed in *Corynebacterium glutamicum* and *Brevibacterium flavum* via pretyrosine (1-carboxy-4-hydroxy-2,5-cyclohexadien-1-yl-β-alanine)[102]. Both bacteria possess a transaminase ⑩, which converts prephenic acid into pretyrosine, and a NADP-dependent pretyrosine-dehydrogenase ⑪, which catalyzes the synthesis of L-tyrosine. The 4-hydroxyphenylpyruvate route of L-tyrosine biosynthesis cannot be detected in either organism.

### 1.2.2 Route of Synthesis of L-Tryptophan

The first enzyme reaction in the biosynthesis of L-tryptophan is the formation of anthranilic acid from chorismic acid by a glutamineaminotransferase: anthranilate synthetase ⑫[209, 305, 334, 422]. Isotope labelling experiments showed that the carbon skeleton of anthranilic acid is identical with that of shikimic acid and that the amino group of anthranilic acid is introduced at C-2 of chorismic acid[484]. The enolpyruvyl-group of chorismic acid is eliminated after protonation of the enolmethylene group[503].

Two types of anthranilate synthetase have been described for this enzyme reaction. Type I was isolated as a single component enzyme from *Bacillus subtilis*[231], *Pseudomonas sp.*[416] and *Serratia marcescens*[199, 428]. Type II is always associated with the second enzyme of L-tryptophan biosynthesis, anthranilate-5-phosphoribosylpyrophosphate transferase (PR-transferase) ⑫–⑬, and has been isolated from *Escherichia coli*[18, 207, 209, 212], *Salmonella typhimurium*[20, 381, 579] and *Aerobacter aerogenes*[123]. The type II enzyme complex can use both L-glutamine and ammonium as amino donor. However, if component I, anthranilate synthetase ⑫, is separated from component II, PR-transferase ⑬, L-glutamine cannot function as an amino donor. Since the PR-transferase stimulates the anthranilate synthetase, it can be assumed that component II is responsible for the binding of L-glutamine. The molecular weight of the multifunctional anthranilate synthetase-PR-transferase aggregate of *Salmonella typhimurium* is 280,000. A molecular weight of about 62,000 was determined for the two non-identical subunits anthranilate synthetase (component I) ⑫ and PR-transferase (component II) ⑬. It was concluded from these studies that each component itself consists of two subunits.

Kinetic studies show that chorismic acid and L-tryptophan bind to component I and that the end product inhibition of anthranilate synthetase depends on antagonism with chorismic acid binding[579].

The individual enzyme, anthranilate synthetase from *Serratia marcescens*, possesses kinetic properties comparable to those of component I from *Salmonella typhimurium*. The molecular weight of this enzyme is 141,000; it consists of a tetrameric complex of non-identical subunits with molecular weights of 60,000 and 21,000. The smaller subunit of this enzyme is responsible for binding of L-glutamine[172, 199, 428].

An enzyme complex, with a molecular weight of 240,000, has been isolated from *Neurospora crassa;* it contains N-5′-phosphoribosyl-anthranilate-isomerase ⑭ and indole-3-glycerophosphate-synthetase ⑮ in addition to anthranilate synthetase ⑫, although the three reactions catalyzed by this complex are not sequential in the biosynthesis[100, 118, 196, 335, 373, 513]. A similar threefold complex, consisting of anthranilate synthetase ⑫, N-5′-phosphoribosylanthranilate isomerase ⑭ and indole-3-glycerophosphate synthetase ⑮, was isolated from *Claviceps spec.*, SD 58. A molecular weight of 200,000 was determined for this enzyme complex in a sucrose density gradient[334]. If this complex is dissociated into a 35,000 dalton and a 165,000 dalton unit, the larger subunit no longer possesses anthranilate synthetase activity. In *Bacillus subtilis*, anthranilate synthetase ⑫ is the enzymological basis for folic acid synthesis[230] and for histidine synthesis[231]. The glutamine-binding-protein X in anthranilate synthetase ⑫ is identical with that in p-aminobenzoic acid synthetase.

Histidine stimulates L-tryptophan synthesis by increasing the activity of $Mn^{2+}$-dependent glutamine aminotransferase.

Anthranilic acid and 5-phosphoribosyl-pyrophosphate, is catalyzed by PR-transferase ⑬ to yield N-(5′-phosphoribosyl)-anthranilic acid, which is further transformed by phosphoribosyl-anthranilic acid isomerase ⑭ into 1-o-carboxyphenyl-D-ribosylamine-5-phosphate[88, 301, 534)]. The next step is an Amadori rearrangement to the desoxyribulose-derivative. This intermediate undergoes cyclization to indole-glycerophosphate with indole-3-glycerol-phosphate-synthetase ⑮. The last step in the synthesis of L-tryptophan consists in separating the glycerophosphate side chain of indole-glycerophosphate and substituting by the alanine moiety of serine in a multienzyme complex of tryptophan synthetase ⑯[79, 118, 211, 392, 561, 571)]. This $\alpha\beta_2\alpha$-multienzyme complex consists of four non-covalently bound polypeptides: two α-chains, which catalyze the aldolysis of indole-glycerophosphate by liberating glyceraldehyde[50, 174, 210)], and a dimer of two β-chains which bind pyridoxal-5′-phosphate and permit the synthesis of L-tryptophan from indole and serine[551)].

Indole, which is both a product and a substrate, normally remains bound to the enzyme in this reaction. However, under appropriate conditions, only one enzyme activity is obtained in good yield, tryptophan aldolase (TS-A) or tryptophan-synthetase (TS-B). Thus, L-tryptophan auxotrophic-mutants, lacking TS-A activity but possessing TS-B activity, can grow in the presence of indole[126, 571)]. The reaction proceeds more specifically in the complex than with the individual subunits. Thus, the $\alpha\beta_2\alpha$-complex is 100 times more active in the aldolase reaction than the α-subunit and 30 times more active in the synthesis reaction than the $\beta_2$-subunit[65, 118)].

Overall, excepting the pretyrosine pathway, the biosynthesis of the aromatic amino acids is uniform in all organisms so far investigated. The evolution reveals itself only in nuances: with anthranilate synthetase ⑫ the use of glutamine as donor of an amino group at physiological pH or ammonium at higher pH[25, 55)], or conversion of indole and indole-3-glycerophosphate by tryptophan synthetase ⑯ although no free indole is recorded under physiological conditions.

## 2 Regulation of the Biosynthesis of Aromatic Amino Acids

Regulation of metabolic processes proceeds by two enzymatic mechanisms[520)]: by control of enzyme synthesis or by regulation of enzyme activity. In branched metabolic pathways, the end product is usually controlled by repression of enzyme synthesis or by feedback inhibition of specific biosynthesis. In the common metabolic pathway, the key enzymes often exist in different forms and the end product only inhibits one of these proteins. Only a concerted inhibition by the end products finishes the biosynthesis.

The biosynthesis of aromatic amino acids is economically, but variously regulated in the individual organisms. This is demonstrated by only a small pool of free amino acids. In *Saccharomyces cerevisiae,* the tryptophan pool in the minimal medium amounts to 0.07 μM per g dry weight[101)]. This is relatively low compared with other amino acids in yeasts[476, 481)].

## 2.1 Regulation of the Common Biosynthesis of Aromatic Amino Acids

Studies on different microorganisms have shown that the 3-desoxy-D-arabinoheptulonic acid-7-phosphate synthetase (DAHP), catalyzing the condensation of D-erythrose-4-phosphate with phosphoenolpyruvate, has the greatest significance in the control of overall biosynthesis[39, 81, 123, 203, 413, 490].

In enterobacteriaceae, this enzyme occurs in three isoenzymes which are always inhibited by feedback with one specific end product[201, 463, 489]. Thus in *Escherichia coli*[40, 41, 85, 216, 217, 529–531], *Salmonella typhimurium*[131], *Neurospora crassa*[85, 87, 90, 168, 311, 405], *Claviceps paspali*[303] and *Corynebacterium glutamicum*[156, 160], the formation and activity of DAHP-synthetase (Tyr) is controlled by L-tyrosine, that of DAHP-synthetase (Phe) by L-phenylalanine and that of DAHP-synthetase (Trp) by L-tryptophan. Nevertheless, there are some exceptions to this scheme of regulation. In *Brevibacterium flavum*, DAHP-synthetase is inhibited synergistically only by L-tyrosine and L-phenylalanine. L-tryptophan has no effect[453]. The same two end products inhibit DAHP-synthetase in *Saccharomyces cerevisiae*[83, 86, 302, 304]. In *Bacillus subtilis*[218–220], *Staphylococcus epidermidis* and *Bacillus licheniformis*[377], there is but a single DAHP-synthetase which is strongly inhibited by chorismic acid or prephenic acid. This is because *Bacillus subtilis* possesses a tri-functional enzyme complex consisting of DAHP-synthetase, chorismate mutase and shikimate kinase[385, 390]. This shows that a scheme of regulation once established cannot be directly transferred to another organism and that there is a certain tolerance here in evolution.

## 2.2 Regulation of L-Phenylalanine and L-Tyrosine Pathway

The first reaction in *Escherichia coli, Aerobacter aerogenes, Salmonella typhimurium* and *Pseudomonas aeruginosa* to synthesize L-phenyl alanine and L-tyrosine from chorismic acid, is catalyzed by a bifunctional enzyme complex possessing both chorismate mutase and prephenate dehydrogenase activity[187]. In cell-free extracts of these bacteria, both L-phenylalanine and L-tyrosine display a typical feedback inhibition on this enzyme complex. The prephenate dehydrogenase is more strongly affected by both amino acids than chorismate mutase. Thus, L-phenylalanine inhibits the prephenate dehydrogenase of all strains, and, in addition, the chorismate mutase of *Aerobacter aerogenes* and *Salmonella typhimurium*[61, 62, 267, 268, 447, 449, 494, 507].

With *Salmonella typhimurium*, the sedimentation constant of the bifunctional enzyme complex increases from 5.6 S to 8.0 S on inhibition and the molecular weight increases from 109,000 to 220,000. This apparent dimerisation is dependent on the concentration of L-phenylalanine and that of the enzyme complex. At a low enzyme concentration, no dimerisation occurs in the presence of L-phenylalanine although the enzyme is inhibited. Presumably, L-phenylalanine binds and inactivates the monomeric enzyme and at a suitable enzyme concentration it dimerises with a further monomeric enzyme[448].

With *Corynebacterium glutamicum*, the prephenate dehydrogenase is not inhibited by the end product; only L-phenylalanine completely inhibits and even L-tryptophan affects the reaction. On the other hand L-tyrosine stimulates the prephenate-dehydrogenase activity and competitively overcomes the inhibition of L-phenylalanine and L-tryptophan[157]. The chorismate-mutase is inhibited by L-phenylalanine by a feedback mechanism and L-tyrosine potentiates the activity. L-tryptophan stimulates this reaction and abolishes the inhibition of the two other amino acids.

The chorismate-mutase is liable to repression by L-phenylalanine[159]. This relatively complicated scheme of regulation has been confirmed in vivo from the corresponding auxotrophic mutants[160].

With *Brevibacterium flavum*, the prephenate dehydratase is inhibited by L-phenylalanine and stimulated by L-tyrosine. The regulation of L-tyrosine biosynthesis, therefore, lies primarily in the synergistic inhibition of DAHP-synthetase by L-phenylalanine and L-tyrosine[453, 455, 498].

Besides this feedback inhibition of key enzymes in the biosynthesis of L-phenylalanine and L-tyrosine, the synthesis of these enzymes is additionally repressed in some microorganisms. Repression of chorismate mutase by L-phenylalanine and repression of prephenate dehydratase by L-tyrosine occurs in *Aerobacter aerogenes* and *Escherichia coli*[208]. With *Salmonella typhimurium*, DAHP-synthetase (tyr) and prephenate dehydratase are coupled genetically to aro F and tyr A. It has been shown with constitutive mutants that the regulator gene tyr R is coupled genetically with pyr F. The synthesis of DAHP-synthetase (phe), determined by aro G, is not regulated by the aporepressor phe R but also by tyr R. Thus, in *Salmonella typhimurium*, the same aporepressor, tyr R, represses the synthesis of both DAHP-synthetases phe and tyr[43, 128–130, 360, 478, 531].

*Bacillus subtilis* contains two chorismate mutases, neither of which are sensitive to feedback inhibition but both are subject to repression in their enzyme synthesis. In this organism, the enzymes which convert prephenic acid are liable to feedback inhibition[216, 386, 390, 425].

By contrast, with *Pseudomonas*, repression mechanisms seem to play a subordinate role in the regulation of amino acid synthesis. Thus the isoleucine and valine synthesis is not repressed in *Pseudomonas aeruginosa*[337] nor the tryptophan synthesis in *Pseudomonas putida*[63]. With regard to the synthesis of L-phenylalanine and L-tyrosine, the two end products have no effect on the synthesis of chorismate mutase, prephenate dehydratase or phenylalanine transaminase[532].

## *2.3 Regulation of the L-Tryptophan Pathway*

Like many mechanisms, the regulation of L-tryptophan biosynthesis has been best studied in *Escherichia coli*[18, 20, 81, 118, 207, 209, 511, 520] (Fig. 2). The sequence of chemical reactions is the same in other microorganisms but differences exist in the enzyme structure[118, 196, 335, 499, 513]. However, with a few exceptions such as *Claviceps paspali*[303, 334] and *Saccharomyces cerevisiae* (Fig. 3), in all cases studied, the first enzyme reaction in the synthesis of L-tryptophan from chorismic acid – anthranilate

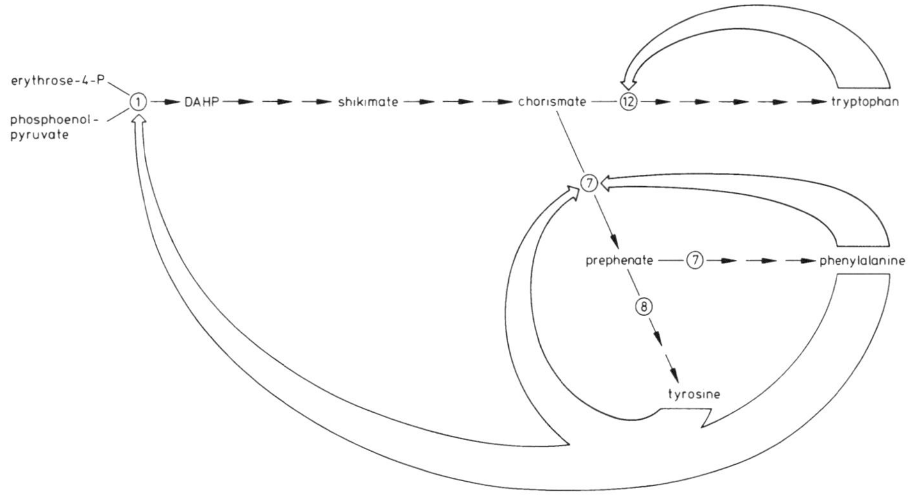


**Fig, 2.** Feedback inhibition in the biosynthesis of aromatic amino acids in *Escherichia coli.* ⇨ Inhibition. ① 3-Desoxy-D-arabinoheptulonic acid-7-phosphate synthetase, ⑦ prephenate dehydratase, ⑫ anthranilate synthetase

synthetase – is the point of attack of the feedback inhibition by L-tryptophan[18, 63, 86, 122, 304, 389, 396, 471, 578] (Figs. 4–6). This anthranilate synthetase consists of two nonidentical subunits.

In *Serratia marcescens,* one subunit catalyzes the reaction of chorismic acid and ammonium ions and possesses a binding site for L-tryptophan. The other subunit has a binding site for L-glutamine and transfers the amino group. L-tryptophan inhibits this enzyme by binding of the amino acid to the regulatory centre; a consequential

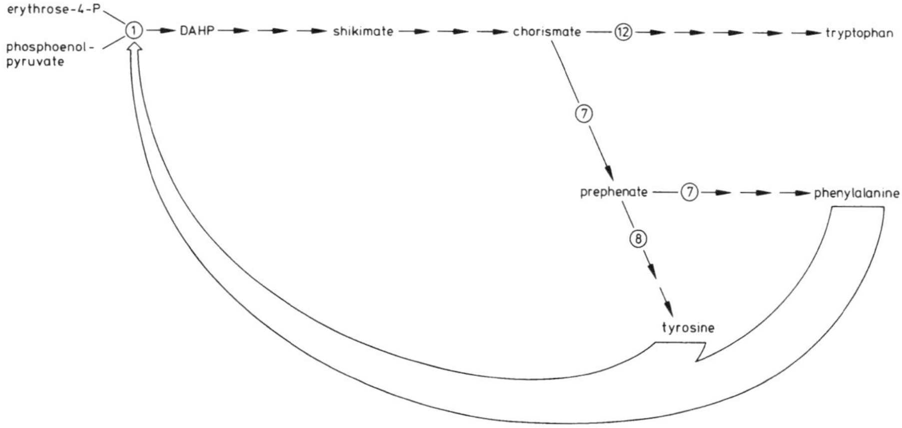


**Fig. 3.** Feedback inhibition in the biosynthesis of aromatic amino acids in *Saccharomyces cerevisiae.* ⇨ Inhibition. ① 3-Desoxy-D-arabinoheptulonic acid-7-phosphate synthetase

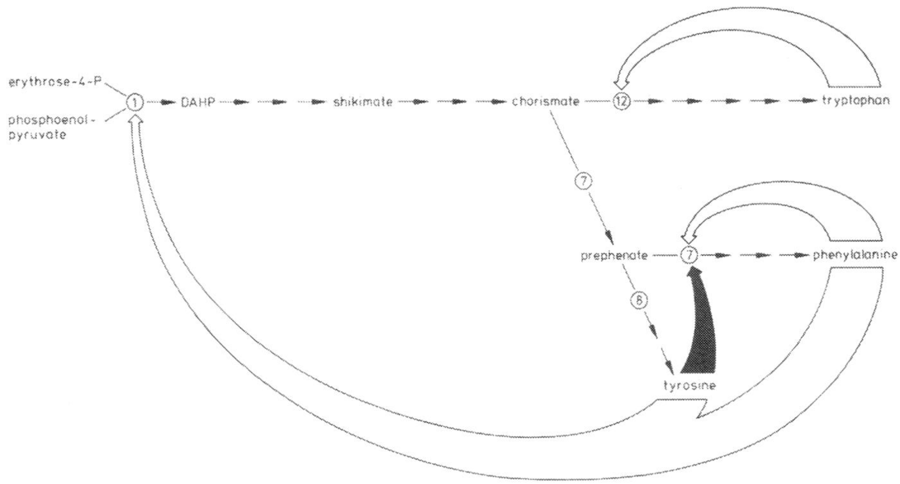


**Fig. 4.** Regulation of the biosynthesis of aromatic amino acids in *Brevibacterium flavum*. ⇨ Inhibition, ➡ activation. ① 3-Desoxy-D-arabinoheptulonic acid-7-phosphate synthetase, ⑦ prephenate dehydratase, ⑫ anthranilate synthetase

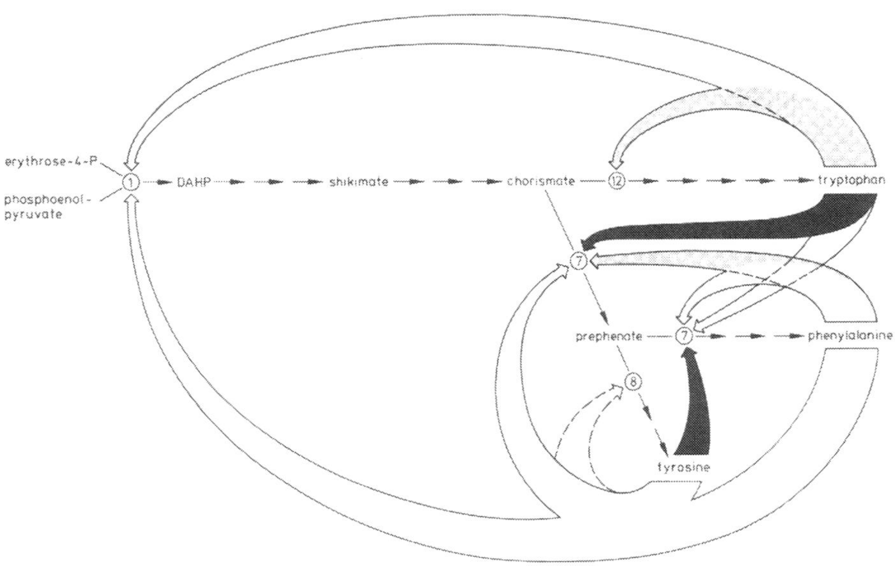


**Fig. 5.** Control of the biosynthesis of aromatic amino acids in *Corynebacterium glutamicum*. ⇨ Inhibition, ⇒ partial inhibition, ➡ activation, ⇸ repression. ① 3-Desoxy-D-arabinoheptulonic acid-7-phosphate synthetase, ⑦ prephenate dehydratase, ⑧ prephenate dehydrogenase, ⑫ anthranilate synthetase

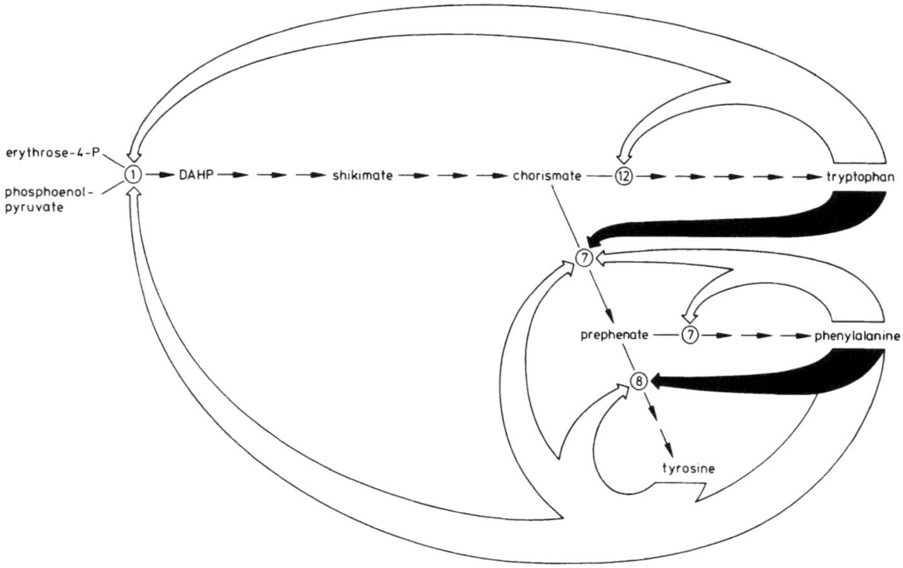


**Fig. 6.** Regulation of the biosynthesis of aromatic amino acids in *Neurospora crassa.* ⇨ Inhibition, ➞ activation. ① 3-Desoxy-D-arabinoheptulonic acid-7-phosphate synthetase, ⑦ prephenate dehydratase, ⑧ prephenate dehydrogenase, ⑫ anthranilate synthetase

conformational change has the effect that both substrates, chorismic acid and glutamine, have a low affinity[23, 578]. As far as is known, this has been maintained throughout evolution.

The conversion of chorismic acid to L-tryptophan takes place by five enzymatically controlled reactions. Again, in *Escherichia coli,* the enzymes are determined by five structural genes within the tryptophan operon[145, 204, 320, 336, 572]. The arrangement of the structural genes and enzymes is summarized in the following Table.

| Structural gene | Enzyme |
|---|---|
| trp E | Anthranilate synthetase component I |
| trp D | Anthranilate synthetase component II<br>Phosphoribosylanthranilic acid transferase |
| trp C | Phosphoribosylanthranilic acid isomerase<br>Indolylglycerophosphate synthetase |
| trp B | Tryptophan synthetase β |
| trp A | Tryptophan synthetase α |

A transcription leader region precedes this structural gene and is 160 base pairs in length[37, 479]. The promotor operator region is before the transcription leader region[28, 292]. Transcription of the operon is started by reaction of the operator with a repressor protein-L-tryptophan complex[379, 432, 433, 458, 480, 481, 584]. A DNA segment of 30 base pairs, the attenuator, or the a-site of the leader region, which is distal to the trp promotor operator, codes for termination of the trp E-polypeptide[28, 292]. In this way, a new type of regulation was discovered in *Escherichia coli* in the trp operon. In contrast to other regulation mechanisms, which are controlled by mRNA-polymerase binding, this attenuator regulates the termination of transcription in a region of the operon which precedes the structural gene[27]. However, a "metabolic regulation" is present only in the absence of the trp attenuator and of the trp repressor depending on the conditions of growth; its mechanism has not been elucidated[414, 433, 573]. These phenomena can be compared with the ribosomal synthesis. Thus L-tryptophan can suppress, in a coordinated manner, the synthesis of all enzymes of the L-tryptophan specific synthetic pathway starting from chorismic acid. Contrary to *Escherichia coli*[205], *Salmonella typhimurium*[33], *Aeromonas formicus*[65], *Bacillus subtilis*[175], *Serratia marcescens*[194] and *Staphylococcus aureus*[405], in *Brevibacterium flavum* repression by L-tryptophan is not coordinated for all the enzymes[489].

# 3 Metabolism of Aromatic Amino Acids

In this section, particular consideration will be given to the catabolic reactions from which metabolites arise, which in turn act as the starting points for biosynthesis of antibiotics.

Since a series of diseases depend on an inborn error in the metabolism of these amino acids, the catabolism of aromatic amino acids has often been studied more in higher organisms than in microorganisms.

## 3.1 Metabolism of L-Phenylalanine

### 3.1.1 Conversion of L-Phenylalanine to L-Tyrosine

In higher animals, a large part of L-phenylalanine is converted into L-tyrosine in the liver[242–245, 517, 540]. Some microorganisms possess a phenylalanine hydroxylase ⑰ for this reaction step[150, 152, 369]. This phenylalanine hydroxylase system requires atmospheric oxygen; the incorporation of $^{18}O$ into the phenyl ring of tyrosine infers an oxygenase mechanism[245, 502]. An inducible phenylalanine hydroxylase was demonstrated in *Pseudomonas* when the bacteria were grown in media containing phenylalanine or tyrosine but no asparagine[151]. Usually, phenylalanine and tyrosine are formed in microorganisms via separate biosynthetic pathways.

### 3.1.2 Transamination of L-Phenylalanine

The first step in the metabolism of L-phenylalanine is the transamination to phenylpyruvate. Phenylacetyl-L-glutamine is produced from phenylacetic acid and glutamine via the intermediates phenylacetyl-coenzyme A and phenylacetyl-adenylate[493, 562].

Phenylacetaldehyde is formed by decarboxylation of phenylpyruvate[452]. Benzoic acid arises through a transamination of phenylpyruvate, followed by an oxidative cleavage to oxalate and benzaldehyde, and oxidation of the latter. Above pH 8, non-enzymatic cleavage of p-hydroxyphenylpyruvate and phenylpyruvate occurs to yield the corresponding aldehyde and oxalate[412].

### 3.1.3 Precursors of Alkaloids

Phenylalanine and tyrosine are precursors of a large number of alkaloids such as pellotine[19], papaverine, morphine, codeine and thebaine[380]. It is assumed that norlaudanosine is condensed from 3,4-dihydroxyphenylethylamine and 3,4-dihydroxyphenylacetaldehyde. Norlaudanosine is a precursor of papaverine and of morphine alkaloids[146].

## 3.2 Metabolism of L-Tyrosine

### 3.2.1 Oxidative Breakdown of L-Tyrosine

L-tyrosine is transaminated into p-hydroxyphenylpyruvate by glutamate-L-tyrosine transaminase ⑱, an obligatory step in tyrosine oxidation[286, 287, 296, 382, 445]. This transaminase is induced by tyrosine[265, 580]. The copper-containing p-hydroxyphenylpyruvate oxidase requires ascorbic acid to activate and hydroxylate the phenyl ring; by migration of the aliphatic side chain to the vicinal position in the ring and oxidative decarboxylation of the pyruvate moiety homogentisic acid is synthesized[155, 261, 285, 287, 296, 451, 515, 548–550]. 1 Mole molecular oxygen is required for hydroxylation of the aromatic ring. The mechanism of this monooxygenase[574] probably proceeds by reaction between the substrate and the oxygen-copper-ion-enzyme complex giving a cyclic peroxide intermediate[557]. This produces homogentisic acid by decarboxylation and migration of the sidechain[57]. Further conversion of homogentisic acid, with fission of the aromatic ring, results in maleylacetoacetic acid, which is transformed into the corresponding trans-isomer by maleylacetoacetic acid isomerase ⑲.

This isomerase is activated by glutathione, γ-glutamylcysteine, cysteinylglycine and cysteine[91, 263, 264, 283, 284, 500, 506]. Fumaric acid and acetoacetic acid are produced by cleavage of fumarylacetoacetic acid with an acylpyruvase ⑳[423].

### 3.2.2 Biosynthesis of the Catecholamines

Catecholamines enhance the reactivity or amount of specific enzymes which affect the activity of a large number of human cells in the liver, heart, brain and smooth muscles[7, 374].

The biogenesis of the catecholamines begins with L-tyrosine, which is oxidized to L-3,4-dihydroxyphenylalanine (L-dopa) by L-tyrosinase ㉑, the rate limiting step in catecholamine metabolism[132, 255, 293, 383, 518, 519]. This enzyme is activated by tetrahydropteridine, iron ions and oxygen and possesses a high specificity for L-tyrosine[69, 149].

The mechanism is that of a mixed function oxidase. First, an oxidized form of the enzyme is reduced with 2-amino-4-hydroxy-6,7-dimethyltetrahydropteridine; then the oxidized pteridine is dissociated followed by aerobically oxidation of L-tyrosine to L-dopa and the oxidized form of the enzyme[201]. Formation of dopamine from L-dopa is controlled by L-dopa decarboxylase ㉒. This enzyme requires pyridoxal phosphate and has broad substrate specificity although it has only slight specificity with regard to functional groups.

Because of the broad profile of action with respect to decarboxylation of L-dopa, L-5-hydroxytryptophan, L-phenylalanine, L-tyrosine, L-tryptophan and L-histidine, this enzyme is generally known as decarboxylase of aromatic amino acids.

The further β-hydroxylation of dopamine to noradrenaline is catalyzed by dopamine-β-hydroxylase ㉓[115]. This enzyme, also a mixed function oxidase, catalyzes the β-hydroxylation of a few other phenylethylamines[66, 127, 298]. The necessary two moles of copper ions per mole enzyme undergo a cyclic reduction and oxidation during the β-hydroxylation[245, 246]. As the last step in the biosynthesis of adrenaline, phenylethanolamine-N-methyltransferase ㉔ catalyzes the transfer of the methyl group from S-adenosyl-methionine to noradrenaline[255].

Because of the low specificity of some enzymes participating in this biosynthesis, alternative biosynthetic pathways are known for catecholamines[374].

### *3.2.3 Phenol Formation*

L-Tyrosine can be broken down to phenol, by fission of the side chain; in microorganisms the enzyme L-tyrosine-phenol lyase ㉕ depends on pyridoxal phosphate[38, 279] to form pyruvate and ammonia. Tyrosine-phenol lyase is an enzyme with broad substrate specificity and, besides this reaction, it catalyzes a series of α,β-elimination, β-replacement and racemisation reactions[278, 279]. In fact, L-tyrosine can be synthesized in the reverse reaction from ammonia, pyruvate and phenol[564].

### *3.2.4 Melanogenesis from L-Tyrosine*

The first step in the biosynthesis of melanin is the hydroxylation of L-tyrosine to L-3,4-dihydroxyphenylalanine (L-dopa) by the copper-containing enzyme tyrosinase ㉑. L-Dopa is further oxidized to L-dopaquinone. In fungi, L-dopa does not seem to be a free metabolite. L-Cyclodopa is formed by spontaneous cyclization of L-dopaquinone[563] and is then rearranged to the intermediate 5,6-dihydroxyindole by oxidation and decarboxylation. This dopamelanin polymerises to a polyquinone. 4,7-linked structures[21, 22, 357] and 3,7-linked polymers[45] can arise in this way. No uniform polymerisation has been demonstrated[254]. The only enzyme known so far in this biosynthesis is tyrosinase which triggers the starting reaction in melanin formation[251]. The later, spontaneous reactions proceed in a relatively non-specific manner.

A suitable scheme of biosynthesis has been summarized by various authors[355–358, 420, 421].

Total or partial absence of tyrosinase is responsible for albinism in man and animals.

The metabolism of phenylalanine and tyrosine is summarized in Fig. 7.

## *3.3 Metabolism of* L-*Tryptophan*

Important metabolites for the organism are produced from L-tryptophan. These include nicotinic acid and various nicotinamide coenzymes, serotonin – a potent vasoconstrictor in nerve tissue – as well as ommochromes, eye pigments in certain insects[68]. The metabolism of L-tryptophan is comparable in various organisms. So far, the oxidative breakdown of L-tryptophan to kynurenine and 5-hydroxytryptophan, both central intermediates for essential metabolites in individual organisms has been found without variations.

### *3.3.1 Oxidative Breakdown of* L-*Tryptophan*

The first step in the oxidative breakdown of L-tryptophan is catalyzed by 1-tryptophanpyrrolase ㉖ and leads to N-formyl-kynurenine by introducing two oxygen atoms into the pyrrole ring[266, 275]. Further hydrolysis of this intermediate by kynurenine formylase ㉗ leads to the key intermediate, kynurenine[35, 171, 214, 274, 361, 362], from which alternative metabolic pathways diverge.

#### 3.3.1.1 Anthranilic Acid Route

The side chain of L-kynurenine is split off by the action of kynureninase ㉘ to produce L-alanine and anthranilic acid. This reaction depends on pyridoxal phosphate. As well as in *Escherichia coli* and *Bacillus subtilis,* kynureninase ㉘ is widely distributed[67, 262, 554–556]. Besides kynurenine, kynureninase ㉘ splits L-formylkynurenine to L-alanine and L-formylanthranilic acid, and 3-hydroxykynurenine to L-alanine and 3-hydroxyanthranilic acid[215]. In *Neurospora,* tryptophan is preferably converted to anthranilic acid via formylkynurenine[213].

#### 3.3.1.2 Nicotinamide-Nucleotide Route

Another important metabolic pathway starting from tryptophan also branches from L-kynurenine. This is first hydroxylated by a mixed function enzyme, L-kynurenine-3-hydroxylase ㉙[73]. Atmospheric oxygen is employed for the oxidation[438]. 3-Hydroxykynurenine is split into 3-hydroxyanthranilic acid and alanine by kynureninase ㉘.

The complete oxidation of the aromatic ring of 3-hydroxyanthranilic acid proceeds via α-amino-β-carboxy-*cis-cis*-muconic acid semialdehyde and glutaryl-coenzyme A.

The aldehyde of 3-hydroxyanthranilic acid is formed by ring fission next to the hydroxyl group. The reaction is catalyzed by 3-hydroxyanthranilic acid oxidase ㉚, a mixed function enzyme requiring iron for oxidation[170, 495].

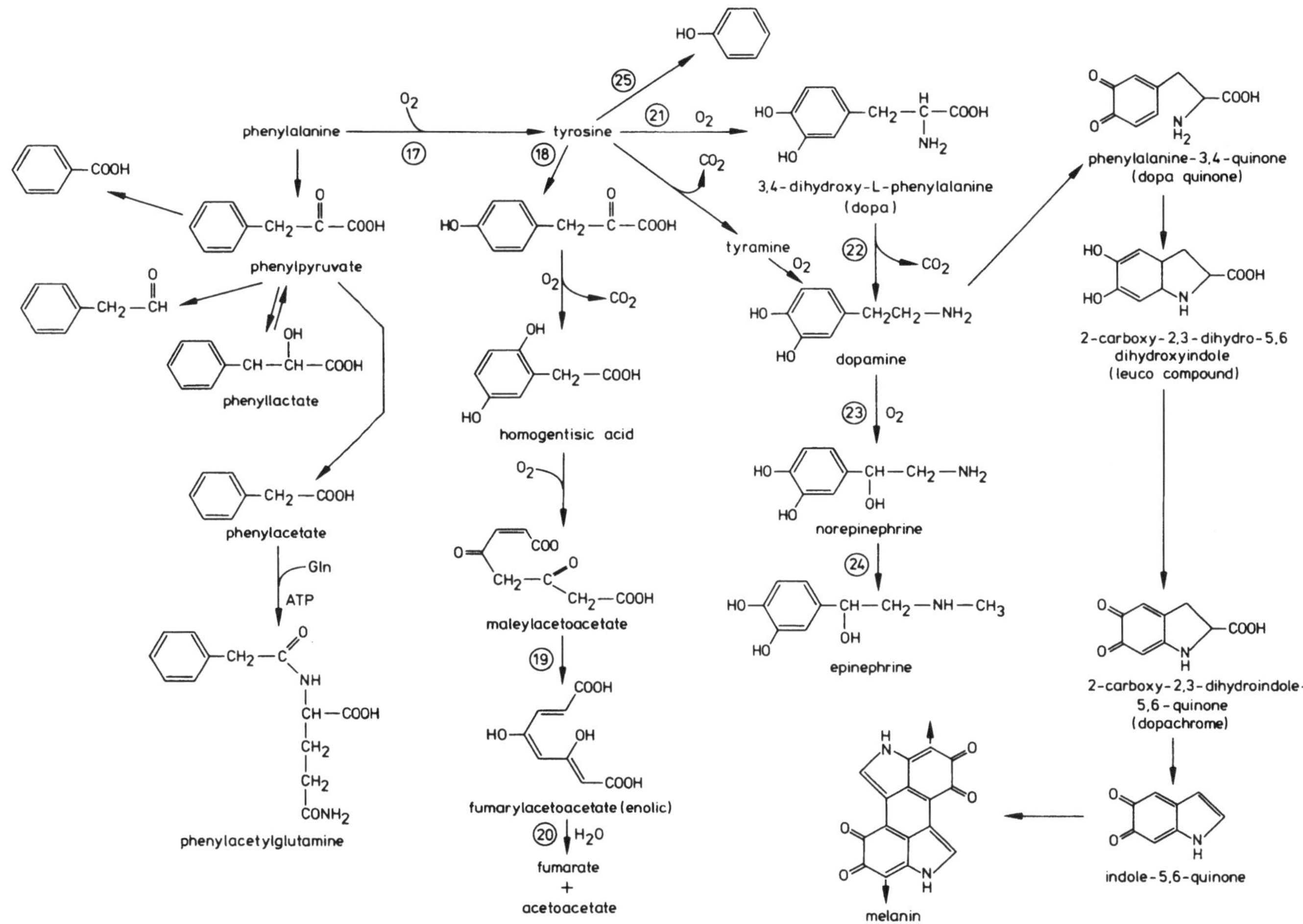


**Fig. 7.** Metabolism of phenylalanine and tyrosine, ⑰ Phenylalanine hydroxylase, ⑱ glutamate-L-tyrosine transaminase, ⑲ maleyacetoacetic acid isomerase, ⑳ acylpyruvase, ㉑ L-tyrosinase, ㉒ L-dopa-decarboxylase, ㉓ dopamine-β- hydroxylase, ㉔ phenylethanolamine-N-methyltransferase, ㉕ L-tyrosine-phenol lyase

α-Amino-β-carboxy-*cis-cis*-muconic acid semialdehyde is the starting material for synthesis of the pyridine ring of nicotinamide nucleotides[57, 202, 394)]. Quinolinic acid is synthesized in this way by non-enzymatic decarboxylation and recyclisation, and then picolinic acid by enzymatic reaction. Picolinic acid is also formed via α-amino-β-carboxy-*cis-cis*-muconic acid semialdehyde. This means that the synthesis of niacin is competitive with two other reactions[363)]. Quinolinic acid transphosphoribosylase ㉛ cataylzes the synthesis of niacin ribonucleotide from quinolinic acid and 5-phosphoribosyl-1-phosphate[58, 120)]. The intermediate, quinolinic acid ribonucleotide, seems to be enzyme-bound[384)].

The biosynthesis of nicotinic acid is not entirely uniform. Thus, *Escherichia coli* and *Bacillus subtilis* possess no kynureninase activity. Therefore, radioactive indole or tryptophan is incorporated only very slowly into nicotinic acid in these bacteria[569, 570)]. In the corresponding auxotrophic mutants, neither kynurenine nor 5-hydroxyanthranilic acid acts as a growth factor[526)]. Glycerol and succinic acid are much more efficiently incorporated into nicotinic acid in these strains. The carboxyl group of nicotinic acid is derived from succinic acid; the methylene carbon atom of succinic acid and C-1 and C-3 of glycerol are incorporated into the pyridine ring[401)].

#### 3.3.1.3 Further Metabolites from Oxidative Breakdown

Besides catabolism of tryptophan via kynurenine and anthranilic acid to catechol, *cis-cis*-muconic acid and β-ketoadipic acid, from which succinic acid and acetyl-coenzyme A are formed finally[200, 202, 488, 497, 507)], the transamination of kynurenine widely occurs. The corresponding ketoacid is formed by the action of kynurenine transaminase, and cyclizes spontaneously to kynureninic acid. Analogous transformation of hydroxykynurenine yields the corresponding xanthurenic acid. Pseudomonas strains are able to metabolize kynureninic acid via oxidative breakdown to α-ketoglutarate, oxalacetate and ammonia.

### *3.3.2 Serotonin Pathway*

Hydroxylation of L-tryptophan to L-5′-hydroxytryptophan is the first and probably rate determining step in serotonin biosynthesis. L-Tryptophan hydroxylase ㉜ is analogous to L-phenylalanine hydroxylase and requires pteridine and molecular oxygen[69, 149)]. L-5′-hydroxytryptophan is converted by L-5′-hydroxytryptophan decarboxylase ㉝ to 1,5′-hydroxytryptamine, serotonin[516)]. In *Chromobacterium violaceum,* a characteristic pigment, violacein, is formed from L-5′-hydroxytryptophan[22, 370)].

### *3.3.3 Tryptophanase Reaction*

In a few bacteria, L-tryptophan is broken down by the tryptophanase reaction ㉞ to indole, pyruvate and ammonium ions[227, 228, 313, 376–378)]. Pyridoxal phosphate serves as the coenzyme. This pyridoxal phosphate forms a pyridoxal phosphate-tryptophan-metalloenzyme complex with the substrate[365)], analogous to the tyrosine-phenol lyase reaction.

Indole and pyruvate are formed by α,β-elimination and fission of the pyridoxal phosphate and ammonium ion. This enzyme has a low substrate specificity and catalyzes a series of other α,β-eliminations and β-replacement reactions[391, 392]. Indigo, indirubin, indoxyl and indican are further metabolites of indole[391].

The metabolism of tryptophan is summarized in Fig. 8.

Studies with different 5-fluorotryptophan-resistant mutants of *Brevibacterium flavum* have shown that the five enzymes, anthranilate-synthetase, anthranilate-phosphoribosylpyrophosphate transferase, N-5-phosphoribosylanthranilate isomerase, indole-3-glycerophosphate synthetase and tryptophan synthetase A, B can be repressed independently[456].

The tryptophan enzymes of *Acinetobacter calcoaceticus* can be split into three independently repressible groups[56].

*Chromobacterium violaceum* seems to have no effect on the synthesis of tryptophan-specific enzymes[535].

In *Streptomyces coelicolor,* the synthesis of all L-tryptophan-specific enzymes is regulated with the exception of indole-3-glycerophosphate synthetase[466].

*Corynebacterium glutamicum* possesses an enzyme repression for anthranilate synthetase[158].

Such regulation mechanisms, depending on repression, can be altered by mutation, which can be inferred from the example of *S. dysenteriae. Escherichia coli, Salmonella typhimurium* and *Serratia marcescens* are completely homologous with respect to tryptophan synthetase enzyme, the RNA-polymerase and tryptophan repressor are interchangeable[336]. In the nucleotide sequence of the promoter operator region of the tryptophan operon, *S. dysenteriae* displays a difference of only two base pairs. This results in a 10-fold reduction in tryptophan promoter function so that the remaining transcription proceeds practically constitutively[368].

The regulation of aromatic amino acids illustrates that, with few exceptions, the sequential build-up of L-phenylalanine, L-tyrosine and L-tryptophan is common in the organisms but, in complete contrast, the regulation of the synthetic pathways is very heterogenous. This heterogeneity – with difficulty demonstrable in the poorly studied actinomycetes – in producers of secondary metabolites may permit a partial deregulation of primary metabolism, thus forming excess intermediates for secondary metabolism.

# 4 Metabolic Products with Antibiotic Activity

Besides the aromatic amino acids and their breakdown products with their central role for maintenance of bacterial metabolism, a series of metabolites with antibiotic activity is known; their biogenesis originates from these amino acids, although their function for the cell is not always obvious.

The biosynthesis of these antibiotics is considered below according to their origin from the appropriate amino acid, the end product or their metabolites.

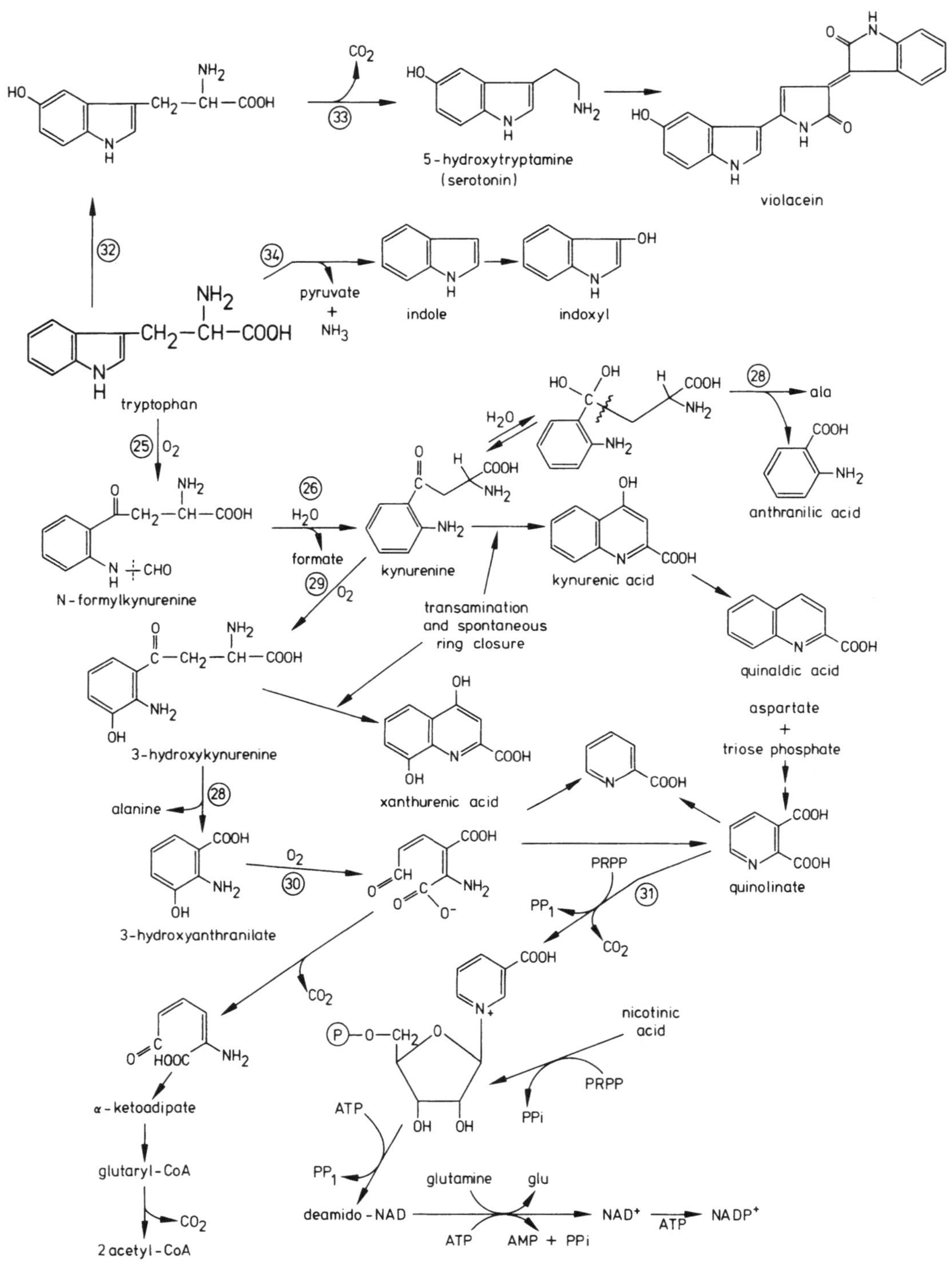


**Fig. 8.** Metabolism of tryptophan. ㉖ Tryptophan pyrrolase, ㉗ kynureneineformylase, ㉘ kynureninase, ㉙ L-kynurenine-3-hydroxylase, ㉚ 3-hydroxyanthranilic acid oxidase, ㉛ quinolinic acid transphosphoribosylase, ㉜ L-tryptophan hydroxylase, ㉝ L-5'-hydroxytryptophan decarboxylase, ㉞ tryptophanase

## 4.1 Antibiotics from Shikimic Acid

### 4.1.1 Biosynthesis of 2,5-Dihydrophenylalanine

L-2,5-Dihydrophenylalanine (DHPA) is a relatively widely distributed metabolic product in actinomycetes and has been isolated by different groups[103, 104, 443, 566]. DHPA acts as an antagonist to phenylalanine both in rats[468] and in microorganisms[443, 566] and is also an inhibitor of tryptophan-hydroxylase. The antibacterial activity of DHPA in *Escherichia coli* is the result of a false feedback inhibition of prephenate dehydratase and the phenylalanine-sensitive DAHP synthetase[104]. The inhibition can be abolished competitively by tyrosine and non-competitively by phenylalanine[103, 104].

The poor incorporation of phenylalanine (0.02%) and the good incorporation of (U-$^{14}$C)-shikimic acid (2.5%) signifies that biosynthesis does not proceed by reduction of phenylalanine but is a new variant of shikimic acid metabolism[443]. Incorporation experiments in the biosynthesis of 2,5-dihydrophenylalanine in *Streptomyces arenae* TÜ 109 show that shikimic acid only provides the ring carbon atoms of DHPA, that the side-chain is substituted at C-1 of shikimic acid, and that the asymmetry of the ring is maintained. Labelled chorismic acid and prephenic acid, but not L-(3-$^{14}$C)-serine, are incorporated into DHPA. 5,6-Dihydro-(4-$^{3}$H)-prephenic acid is not an intermediate in DHPA biosynthesis. This indicates that DHPA is biosynthesized from shikimic acid via chorismic acid and prephenic acid. The reaction sequence for transforming prephenic acid to DHPA is assumed to be an allyl rearrangement, 1,4-reduction of a conjugated dione and a combined decarboxylation/dehydration. The biosynthesis can, therefore, be shown as follows[108, 457].

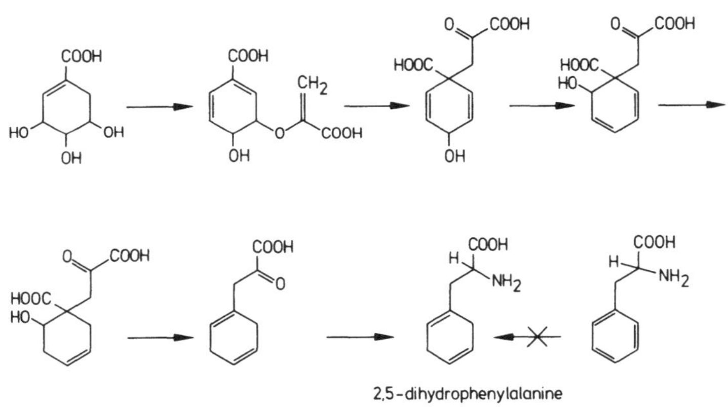


**Fig. 9.** Biosynthesis of 2,5-dihydrophenylalanine

### 4.1.2 Biosynthesis of Bacilysin

Bacilysin is a dipeptide produced by *Bacillus subtilis* A 14 and decomposes on acid hydrolysis to give an N-terminal L-alanine and tyrosine[429, 430]. This antibiotic acts by lysis of staphylococci.

Growth is biphasic in a chemically defined medium; after consumption of 50% of the added glucose in the second phase, bacilysin is produced during reduced growth. Although very little antibiotic is produced in the stationary phase, biosynthesis is not coupled with protein synthesis. This can be proved by the fact that synthesis of bacilysin is not affected by chloramphenicol.

Only if chloramphenicol is added before the start of bacilysin synthesis the production of the necessary enzymes is inhibited. The statement that mechanisms differing from those for protein synthesis are responsible for synthesis of peptide antibiotics is true for most peptide antibiotics.

As expected, $^{14}$C-alanine is incorporated into the N-terminus of the molecule. However, under the same conditions, DL-(2-$^{14}$C)-tyrosine, 1-(U-$^{14}$C)-tyrosine and (1-$^{14}$C)-acetate are not incorporated into the antibiotic although the labelled compounds penetrate into the bacterial cells. This proves that neither acetate nor tyrosine are direct precursors of the tyrosine moiety of bacilysin. However, since (1,6 ring-$^{14}$C)-shikimic acid is significantly incorporated into the tyrosine fragment, the branching of bacilysin biosynthesis from the metabolism of aromatic amino acids seems to lie between shikimic acid and tyrosine[431]. It is not yet known whether 3-enolpyruvyl-shikimate-5-phosphate, chorismic acid and prephenic acid, intermediates in tyrosine biosynthesis, are also incorporated.

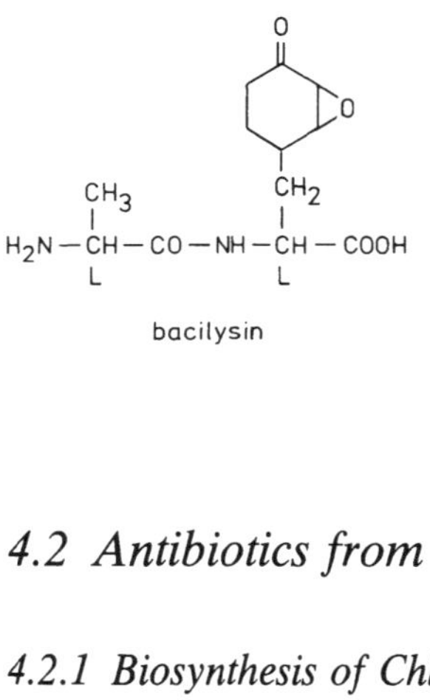


**Fig. 10.** Bacilysin, a dipeptide antibiotic consisting of L-alanine and the active molecule anticapsin

## 4.2 Antibiotics from Chorismic Acid

### 4.2.1 Biosynthesis of Chloramphenicol

Chloramphenicol is produced by a series of actinomycetes, such as *Streptomyces venezuelae*[93, 137], *Streptomyces sp.* 3022, *Streptomyces phaeochromogenes* var. *chloromyceticus, Streptomyces omiyaensis*[522] and *Streptosporangium viridogriseum* var. *kofuense*[504]. Corynecines, antibiotics closely related to chloramphenicol, are produced by *Corynebacterium hydrocarboblastus*[501, 459].

Chloramphenicol contains two asymmetric carbon atoms and therefore four stereoisomers are possible. The D(-)threo form is the active component, L(+)threo-chloramphenicol has only weak activity; the two erythro-isomers are inactive.

Chloramphenicol is a broad spectrum antibiotic with bacteriostatic activity, its mode of action is based on inhibition of protein synthesis[161, 162]. At a concentration of 50 μg/ml, it inhibits protein synthesis in *Streptomyces griseus*[344], but even at higher concentrations, it has no effect on the growth of chloramphenicol producing Streptomycetes during the production phase[333].

Chloramphenicol binds to the 50 S subunit of the ribosome, prevents the binding of amino-acyl-tRNA by a conformational change of the protein and inhibits the activity of peptidyltransferase[328, 409, 410].

The biosynthesis of chloramphenicol by *Streptomyces venezuelae* in a chemically defined medium is stimulated by glucose as a good carbon source for growth and antibiotic formation, as well as by glycerol and lactic acid, which promote a higher yield in some media[140], by nitrate and particularly by phenylalanine[138, 139, 464].

In complex nutrient media, chloramphenicol production is biphasic while it is more associated with the growth phase in defined media[330]. The basis for depression of antibiotic production in a defined medium may be growth limitation or deficiency of an antibiotic synthetase repressor at the start of fermentation.

Despite the structural similarity between phenylalanine and the p-nitrophenylserinol part of chloramphenicol, the stimulating effect of L-phenylalanine does not depend on incorporation of phenylalanine into the p-nitrophenylserinol moiety of the antibiotic[139]. Similarly, phenyllactic acid and p-hydroxyphenyllactic acid are not incorporated into chloramphenicol. By contrast, p-aminophenylalanine[222, 460] and threo-p-aminophenylserine[321] are incorporated selectively into chloramphenicol. p-Aminophenylalanine was also isolated from the culture filtrate of *Streptomyces sp.* 3022, which points to its significance for chloramphenicol synthesis[322].

Although labelled chorismic acid and prephenic acid are not accepted by the producer strains for chloramphenicol synthesis, (U-$^{14}$C)-shikimic acid, without previous breakdown, is incorporated into the aromatic ring of chloramphenicol[525]. Glycerol, the main carbon source in the medium, is incorporated into all parts of the molecule. In fact, three neighboring C-atoms from glycerol can be demonstrated in the serinol moiety[400].

Because of the disadvantage that chorismic acid and prephenic acid do not penetrate into the mycelium, a statement about the branch point in aromatic metabolism in vivo is difficult[99]. After isolating the first enzyme in chloramphenicol biosynthesis from *Streptomyces sp.* 3022 a, arylamine-synthetase ㉜, which converts chorismic acid into L-p-aminophenylalanine, chorismic acid can be defined as a starting substrate for synthesis of chloramphenicol[223]. Glutamine acts as an amino-donor for this enzyme. $NAD^+$ increases its activity.

Arylamine-synthetase ㉜ probably includes two enzyme activities by which p-aminophenylpyruvate is first produced as an intermediate; then further transformation is achieved by the aminotransferase activity ㉝. An enzyme for such a reaction has been isolated from *Streptomyces sp.* 3022 a[322, 329, 514]. This enzyme, however, also possesses aminotransferase activity for tyrosine and phenylalanine; this indicates that a single, non-specific aminotransferase can perform various transaminations.

The following reaction steps convert p-aminophenylalanine into chloramphenicol: oxidation of the β-carbon, oxidation of the p-amino group, reduction of the carboxyl group to the alcohol, acylation and dichlorination in the serinol moiety[329].

The specific incorporation of DL-threo(carboxyl-$^{14}$C)-p-aminophenylserine into chloramphenicol proves the oxidation of the β-carbon of p-aminophenylalanine to p-aminophenylserine. On the other hand, the lack of incorporation of p-nitrophenylalanine into chloramphenicol excludes oxidation of the p-amino group from this early stage in the synthesis. A metabolite of *Streptomyces venezuelae,* p-aminophenylserinol, is incorporated into chloramphenicol[496, 533, 544]; this suggests that the p-amino group of p-aminophenylserine is substituted into N-malonyl-p-aminophenylserine and that the two chlorine atoms are then introduced from chloride ions in the nutrient medium[462].

The last step is oxidation of the p-amino group. If chlorine in the medium is replaced by bromine, N-bromochloracetyl- and N-dibromoacetyl-chloramphenicol are obtained[462, 496]. While the biosynthesis of aromatic amino acids in different microorganisms is well regulated many species of streptomyces and micromonospora possess no feedback regulation of DAHP-synthetase[221]. This is also the case with the chloramphenicol producer, *Streptomyces sp.* 3022 a[308].

The consequential enrichment of chorismic acid (enhanced by inhibition of anthranilate-synthetase and prephenate-dehydratase by tryptophan and phenylalanine respectively) results in a chorismic acid pool at the end of the growth phase; then aromatic amino acids are no longer required to a full extent for protein synthesis. This accumulation of chorismic acid may be the reason of the induction of chloramphenicol synthesis via the arylamine-synthetase ㉜. This first enzyme is liable to feedback inhibition by p-aminophenylalanine and p-aminophenylpyruvate[112].

Chloramphenicol itself represses its own biosynthesis[332]. However, exogenous chloramphenicol does not appear to affect the endogenous synthesis of the antibiotic, rather, a high extracellular concentration appears to hinder the excretion of chloramphenicol. The resulting intracellular pool then represses the arylamine-synthetase ㉜. Since chloramphenicol is rapidly broken down to p-nitrophenylserinol by the producer strain[331], this compound seems to be the true repressor.

Under the action of curing agents, such as acriflavin, or at high incubation temperatures, the ability of *Streptomyces venezuelae* to produce chloramphenicol can be lost with a frequency of 2–55%[5, 6]. Furthermore, plasmid participating in chloramphenicol production can be proved by crossing auxothropic non-producers and producer strains[6]. By contrast, absolute non-producers are obtained by UV-irradiation or with nitrosoguanidine; their mutation is in the structural gene, coding the enzyme which hydroxylates p-aminophenylalanine in 3-position. Plasmid-free strains produce a small amount of the antibiotic. Thus, the structural gene for all, or at least for most steps in the biosynthesis, seems to be located on the chromosome. By contrast, the plasmid gene seems to be responsible for regulating the synthesis[4, 6].

Gene mapping of the 18 megadalton size plasmid using 8 markers as well as chloramphenicol production and melanin pigment formation demonstrated that chloramphenicol production and melanin formation are controlled by a plasmid, which does not, however, seem to be transferable by conjugation[5].

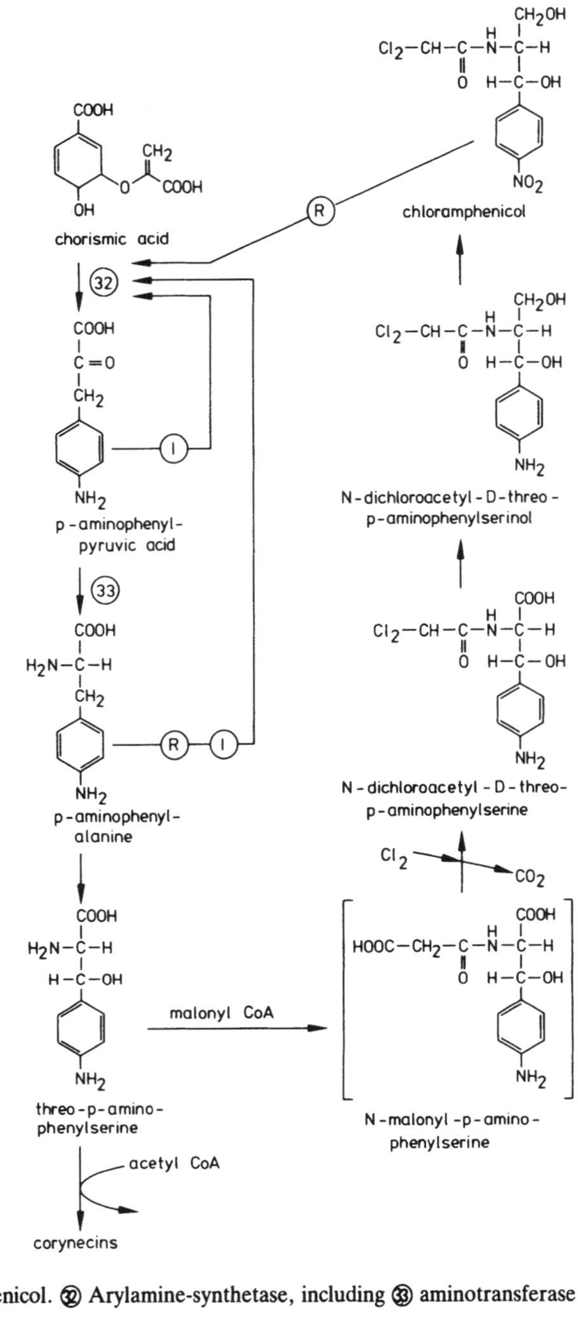


**Fig. 11.** Biosynthesis of chloramphenicol. (32) Arylamine-synthetase, including (33) aminotransferase activity, I induction, R repression

### *4.2.2 Biosynthesis of Candicidin*

Candicidin, a heptaene macrolide antibiotic, is produced by *Streptomyces griseus* IMRU 3570[290, 528]. Its antifungal action depends on binding to sterols in the membrane of yeasts and fungi. Because of a change in permeability, essential cytoplasmic constituents are lost[166, 253, 289], particularly potassium and magnesium ions; this reduces the protein and RNA synthesis[306, 307]. It is not clear whether the polyene antibiotics form pores in the membrane by this way[107].

Although the structure of candicidin is still not completely known, the partial structure reveals that the antibiotic molecule contains an amino sugar, mycosamine, and an aromatic moiety, p-aminoacetophenone, in addition to a macrolide lactone ring[289]. As in the case of nystatin[31], the aglycone seems to be constructed from acetate and propionate units. Mycosamine arises from the metabolism of D-glucose

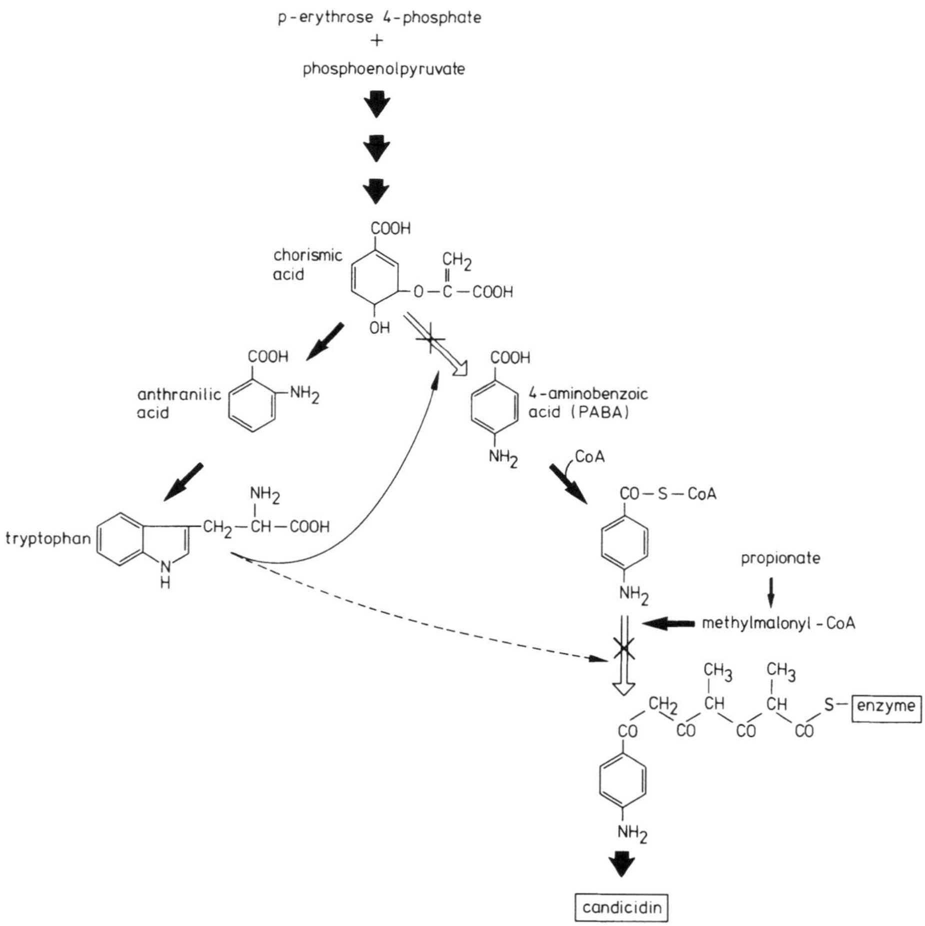


**Fig. 12.** Biosynthesis of the aromatic starter molecule for candicidin and inhibition by tryptophan, → complete, --→ partial

without an alteration in the carbon skeleton. It has been shown by feeding various, potential, $^{14}C$-labelled precursors, such as (UL-$^{14}C$)-shikimic acid, (1-$^{14}C$)-sodium acetate, (2-$^{14}C$)-sodium acetate, ($^{14}C$-methyl)-methionine, (UL-$^{14}C$)-D-glucose, (UL-$^{14}C$-ring)-p-aminobenzoic acid and (7-$^{14}C$)-p-aminobenzoic acid, that the aromatic moiety does not originate from the aromatisation of the polyketide condensation product, as in tetracyclines, but from the shikimic acid pathway. Thus, p-aminobenzoic acid (PABA) has been shown to be a precursor of the aromatic part of candicidin. It has been proved with radioactive (ring UL-$^{14}C$) PABA and (7-$^{14}C$) PABA, labelled in the carboxyl group, that the intact ring and the carboxyl group of PABA are incorporated into candicidin[310)]. PABA has also been incorporated into phosphate limited resting cells of *Streptomyces griseus*[349)].

Precursors in the biosynthesis of aromatic amino acids, such as shikimic acid, are also incorporated into the aromatic moiety of candicidin but with a significantly lower yield than with PABA. In comparison with other metabolic products investigated, PABA seems to be the direct precursor of candicidin.

Thus chorismic acid is the branch in the metabolism of aromatic amino acids[123)], which, in the presence of glutamine, is further converted into PABA[121)].

Biosynthetic studies on aromatic polyketide metabolic products indicate that the aromatic ring functions as a starter for polyketide formation. Rifamycin, a macrolactam antibiotic is an example. In its biosynthesis, the aromatic C 7-unit acts as the starter and the polyketide chain is built up by linear condensation by eight methylmalonate and two malonate molecules[234, 546, 547)]. Studies on the biogenesis and origin of the C 7-primer prove that it is formed by condensation from phosphoenolpyruvate and erythrose-4-phosphate in the general biosynthetic pathway for aromatic amino acids[545)].

Cerulenine, a specific inhibitor of polyketide synthesis, inhibits the synthesis of candicidin starting from PABA. This may be an indication that the aromatic moiety of the polyene macrolide is the starter molecule for polyketide synthesis[340, 352)]. Based on this assumption, the synthetic scheme shown in Fig. 13 is postulated for the construction of candicidin with the starter molecule p-aminobenzoyl-CoA.

Tryptophan inhibits the biosynthesis of candicidin. Because feedback inhibition of DAHP-synthetase is lacking in streptomycetes and micromonospora[289)], but some protein subunits of anthranilate-synthetase and PABA-synthetase are identical, the first of which is controlled by feedback in streptomycetes, it can be assumed that the inhibition of candicidin biosynthesis by tryptophan depends on inhibition of PABA-synthetase.

The biosynthesis of many antibiotics is inhibited by nitrogen[1)] or by inorganic phosphate[341, 536)]. The latter is also true for candicidin[311)]. In a resting cell system of *Streptomyces griseus,* addition of more than 1 mM phosphate leads to a concentration-dependent inhibition of candicidin synthesis[345)]. Phosphate inhibits candicidin synthesis within 15 minutes of its addition without any inhibitory effect on protein or RNA synthesis. Addition of phosphate causes an immediate increase in the intracellular ATP pool. It is therefore assumed that the intracellular ATP concentration is the only phosphate effect[341)] which causes repression and inhibition of candicidin-synthetase[308, 350)]. On this basis, fermentation of many antibiotics must be performed under limitation of phosphate[342, 353)]. Mutants which have no phosphate-controlled candicidin biosynthesis are therefore better candicidin producers than the wild type.

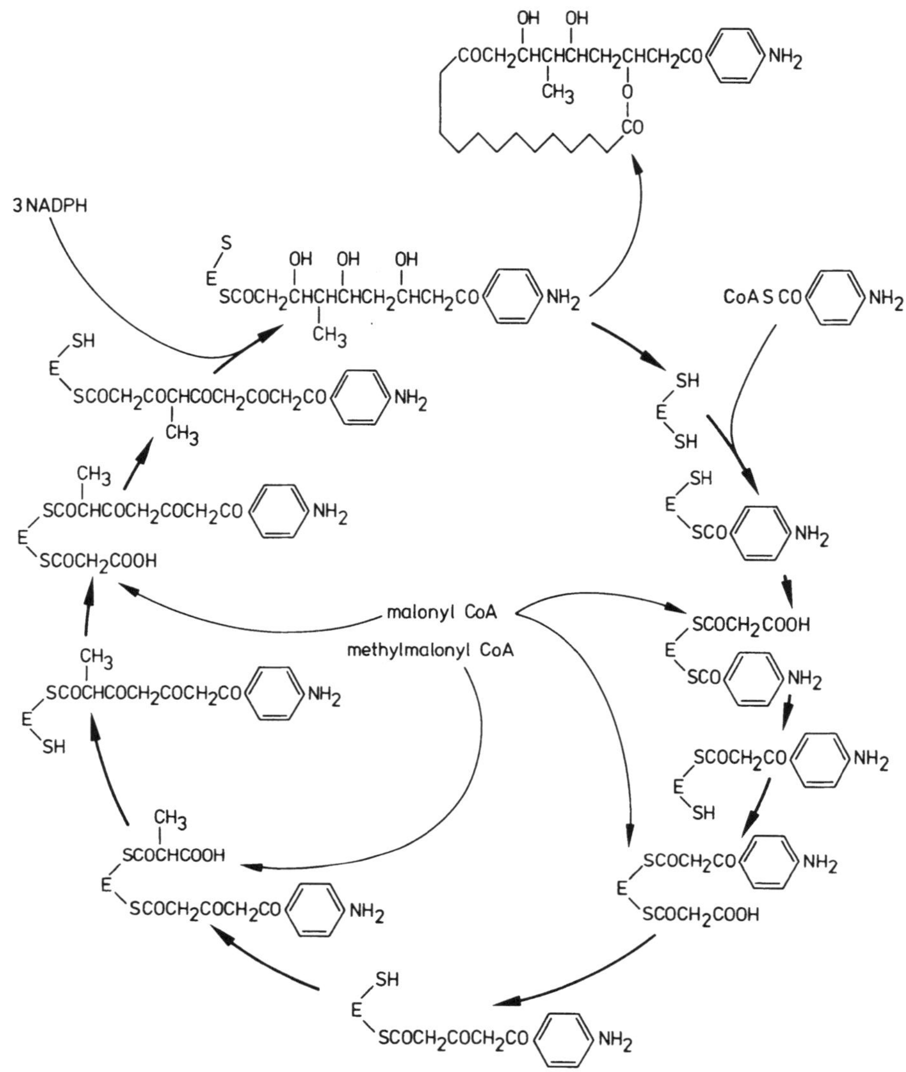


**Fig. 13.** Biosynthesis of the aliphatic portion of candicidin. Beginning with aromatic starter molecule, p-aminoenzoyl-CoA, malonyl-CoA- and methylmalonyl-CoA-elongation units are added by condensation

The high yield of candicidin from these mutants is the result of a higher rate of synthesis and a longer production phase[354].

## 4.3 *Antibiotics from* L-*Phenylalanine*

### 4.3.1 *Biosynthesis of Neoantimycin*

The cyclic polyester antibiotic, neoantimycin, is produced by *Streptoverticillium orinoci*[47]. 3,4-Dihydroxy-2,2-dimethyl-5-phenylvaleric acid is a constituent unit in

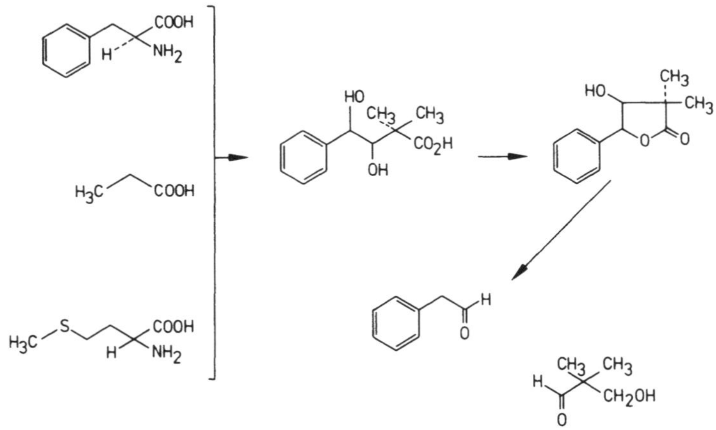


**Fig. 14.** Biosynthesis of 3,4-dihydroxy-2,2-dimethyl-5-phenyl-valeric acid, a constituent unit of neoantimycin, and its chemical breakdown with $LiAlH_4$

this antibiotic, and is synthesized from phenylalanine, propionic acid and methylation with L-methionine[46]. Its structure is thus the first natural product resulting from condensation of a $C_6C_3$- with a $C_3$-unit.

## *4.4 Antibiotics from L-Tyrosine*

### *4.4.1 Biosynthesis of Edeine*

The linear oligopeptide antibiotic, edeine, is produced by *Bacillus subtilis.* β-L-Tyrosine is a component in this antibiotic and is formed from α-L-tyrosine by L-tyrosine α,β-aminomutase[281]. This enzyme catalyzes the direct transfer of the amino group from the α- to the β-position at a pH optimum of 8.5 in the presence of ATP.

In requiring ATP, this aminomutase differs from enzymes which require pyridoxal phosphate to carry out the same reaction[53, 54]. The constituent units in the antibiotic are shown in the structural formula.

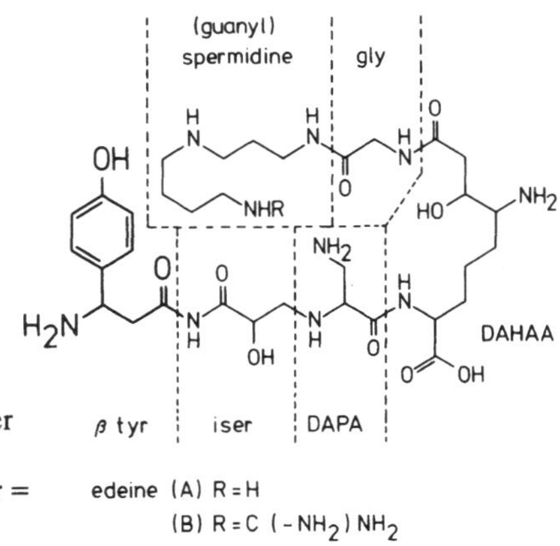


**Fig. 15.** Structural constitution of edeine A and B. iser = Isoserine, DAPA = α,β-diaminopropionic acid, DAHAA = 2,6-diamino-7-hydroxyazelaic acid, β-tyr = β-tyrosine, gly = glycine

### *4.4.2 Biosynthesis of Novobiocin*

Novobiocin is produced by *Streptomyces niveus* and *Streptomyces spheroides*. The sugar components are synthesized from glucose[29, 30, 270]. The aminocoumarin unit and the substituted benzoic acid originate from tyrosine[51, 247]. The heterocyclic oxygen of the aminocoumarin arises from the carboxyl oxygen of tyrosine[248]. Incorporation experiments with $^{15}N$-labelled tyrosine demonstrate that the amino group of the aminocoumarin arises similarly from the α-amino group of this amino acid[48]. Therefore, the aminocoumarin unit presumably leads via a new biosynthetic pathway to 7-hydroxycoumarin by oxidative cyclization from tyrosine with retention of the amino group[48, 248]. By contrast, $^{14}C$-phenylalanine is a major precursor of 7-hydroxycoumarin in plants but not for the biosynthesis of novobiocin. The reason for this is a difference in the biosynthesis of 7-hydroxycoumarin. In plants, the first reaction is deamination of L-phenylalanine to cinnamic acid and then via p-coumaric acid to 7-hydroxycoumarin[388]. This enzyme is not present in novobiocin-producing strains of *Streptomyces niveus*.

In microorganisms two main reactions are possible for conversion of tyrosine into 4-hydroxybenzoic acid.

1. Tyrosine → 4′-hydroxycinnamic acid → β-oxidation → 4-hydroxybenzoic acid[375].
2. Tyrosine → 4′-hydroxyphenylpyruvate → 3-(4′-hydroxyphenyl)-3-hydroxypropionic acid → 4′-hydroxycinnamic acid → β-oxidation 4-hydroxybenzoic acid[565].

Since no non-oxidative deamination has been demonstrated in *Streptomyces niveus*, and because $^{14}C$-4′-hydroxyphenylpyruvate is an efficient precursor of hydroxybenzoic acid, the 2nd reaction is assumed to apply to the biosynthesis of the benzoic acid unit in novobiocin.

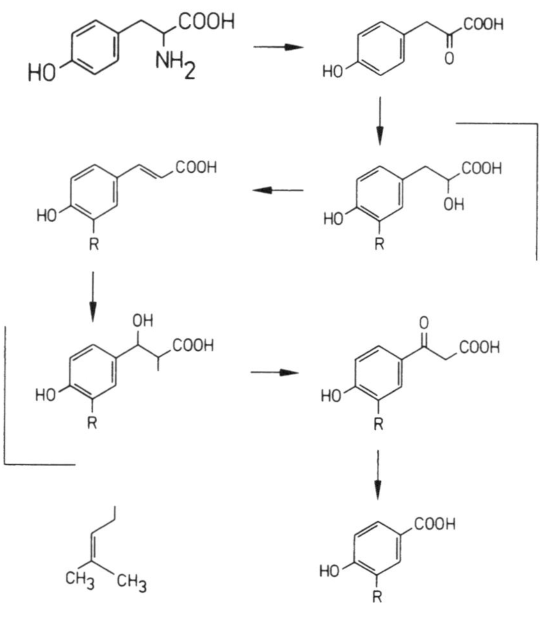


**Fig. 16.** Biosynthesis of the benzoic acid unit in novobiocin, starting from tyrosine

## 4.5 Antibiotics from L-Tryptophan

### 4.5.1 Biosynthesis of Indolmycin

Indolmycin is produced by *Streptomyces griseus*[339, 418] and is highly active against resistant Staphylococci.

Besides weak activity against gram-negative bacteria, grampositive bacteria and *Mycobacterium tuberculosis* are particularly inhibited[404].

In addition to indolmycin, N-demethyl-D-demethylindolmycin can be isolated[177]. 5-Hydroxy- and 5-methoxyindolmycin derivatives obtained by biotransformation possess enhanced activity against both gram-positive and gram-negative organisms[541]. Such derivatives should be obtained in higher yields if the corresponding mutants, which are blocked in antibiotic biosynthesis or possess a tryptophan auxothrophy, are used[540]. Among the chemically synthesized derivatives, the thio-analogue of indolmycin[167] has a good antiviral and antibacterial activity.

The mode of action of indolmycin depends on bacterio-specific inhibition of tryptophanyl-tRNA-ligase in prokaryotes but not in eukaryotic cells[543]. The penetration of the antibiotic into gram-positive bacteria is catalyzed by tryptophan-permease[539]. Indolmycin has no significant effect on the metabolism of tryptophan in rat liver[542].

Since the chemical structure of indolmycin contains the tryptophan nucleus, radiolabelled precursors were incorporated into *Streptomyces griseus* ATCC 12648 to support this theory of biogenesis[183]. The studies revealed good incorporation of ring-labelled anthranilic acid, ($^{14}$C-guanido)-arginine, indole, ($^{14}$C-methyl)-methionine and ring- or sidechain-labelled tryptophan, but not of acetate, alanine, formic acid, glucose, L-threonine and urea. Further breakdown experiments with the labelled indolmycin compounds showed that the radioactivity of the guanido group is quantitatively incorporated into the $CO_2$ produced by alkaline hydrolysis; furthermore ($^{14}$C-3-alanine)-tryptophan preferably labels C-6 of the molecule, whereas the methyl group of methionine is distributed by 31% in the C-methyl group and by 55% in the N-methyl group. The building units of indolmycin can be summarized from these labelling experiments (Fig. 17).

Since α-indolmycinic acid is a precursor in the fermentation and the radiolabelled analogue with the natural configuration (2 S, 3 R)[182, 183] is incorporated with good yield (17%) into the antibiotic, C-methylation must occur before the oxazolinone ring is closed[185]. This methylation occurs at the indolepyruvate stage since the C-methylase only reacts with indolepyruvate and not with tryptophan. The synthesis of indolmycinic acid and initiation of indolmycin biosynthesis must, therefore, proceed by transamination of tryptophan to indolepyruvate, its C-methylation to β-methylpyruvate and reduction of this to indolmycinic acid[472]. A comparable C-methylation has been described for indole-isopropionic acid[184].

In the formation of the oxazolinone ring, the intact amidino group of arginine is transferred[581]. This is proved by significant incorporation of labelled L-arginine with $^{14}$C-amidino-carbon and with $^{14}$N in both amidino nitrogens. Whether the amidino group is first bound to the free hydroxyl group or to the free carboxyl group, or whether direct coupling to both functional groups is catalyzed, remains an open

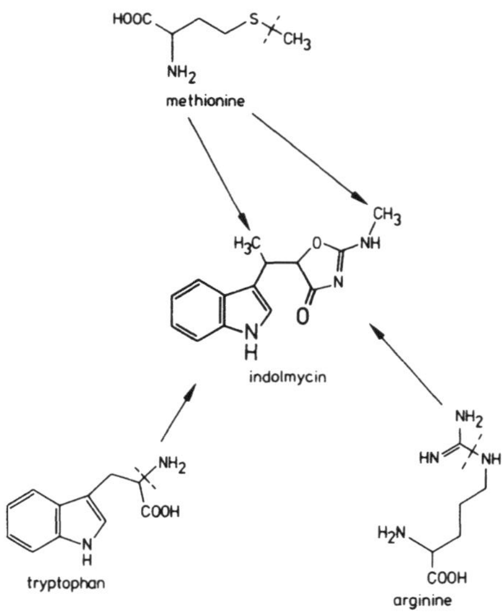


**Fig. 17.** Biogenetic building units of indolmycin

question. Assumably the stereospecific N-methylation leads to indolmycin as the final reaction.

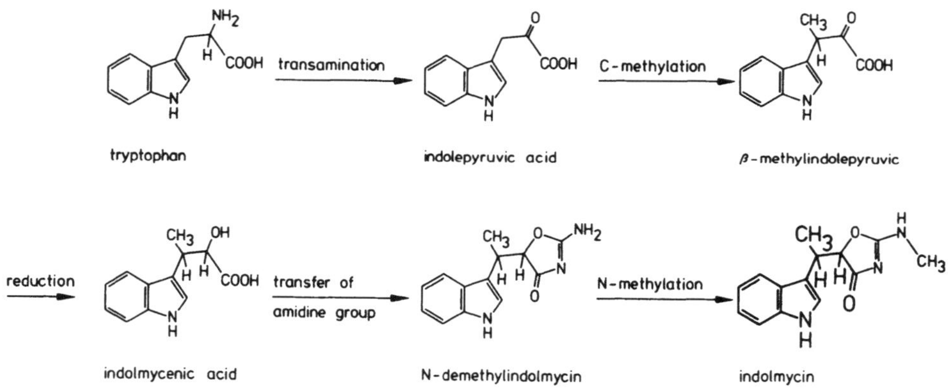


**Fig. 18.** Indolmycin biosynthesis, starting from tryptophan

In indolmycin biosynthesis, only glucose at a concentration of 0.5% inhibits the production of indolmycin whereas, at lower concentrations, 0.005 and 0.01%, it stimulates the synthesis. This effect is smaller with mannose while sucrose, galactose and to some degree fructose are stimulating. This indicates that the biosynthesis of indolmycin is regulated by catabolite inhibition or repression[188].

### *4.5.2 Biosynthesis of Pyrrolnitrin*

Pyrrolnitrin is produced by various strains of Pseudomonas[10, 11, 205] and possesses antifungal activity particularly against dermatophytes, such as Trichophyton species[134, 393]. In addition, naturally occurring analogues of the antibiotic, such as isopyrrolnitrin[169], oxypyrrolnitrin[163], 2-chloropyrrolnitrin[164], deschloropyrrolnitrin[168] and aminopyrrolnitrin[164] have been isolated. The bromo-analogue of pyrrolnitrin is obtained by addition of ammonium bromide to synthetic fermentation media[3].

The action of the antibiotic depends on damage to the cell wall by an exchange reaction with phospholipids[395]. Besides this mechanism of action, inhibition of the respiratory electron transport system can be observed[282, 511, 559, 560].

The biogenesis of pyrrolnitrin originates from tryptophan since both $^{14}$C-D- and $^{14}$C-L-tryptophan are incorporated into the antibiotic[136, 164, 312] although only D-tryptophan enhances the production of pyrrolnitrin[97, 109]. A possible explanation of this effect may be as follows.

Apart from biosynthesis of pyrrolnitrin, two main pathways for degradation of tryptophan are known in Pseudomonas.

One is the quantitatively less important indole-acetic acid route by which D- and L-tryptophan are converted. The other is the kynurenine-anthranilic acid route which is initiated by tryptophan-2,3-dioxygenase. This enzyme is only induced by L-tryptophan and not by the D-isomer. The product of this metabolic route is anthranilic acid, a potent inhibitor of pyrrolnitrin synthesis.

This may be the explanation why L-tryptophan is unable to stimulate pyrrolnitrin synthesis in contrast to D-tryptophan. The D-isomer is only slowly converted into L-tryptophan so that no L-tryptophan is formed at a level to induce the dioxygenase[439]. Mutants which are resistant to 5- and 6-fluoro-tryptophan do not need an addition of D-tryptophan to produce high antibiotic titers in the fermentation[98].

Since aminopyrrolnitrin, a naturally occuring metabolite increases the pyrrolnitrin formation when added to the medium and is incorporated into the molecule, aminopyrrolnitrin is assumed to be the last intermediate step to pyrrolnitrin, the amino group being oxidized to the nitro group.

The following synthetic pathway is suggested for the transformation of tryptophan into pyrrolnitrin[52, 108] (Fig. 19).

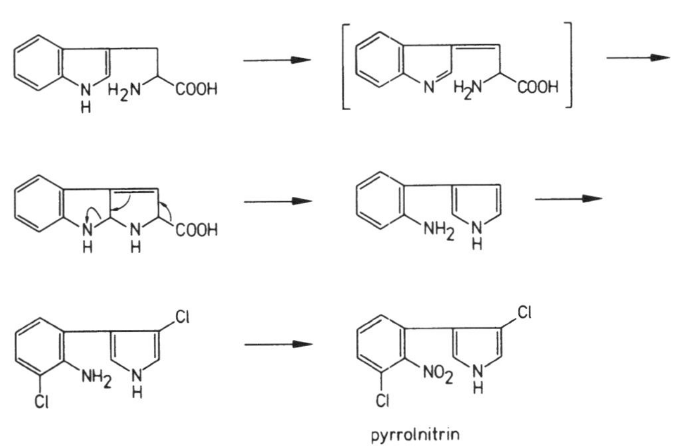


**Fig. 19.** Biosynthesis of pyrrolnitrin

Pyrrolnitrin is a good example of the possibilities of directed biosynthesis in which, by addition of tryptophan analogues such as 6-fluorotryptophan and 7-methyltryptophan, the corresponding pyrrolnitrin analogue is obtained[135]. The fluoro-derivative has a somewhat greater antifungal activity than pyrrolnitrin. By addition of 4-, 5-, 6- and 7-substituted tryptophan derivatives, three different types of derivative are obtained, ring-substituted 3-chloroindole from 4-halogenated tryptophan, benzene ring-substituted pyrrolnitrin analogues and benzene ring-substituted aminopyrrolnitrin analogues[165].

### *4.5.3 Biosynthesis of Streptonigrin*

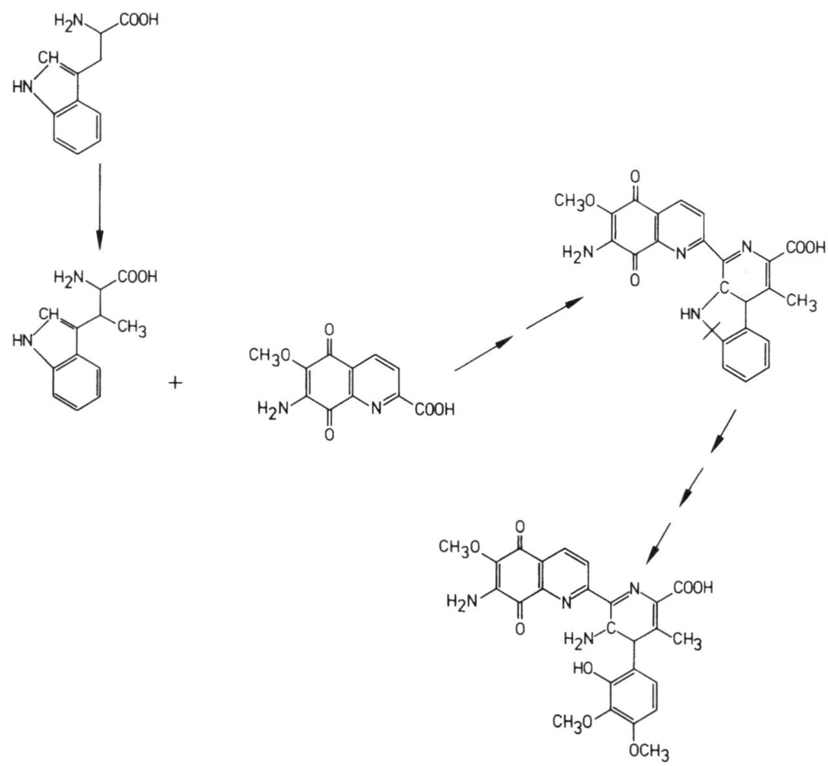


**Fig. 20.** Streptonigrin biosynthesis and synthesis of the precursors

Streptonigrin is an antibiotic produced by *Streptomyces flocculus* ATCC 13257[338, 417, 419] and possesses antitumor activity[318, 397, 552]. The 4-phenylpicolinic acid moiety is derived from the metabolism of tryptophan[141]. This indicates a new metabolic formation of the pyridine ring[474, 505] and so a new metabolism for tryptophan. The biosynthesis starts by cleaving the intact indole ring at the C–N bond. In the biosynthesis of pyrrolnitrin, the $N_b$-$C_2$ bond is cleaved; the obtained aromatic amine finally is oxidized to the nitro group[109]. The incorporation of (2-$^{13}C^{15}N_b$)-DL-tryptophan into the antibiotic helps to investigate the cleavage from tryptophan to streptonigrin[141]. Since both isotopes are present in streptonigrin and the C-5′-signal is a doublet because of $^{15}N$-coupling, this C–N bond must remain intact during biosynthesis[142]. Therefore, the $N_b$-$C_{7a}$ bond is broken, quite unusual for metabolism of tryptophan. The four methyl groups in streptonigrin originate from methionine[235].

Tryptophan first is methylated to β-methyltryptophan[184, 469] and condensed with a quinolinic-carboxylic acid. The pyridine ring then is formed by intramolecular attack from the nucleophilic α-position of the indole on the amide carbonyl function, followed by aromatisation of the resulting dihydropyridine and fission of the indole ring. This is a new metabolic formation of the pyridine ring and so a new metabolism for tryptophan[141].

Radiolabelling experiments exclude the incorporation of the intact tryptophan molecule into the quinoline ring system. Thereby two of three known routes for synthesis of the quinoline ring are excluded[198, 317]. Experiments on the biosynthesis of the quinoline ring are not yet concluded.

### *4.5.4 Biosynthesis of Tryptanthrin*

Tryptanthrin is synthesized in *Candida lipolytica* from 1 mole tryptophan and 1 mole anthranilic acid[446]. In the biosynthesis, tryptophan first is broken down alternately to anthranilic acid and indole-pyruvic acid and then the two molecules are coupled together.

Although many yeasts excrete anthranilic acid and β-3-H-indolylpyruvic acid on feeding tryptophan, the production of tryptanthrin is limited to *Candida lipolytica*.

Because of the relatively non-specific reaction of the enzyme concerned, feeding derivatives of the starting compounds yields analogues of the antibiotic. The halogenated compounds are particularly active[105]. The spectrum of action lies particularly in the gram-positive area, probably because the compound is lipophilic.

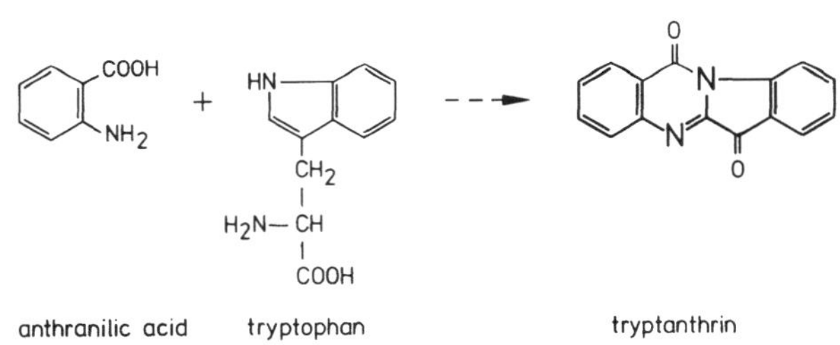


**Fig. 21.** Biosynthetic units and structural formula of tryptanthrin

## *4.6 Antibiotics from End Products of the Biosynthesis of Aromatic Amino Acids*

### *4.6.1 Biosynthesis of Tyrocidin*

Besides the metabolic origin of antibiotics from amino acid metabolism, there is a series of peptide antibiotics which are synthesized by coupling of non-metabolized amino acids[257–259]. The synthesis of these antibiotics differs from ribosomal protein synthesis in the following points.

Inhibitors of ribosomal protein synthesis such as chloramphenicol, puromycin[323], erythromycin, terramycin and lincomycin[49] do not inhibit the biosynthesis of the decapeptide tyrocidin[324]. Contrary to protein synthesis, tyrocidin synthesis is not dependent on continuous synthesis of RNA[325]. Assumably such small polypeptides are synthesized by stepwise coupling of the individual amino acids with specific enzymes or possibly with a multi-enzyme complex. Some of these basic differences in the biosynthesis of proteins and small polypeptides have been confirmed, apart from tyrocidin, for some other antibacterial peptides, such as polymycin[407], gramicidin S[94, 95], mycobacillin[15] and edeine[34]. The three forms of tyrocidin A, B and C differ in the substitution of a single amino acid such as phenylalanine or tryptophan[252].

By analogy with ribosomal protein synthesis, the substitution of amino acids in tyrocidin is under direct genetic control. Since a single culture of *Bacillus brevis* can produce all three derivatives, direct genetic control implies genetic heterogeneity with respect to the synthesis of tyrocidin A, B and C. Another explanation is provided by the presence of different templates which differ from each other in substitution of a single nucleotide sequence and which cause simultaneous production of several tyrocidins. In fact, the synthesis of different tyrocidin derivatives is not dependent on such genetic consequences, but is determined by the concentration of the amino acids necessary for the synthesis of the antibiotic. This phenomenon can be explained by the poor specificity of the enzyme systems responsible for the incorporation of structurally similar amino acids.

Thus, an excess of L-phenylalanine in a culture of a minimal medium leads to almost exclusive synthesis of tyrocidin A (3 Phe). If L-tryptophan is added, the phenylalanine-containing tyrocidins are not synthesized, only tyrocidin D containing three molecules of tryptophan[437]. On addition of both amino acids, L-phenylalanine and L-tryptophan, all 4 derivatives are formed. Feeding of the corresponding D-amino acids has the same effect.

Such flexibility, and particularly the possibility of changing the primary structure of polypeptide chains by changing external factors, can lead to biologically important polypeptides. Other peptides, which may be also synthesized by flexible control of the amino acid sequence, can be specifically taken up by macromolecules. Such polypeptides may play an important role in regulation of different cellular activities[324].

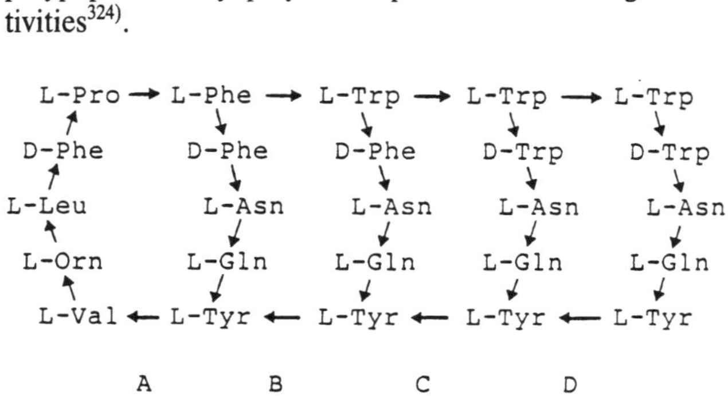


**Fig. 22.** Structures of tyrocidin A, B, C, D

### *4.6.2 Biosynthesis of Gramicidin*

Gramicidin S, a cyclic peptide produced by *Bacillus brevis*, consists of two pentapeptide units having the sequence D-phenylalanine-L-proline-L-valine-L-ornithine-L-leucine. Synthesis of this antibiotic takes place through gramicidin S synthetase after the exponential growth phase.

In a cell-free system, there is a different pH optimum for incorporation of each individual amino acid: leucine 7.5–7.7, proline 7.5–7.7, phenylalanine 7.7–7.9, ornithine 7.7–7.9, valine 8.0–8.2[106]. The pH optimum for valine at 8.0–8.2 reduces the incorporation of leucine by about 25%, whether this has a regulatory effect in vivo as well, is not known.

10–18 μg/ml actinomycin D causes 80–90% inhibition of the incorporation of 1-($^{14}C$)-phenylalanine and 5-($^{14}C$)-ornithine in protein synthesis, while incorporation into the gramicidin S fraction is unaffected[96].

The effect of some amino acids on the biosynthesis of gramicidin S has been studied[582, 583]. Leucine, valine, proline, ornithine, phenylalanine[582], glycine, tyrosine, methionine and aspartic acid stimulate formation of gramicidin S[583]. Phenylalanine, a component of gramicidin S has the greatest stimulating effect. The observation that fluorophenylalanine and β-phenyl-β-alanine inhibit formation of gramicidin S but not growth[527, 553] confirms the significance of phenylalanine. In principle, the significance of this amino acid may lie in the phenylalanine dependence of gramicidin S synthetase formation, in the stabilization of the enzyme or in the antibiotic formation.

Two enzymes, gramicidin S synthetase I and II, are responsible for the synthesis of the antibiotic[359]. If the amino acids present in the antibiotic are added to a fermentation in a defined minimal medium, only L-phenylalanine effects a significant increase in antibiotic production; 1% of the amino acid increases the yield of the antibiotic in the medium 3 fold. This effect causes no induction of gramicidin S synthetase I and II, no stabilization of the enzyme, and no increased growth. L-Phenylalanine apparently stimulates production of the antibiotic through its role as a precursor of the D-phenylalanine in the gramicidin S molecule[77].

However, addition of D-phenylalanine, which also stimulates the production of gramicidin S, inhibits growth and so requires a longer fermentation time.

Consequently, L-phenylalanine overproducer mutants of *Bacillus brevis* should produce increased amounts of gramicidin. Such mutants were of industrial interest for large-scale production.

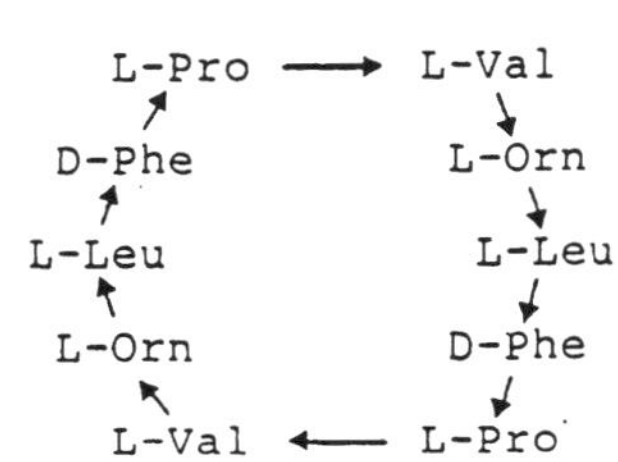


**Fig. 23.** Structure of gramicidin S

## 4.7 Antibiotics from Metabolism of Tyrosine

### 4.7.1 Biosynthesis of Lincomycin

Lincomycin is produced by *Streptomyces lincolnensis*. Biosynthetic studies prove that the pyrrolidone structure in lincomycin originates from tyrosine. These data show that propylproline and ethylproline are formed respectively from L-dopaquinone and 2-carboxy-2,3-dihydro-5,6-dihydroxyindole. These last two compounds are also intermediates in melanin synthesis. This means that monophenol mono-oxygenase effects the biosynthesis of both melanin and lincomycin starting from L-tyrosine[366].

The importance of monophenol mono-oxygenase for the formation of the antibiotic was studied in $mel^+$ and $mel^-$ spontaneous mutants of *Streptomyces lincolnensis* NCIB 9413. Lincomycin is produced at the end of the logarithmic phase. In $mel^+$ strains, the maximum activity of the oxygenase is reached in the middle of the logarithmic phase, correlating with melanin formation and just before antibiotic formation. In lincomycin high producer $mel^-$ mutants, monophenol mono-oxidase activity remains below 5% of that in $mel^+$ strains. In the stationary growth phase, in which lincomycin is produced, oxidase activity in the mycelium is 4–5 times higher than in the $mel^+$ strain. The yield of antibiotic in these $mel^-$ strains is twice of that in $mel^+$ strains. Non-producer mutants possess no monophenol mono-oxygenase. Further evidence for monophenol mono-oxygenase dependence of lincomycin production is the action of inhibitors on this enzyme. Thiourea at a concentration of 10 mM has no effect on inhibition of growth but reduces oxidase activity by half in the logarithmic phase of $mel^+$ $linco^+$ strains and by a factor of 15 in the stationary phase. The yield of the antibiotic is thereby reduced by 90%. Formation of the antibiotic is also inhibited by tryptophan, anthranilic acid, 3-hydroxyanthranilic acid, indolyl pyruvate and indole. Indole, with 70% inhibition, is the most potent. The other compounds only inhibit synthesis by 30–40%.

The tyrosine pool in $mel^-$ $linco^+$ mutants is about 20% while in the $mel^+$ $linco^+$ parent strain it does not exceed 3%. This fact, the lack of melanogenesis, and the higher oxidase activity increase the yields of lincomycin[366].

Propylproline and ethylproline are enriched in the fermentation medium of *Streptomyces lincolnensis* under sulphur limitation, and can be considered as precursors of lincomycin and 4′-depropyl-4′-ethyllincomycin respectively. The production of propylproline and ethylproline is stimulated by adding L-tyrosine and L-dopa; this implies that lincomycin originates in part from the intermediary metabolism of melanin. 1-$^{14}$C-Tyrosine and $^{15}$N-tyrosine are incorporated adequately into the propylproline moiety of lincomycin and into the ethylproline moiety of 4′-depropyl-4′-ethyl-lincomycin. Seven C-atoms of tyrosine are incorporated into the propylproline moiety of lincomycin which indicates that tyrosine is first cyclized and partially degraded before it is incorporated as propylproline into lincomycin[558]. The methylhydroxyl-anthranilic acid part originates from tryptophan via 3-hydroxyanthranilic acid on which the methyl group of methionine is substituted[450].

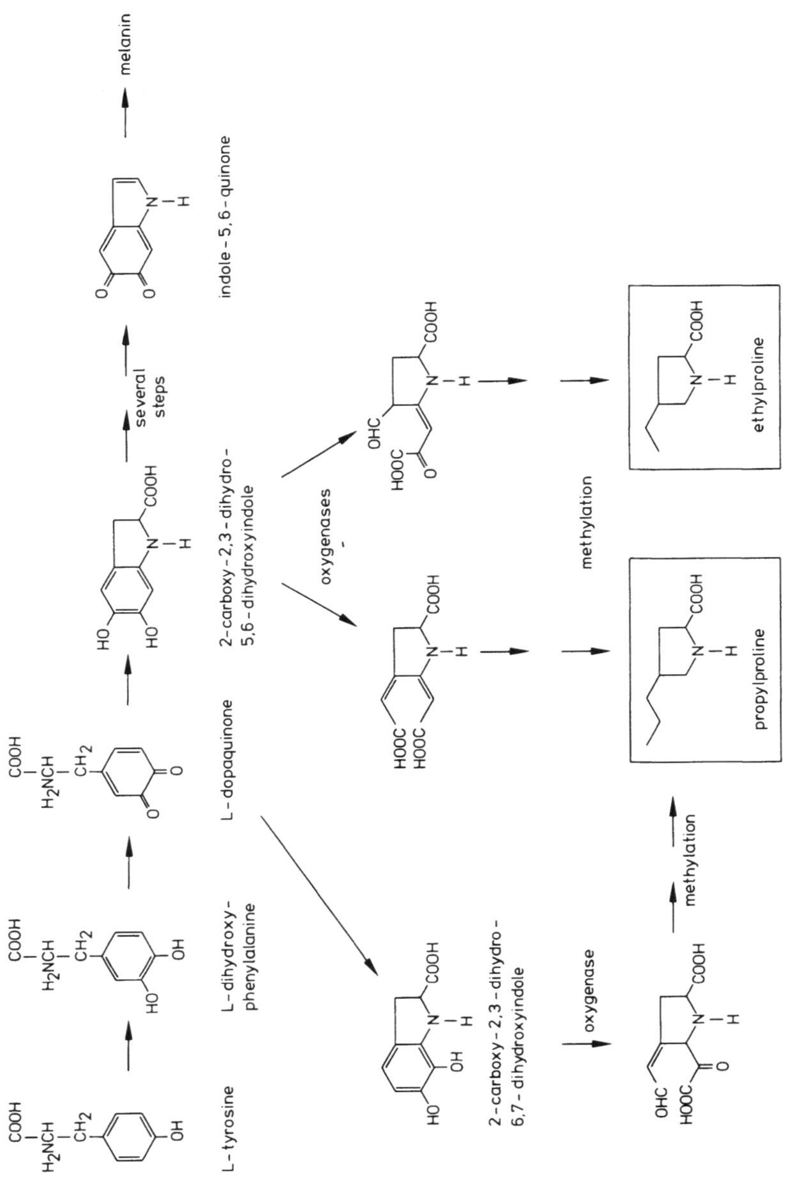


**Fig. 24.** Hypothetical possibilities for the biosynthesis of the lincomycin precursors, propylproline and ethylproline, from melanin metabolism

## 4.8 Antibiotics from Kynurenine Metabolism

A group of antibiotics, which are synthesized from the metabolism of tryptophan via kynurenine, belong to the pyrrolo-(1,4)-benzodiazepine class and include anthramycin, tomaymycin, sibiromycin and neoanthramycin A, B, structurally related antitumor antibiotics[402, 403, 521].

This class of substances is biosynthesized on one hand from tryptophan via kynurenine to anthranilic acid; on the other hand, tyrosine provides the $C_2$ and $C_3$ proline units in the antibiotics via dopa. The yield of antibiotic in a defined medium increases 3-fold when the precursors tryptophan, tyrosine and methionine are added in the range 0.1–0.5 mM. If only tryptophan is given, about the same effect is obtained as with all three amino acids at a concentration of 0.5 mM. The time course of antibiotic synthesis is unaffected. Addition of tyrosine and methionine at concentrations exceeding 0.5 mM, reduces the stimulating effect of tryptophan. Methionine at 0.5 mM produces a 67% inhibition of antibiotic synthesis, and 74% at 2.5 mM. Tyrosine has no inhibitory effect at 0.5 mM, at 2.5 mM it inhibits biosynthesis by 43%. Addition of tryptophan during antibiotic formation is less stimulating on biosynthesis than at the start of fermentation. This may indicate that tryptophan causes enzyme induction of antibiotic biosynthesis, as is known for alkaloids[277], or that it only causes a substrate effect which permits enzyme saturation in a rate limiting step. In particular, the key enzyme in the biosynthesis of pyrrolo-(1,4)-benzodiazepines, tryptophan oxygenase, is a possible candidate for antibiotic regulation (Speedie, personal communication).

The compelling evidence for participation of tryptophan, tyrosine and methionine in the biosynthesis of anthramycin, tomaymycin and sibiromycin has been obtained by experiments with radioisotopes and stable isotopes[189, 190, 192, 193]. The relatively rapid incorporation of the three amino acids in anthramycin compared with tomaymycin and sibiromycin correlates with the more rapid production of anthramycin.

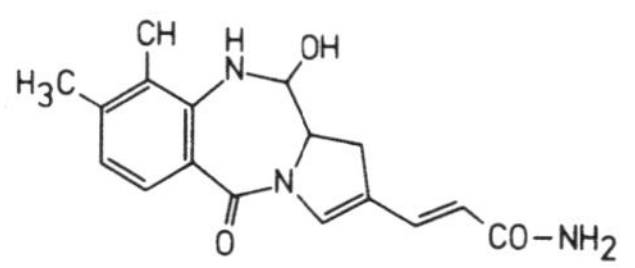


anthramycin

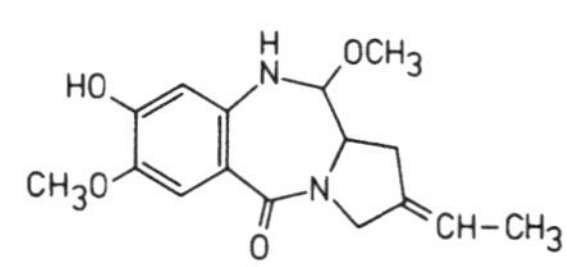


tomaymycin

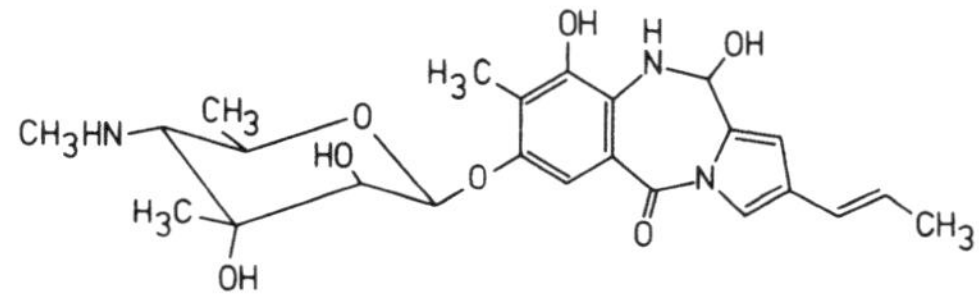


sibiromycin

**Fig. 25.** Pyrrole(1,4)benzodiazepines: anthramycin, tomaymycin, sibiromycin

### *4.8.1 Biosynthesis of Anthramycin*

Anthramycin is produced by *Streptomyces refuineus* var. *thermotolerans* NRRL 3143[294, 295, 508]. Anthramycin possesses antitumor, antibiotic, amoebicidal and chemosterilant properties[186]. All these biological parameters indicate that anthramycin affects the biosynthesis of nucleic acids[125, 148, 269]. Anthramycin inhibits RNA synthesis besides DNA synthesis in bacterial and human cells[492]. In the biosynthesis of anthramycin, tryptophan is metabolized with tryptophanpyrrolase and then converted to anthranilic acid, and tyrosine is degraded via L-dopa with subsequent meta-cleavage. The metabolites are condensed and methylated[194].

The methyl-hydroxyl-anthranilic acid moiety (MHAA) of anthramycin is also an essential structural element of tomaymycin[12, 223], sibiromycin[364] and lincomycin[450]. This structural similarity suggests a common biosynthetic origin. Tyrosine is the biogenetic source of the propylproline group of lincomycin A, the two $C_1$ units deriving from methionine[558]. In the case of tomaymycin, the antibiotic is formed by a synthesis analogous to that of anthramycin[191, 194].

The significance of tryptophan, methionine and tyrosine as precursors in the synthesis of anthramycin was shown by chemical degradation of the antibiotic radiolabelled with the amino acids. Acid hydrolysis yielded 4-methyl-3-hydroxyanthranilic acid (MHAA). The entire radioactivity of tryptophan was demonstrated in MHAA. 42% of the activity of methionine, which is localized in the methyl group, was found in the degradation product but none in that of tyrosine[195]. The origin of MHAA in tryptophan and methionine is analogous to that in the biosynthesis of actinomycin[175, 441] and probably represents a version of the known metabolism of tryptophan to 3-hydroxyanthranilic acid[147]. By analogy with the biosynthesis of the

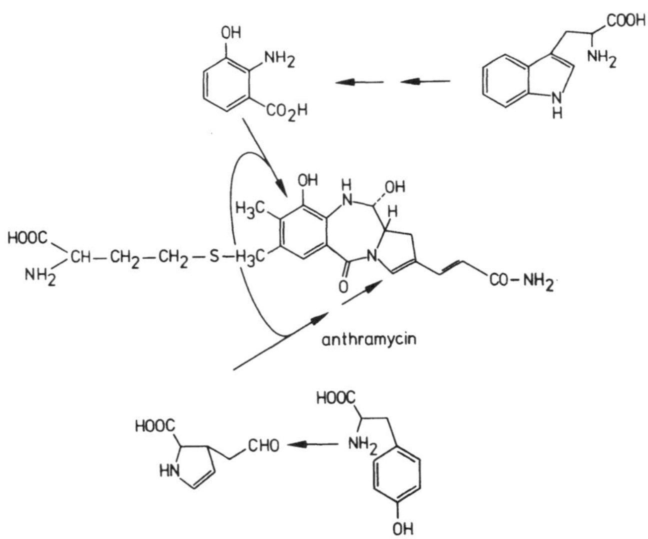


**Fig. 26.** Building units in the biosynthesis of anthramycin

propylproline moiety of lincomycin A[558], the following scheme can be drawn up for the biosynthesis of anthramycin:

The MHAA unit is synthesized from tryptophan via 3-hydroxyanthranilic acid and the methyl group of methionine. In addition, 7 of the 8 C-atoms in the acrylamide-proline unit originate from L-tyrosine, the 8th C-atom is another C-1 from methionine[195]. Also by analogy with the propylproline synthesis of lincomycin A, the amide carbon atom of the acrylamide side chain originates from methionine. Feeding experiments with doubly-labelled (Me-$^{14}$C, Me-$^{3}$H)-methionine confirm this theory and the incorporation of an intact $CH_3$-group into 3-hydroxyanthranilic acid.

Studies of ring cleavage of doubly-labelled (1-$^{14}$C, 3- or 5-$^{3}$H)-tyrosine reveal an extradiol (meta) ring cleavage with both antibiotics, anthramycin and tomaymycin. Similar studies with lincomycin A also resulted in the finding of extradiol cleavage of the aromatic ring. It is assumed, therefore, that extradiol cleavage of 5,6-dihydroxy-cyclo-dopa and 6,7-dihydroxycyclo-dopa (Fig. 27) lead to lincomycin A and sibiromycin respectively.

The unanswered question in the biosynthesis of anthramycin is, whether tyrosine or dopa condense with the anthranilic acid moiety before or after conversion into the

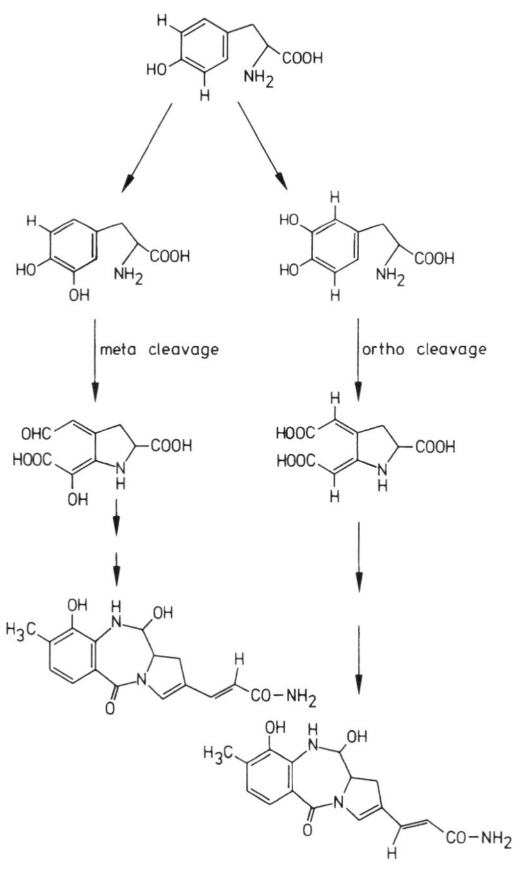


**Fig. 27.** Alternative cleavage of tyrosine in the acrylamide-proline part of anthramycin

$C_3$-proline unit. From labelling experiments and by analogy with cyclopenin, it is concluded that tyrosine condenses with MHAA before it is converted.

Regulation of the biosynthesis of tyrosine in the anthramycin producing *Streptomyces refuineus* var. *thermotolerans* was investigated in the wild type and the corresponding tyrosine auxotrophic strain. The auxotrophic mutant grows in a minimal medium with 50 μg/ml phenylalanine; obviously phenylalanine increases the tyrosine pool.

In fact, the tyrosine content increases proportionally to the added phenylalanine. However, no radioactive tyrosine could be detected when $^{14}C$-phenylalanine was added to the medium. This indicates that no phenylalanine hydroxylase is present in the cells; rather, there is feedback inhibition of prephenate dehydratase by phenylalanine, causing an increase in NAD-dependent prephenate dehydrogenase activity, which initiates the conversion of prephenic acid to tyrosine[473)].

Another explanation for the increase in tyrosine in the presence of phenylalanine may be the induction of an alternative synthetic pathway in the auxotrophic mutant from prephenate via pretyrosine to tyrosine as has been described for some microorganisms[102)].

### *4.8.2 Biosynthesis of 11-Demethyltomaymycin*

11-Demethyltomaymycin is an antiviral antibiotic produced by *Streptomyces achromogenes* var. *tomaymyceticus*[12, 233)]. It is built from tryptophan, tyrosine, and by methylation via methionine. The biosynthesis follows the diagram in Fig. 28.

The metabolism of tyrosine proceeds via dopa. As in the biosynthesis of anthramycin[190)], the ring of dopa is cleaved in the meta-position. Tryptophan is degraded by the tryptophan-pyrrolase reaction to hydroxymethylanthranilic acid. Both metabolites are condensed and methylated with methionine[190, 191)].

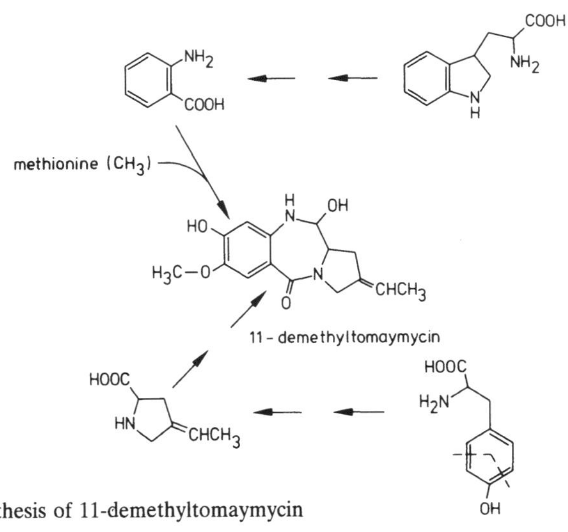


**Fig. 28.** Building units of the biosynthesis of 11-demethyltomaymycin

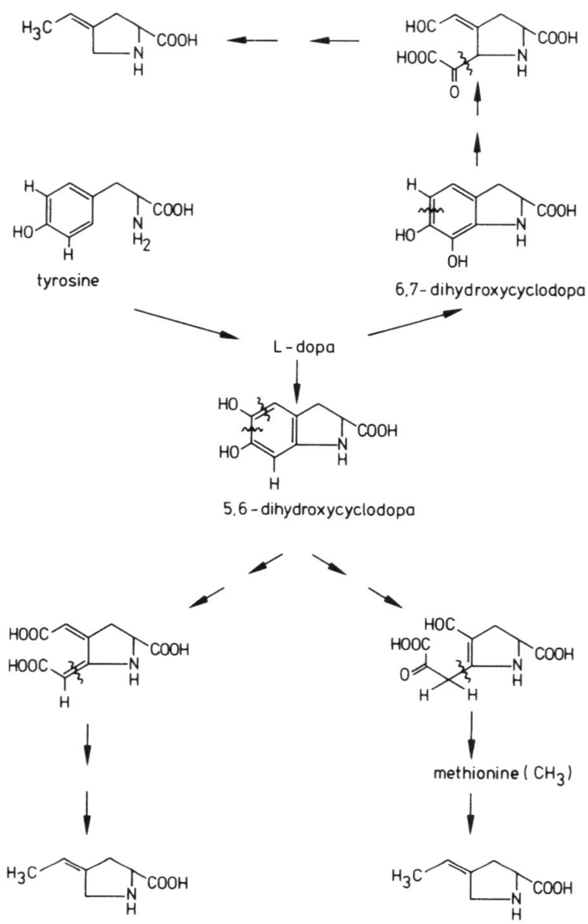


**Fig. 29.** Alternative routes of cleavage of L-dopa for the biosynthesis of the proline part of 11-demethyltomaymycin. The synthetic pathway shown in the upper section of the Figure has been confirmed by labelling experiments

### *4.8.3 Biosynthesis of Antimycin A*

Antimycin A, an antifungal non-polyene antibiotic, is produced by *Streptomyces antibioticus* NRRL 2838[426, 523]. In the biosynthesis of this antibiotic, tryptophan should be considered as a precursor of the aromatic part of the molecule[232]. DL-(ring 2-$^{14}$C)-tryptophan is incorporated to a significant extent into antimycin C. A precursor of antimycin, 3-(N-formylamino)-salicyclic acid (FAS), which can be isolated from the culture medium during fermentation, is comparably labelled by the radioactive amino acid. Thus, FAS represents a new intermediate in the metabolism of kynurenine, within the biosynthesis of antimycin. Practically no pool of free, intracellular tryptophan is present in *Streptomyces antibioticus,* whereas the FAS content depends on the different growth phases. The maximum FAS concentration

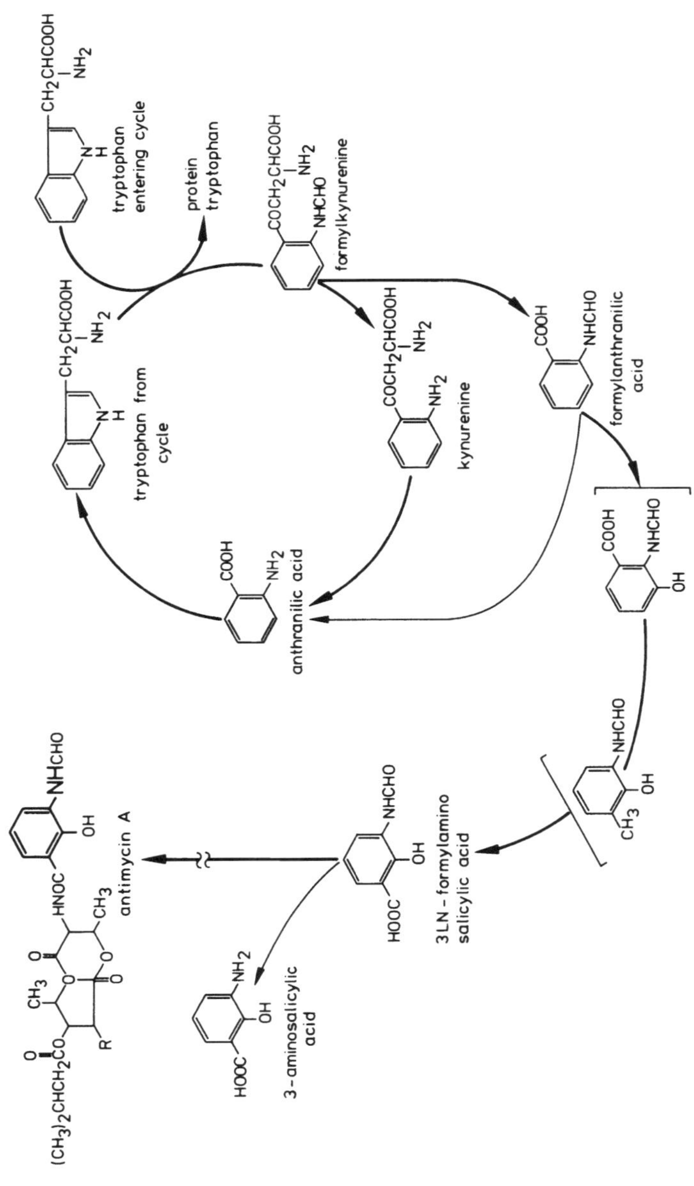


**Fig. 30.** Biosynthesis of antimycin A with particular reference to the aromatic part of the antibiotic

correlates with that of kynurenine and tryptophan synthetase after an incubation time of 40 hours. Displaced in time, antimycin A production persists up to 88 hours. At this time, no tryptophan synthetase activity and no free kynurenine can be detected; the FAS content has fallen by a factor of 6[427]. This indicates that tryptophan is no longer consumed in the primary metabolism at the end of the growth phase; it is then degraded via formylkynurenine to FAS and to antimycin A. This is explained in the biosynthetic scheme of Fig. 30.

### *4.8.4 Biosynthesis of Actinomycin D*

Actinomycin D, a member of the family of chromopeptide antibiotics, is synthesized by *Streptomyces antibioticus*[225, 226, 236, 237]. The actinomycins consist of the chromophore, actinocin (2-amino-4,6-dimethyl-3-phenoxazinone-1,9-dicarboxylic acid) and two cyclic pentapeptides. The phenoxazinone ring of actinomycin D is formed by condensation of two units of 3-hydroxy-4-methylanthranilic acid[147, 240, 408, 441, 461, 538]. 3-Hydroxy-4-methylanthranilic acid is synthesized via the normal metabolism of tryptophan to 3-hydroxyanthranilic acid[147, 537] with subsequent methylation of the aromatic ring. However, alternatively the biosynthesis starts from 3-hydroxykynurenine with subsequent trans-methylation. Feeding experiments demonstrate that 3-hydroxy-4-methylkynurenine increases the yield of actinomycin. This indicates that 3-hydroxy-4-methylkynurenine is an intermediate in the conversion of L-tryptophan; so the latter biosynthetic pathway can be assumed[408].

The synthesis of actinomycin is most probably regulated by the rate-limiting reaction: conversion of L-tryptophan to 3-hydroxymethylanthranilic acid.

The presence of D-amino acids is a characteristic of many peptide antibiotics. However, the synthesis of the antibiotic is usually inhibited specifically by addition of the corresponding D-amino acids to the producer organism. Since the corresponding L-amino acid is able to overcome this antagonism, the L-isomer is assumed to be the precursor of the D-isomer in the peptide molecule. This hypothesis has been confirmed for D-valine in valinomycin[319], actinomycin[241, 440] and penicillin[13], for D-leucine in polymyxin D[80], for D-ornithine in bacitracin[26] and for D-allo-isoleucine in actinomycin[8].

Feeding experiments with D-amino acids which are present in the actinomycin molecule, or act as precursors: D-valine, D-allo-isoleucine, D-isoleucine, D-threonine, D-tryptophan inhibit the actinomycin synthesis by ca. 40% at a concentration of 100 μg/ml, while D-amino acids which probably have no relation to the biosynthesis, stimulate the synthesis by ca. 35% at the same concentration.

L-Tryptophan at a concentration of 100 $\mu g ml^{-1}$ stimulates the synthesis of actinomycin D by 46%. If D-(benzene ring-$^{14}C$) tryptophan is added to the culture medium at concentrations which do not inhibit antibiotic production, the D-amino acid is incorporated about 10% less than the L-isomer (D-form 39.1%, L-form 50.1%). Since the incorporation of both isomers is about the same after 72 hours, this may be because of the slower and later uptake of the D-form into the bacterial cells. The L-form is incorporated at an earlier stage.

An alternative hypothesis is that D-tryptophan is first oxidized to indolepyruvate or anthranilic acid and that 4-methyl-3-hydroxyanthranilic acid is then synthesized from this intermediate and acts as a precursor for the chromophore of actinomycin[537]. This route of synthesis may be comparable with that assumed for the ergot alkaloids[110].

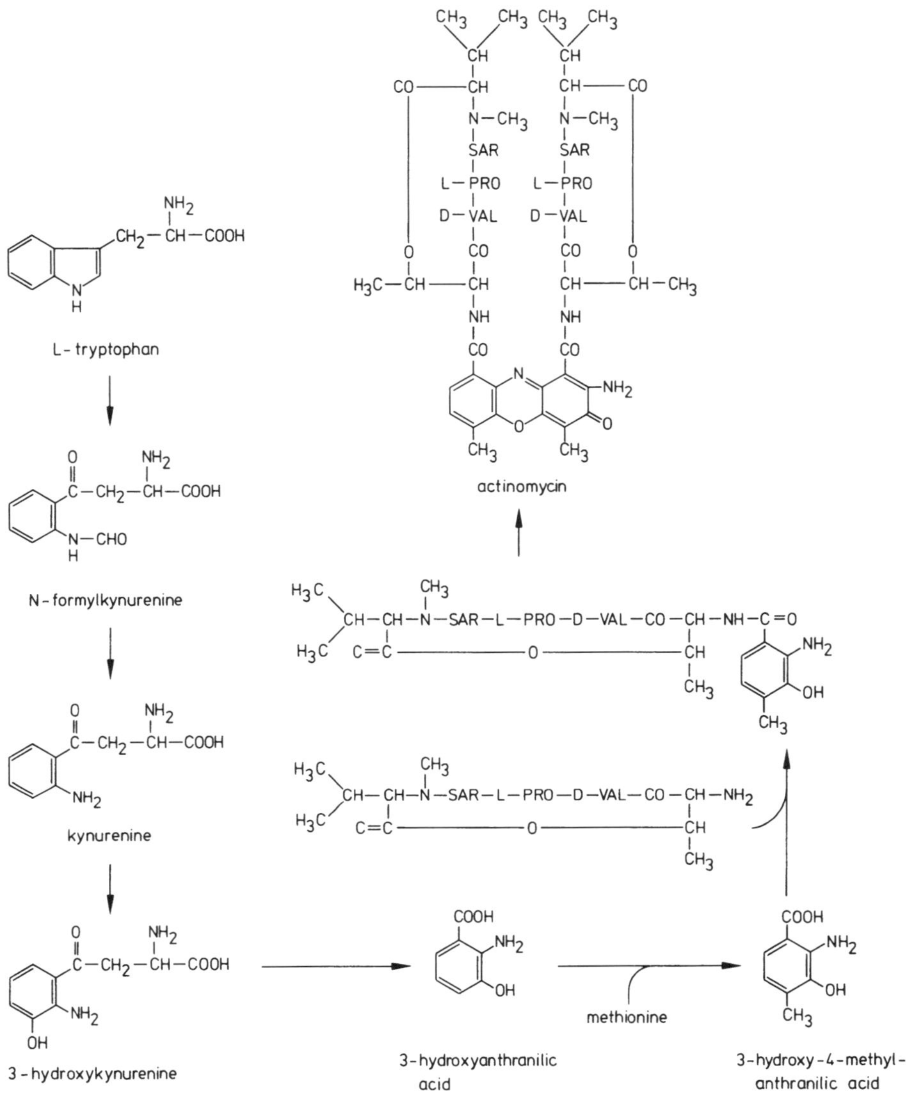


**Fig. 31.** Biosynthesis of actinomycin D (amino acid sequence: L-threonine, D-valine, L-proline, sarcosine, N-methyl-L-valine)

# 5 Regulation of Secondary Metabolites with Antibiotic Activity

Obvious secondary metabolism is more differentiated than primary, yet a regulation mechanism underlies the biosynthesis of secondary metabolites by which their production is controlled. Thus secondary metabolism is controlled both by general regulatory mechanisms such as a nutrient-dependent growth rate, and by specific regulatory mechanisms of the appropriate individual biosynthetic pathway, such as induction, catabolite repression and feedback inhibition[343)].

## *5.1 Initiation of Antibiotic Biosynthesis*

Secondary metabolites with antibiotic activity are mostly formed after the growth phase (trophophase) in the production phase (idiophase). In many fermentations – chloramphenicol, amphenicol, colistin, penicillin and bacitracin – the typical transition from trophophase to idiophase is only observed in complex media and not in a chemically defined medium[153, 154, 173, 206, 330, 411, 467)]. Initiation of antibiotic synthesis can therefore be explained by the deficiency of one or more growth-limiting substances but is not completely understood. The study of antibiotic synthetases, which are significant in the special synthetic pathway for secondary metabolites, casts more light on induction.

The actinocin moiety of actinomycin is formed by phenoxazinone synthetase from two molecules of 4-methyl-3-hydroxyanthranilic acid[236)]. This enzyme is a typical idiophase enzyme in actinocin producing organisms.

While only a low level of enzyme activity is present in the first 20 h of fermentation, in the idiophase it is enriched 12-fold[119)].

Gramicidin S synthetase[280, 359, 509)] and tyrocidin synthetase[117)], synthesizing the corresponding peptide antibiotics, are also first synthesized towards the end of the logarithmic growth phase, shortly before formation of the antibiotic.

Transcription of candicidin synthetase, which is comparable with fatty acid synthetase[434)], does not occur in the first 10 h of fermentation[308)]. The beginning of enzyme synthesis remains repressed until phosphate is exhausted in the medium. Phosphate also inhibits candicidin synthetase when it is formed. Since secondary metabolites have no independent de novo synthesis but arise fom primary metabolism, they often also are subject to the regulatory mechanisms of primary metabolism. Thus, phenylalanine, tyrosine and tryptophan cumulatively inhibit candicidin biosynthesis by 74% while the individual amino acids only inhibit it by 28, 10 and 50% respectively[310, 348)].

Although these regulatory mechanisms starting the antibiotic synthesis are very heterogeneous, they can still be presented in a simple model: a low molecular weight molecule acts as corepressor or inhibitor by blocking the formation or activity of an antibiotic synthetase. To initiate an antibiotic synthesis, this corepressor or inhibitor must attain an intracellular minimum concentration which permits transcription or enzyme activity. Such a model would explain the effects of catabolite-repression, regulation of nitrogen metabolism and phosphate control.

In another model, an inducer is proposed which is formed at a particular time during the fermentation and induces the biosynthesis of the antibiotic.

### *5.1.1 Catabolite-Repression*

Glucose normally is a good carbon source for growth but it affects the biosynthesis of many antibiotics. Poly- or oligosaccharides often are better for antibiotic production[470)]. Combined carbon sources, glucose with a poorly utilized carbon source, have the advantage that a larger cell mass is produced and after the glucose has been consumed, antibiotic production commences with the second carbon source[14, 75, 119)]. This repression of antibiotic biosynthesis is not limited to glucose. The novobiocin-producing organism, *Streptomyces niveus,* grows better with citrate than with glucose. However, novobiocin production is suppressed by citrate[271)].

The molecular mechanism of carbon catabolite regulation may be related to growth rate[44, 347)]. Thus growth-limiting, slow addition of glucose stimulates candidin and candihexin biosynthesis but high doses effect inhibition[351)]. The reduction in biosynthesis of actinomycin by glucose depends on repression of phenoxazinone synthetase catalyzing the formation of the phenoxazinone ring[119)]. Kynurenine formidase I remains unchanged while the synthesis of kynurenine formidase II is completely suppressed[238)]. In some microorganisms, adenosine 3,5-monophosphate (cAMP) is a positive effector for synthesis of inducible, catabolic enzymes[406)]. High doses of glucose also inhibit adenylate cyclase activity and so reduce the intracellular level of cAMP. Then cAMP cannot react with the corresponding receptor protein and this cannot react with the promotor of the operon which activates gene transcription for inducible antibiotic synthetase. This mechanism can be envisaged in particular when the genes are organized within an operon[435, 436)] as in the case of actinorhodin[36)]. The extent to which this is a generally valid mechanism requires confirmation by the appropriate experiments.

The following Table gives some of the antibiotics described in this review together with their corresponding carbon-catabolite repressor molecule as well as the carbon source which is suitable for antibiotic synthesis.

| Antibiotic | Catabolite-repressor | Carbon-source/ antibiotic synthesis |
|---|---|---|
| Actinomycin | Glucose | Galactose |
| Indolmycin | Glucose | Fructose |
| Chloramphenicol | Glucose | Glycerol |
| Novobiocin | Citrate | Glucose |
| Candidin | Glucose | low level of glucose |

### *5.1.2 Regulation of Nitrogen Metabolites*

Many microorganisms possess regulation mechanisms which control the use of the corresponding nitrogen source[90, 326)]. Thus ammonium ions repress enzymes such as nitrite-, nitrate reductase, glutamate dehydrogenase, arginase, ornithine transaminase, extracellular proteases, acetamidase, threonine dehydratase and allantoinase, which have nitrogen available by other reactions than uptake in ionic form.

Although such a regulatory mechanism has not been described for antibiotics from the biosynthesis of aromatic amino acids, it has some relevance for a series of other antibiotics.

The addition of ammonium chloride to erythromycin fermentations drastically suppresses the biosynthesis of the antibiotic[59, 465]. In *Streptomyces noursei,* the producer of nourseothricin, an antibiotic in the streptothricin group, glutamine synthetase is inhibited by high concentrations of ammonium ions[143, 144]. To what extent this is directly connected with simultaneous inhibition of the synthesis of the antibiotic is not completely clear.

Nitrogen metabolite regulation is also involved in cephalosporin synthesis in Streptomycetes[2, 124] where glutamate synthetase and alanine synthetase are concerned in the control mechanism[1].

### *5.1.3 Phosphate Control*

Phosphate is a critical, growth-limiting nutrient and at the same time an inhibitor of antibiotic synthesis in many fermentations. When fermenting *Streptomyces griseus* for candicidin, phosphate has been completely exhausted two h before the start of synthesis[308] and, if added, inhibits the production of candicidin which has already started[311, 345]. If 10 mM phosphate is added to candicidin-producing bacteria, the intracellular level of ATP doubles within 10 min and inhibition of candicidin production commences after 15 min[345]. This indicates that ATP is the active intracellular molecule for the control of antibiotic synthesis. It is still not yet known whether the ATP-concentration or the energy bound by it in the cells is the regulatory parameter.

Besides this ATP regulation mechanism, phosphatase regulation is also significant in control of antibiotic biosynthesis.

In the biosynthesis of streptomycin[367], viomycin[405] and neomycin[327], some of the intermediates, but not the end product, are phosphorylated. Phosphatases which cleave these phosphorylated intermediates are often regulated by feedback-inhibition or repression with inorganic phosphate. Thus the alkaline phosphatase in *Proactinomyces fructiferi* var. *ristomyceticus* is completely repressed in concentrations of inorganic phosphate which inhibit production of ristomycin[510].

### *5.1.4 Induction of Antibiotic Synthesis*

Enzyme induction is a recognized regulation mechanism in primary metabolism. Although this phenomenon has been less well studied in secondary metabolism, it seems to be significant for the synthesis of some antibiotics.

Since it is often difficult to differentiate between an inducer and a precursor of antibiotic synthesis, only those molecules should be designated *inducers* which stimulate antibiotic synthesis if they are added to the fermentation during the growth phase and before the idiophase. *Precursors* can increase the biosynthesis of antibiotics both during the trophophase and during the idiophase.

An interesting induction phenomenon is known in the biosynthesis of streptomycin, in which 2-S-isocapryloyl-3-R-hydroxymethyl-butyrolactone (A-factor) stimulates synthesis[256]. This A-factor can be isolated from the streptomycin producers *Streptomyces griseus* and *Streptomyces bikiniensis* and, if added at the beginning of fermentation of streptomycin "blocked mutants", induces synthesis of this antibiotic by a factor of $10^6$ [250]. However, it cannot be detected in other species which also synthesize streptomycin[249, 250].

Besides induction of streptomycin, the A-factor has a role in the differentiation of bacterial cells. It stimulates both spore formation and intracellular membrane formation. Its mode of action seems to depend on induction of another compound, possibly ADP, which inhibits glucose-6-phosphate dehydrogenase[250].

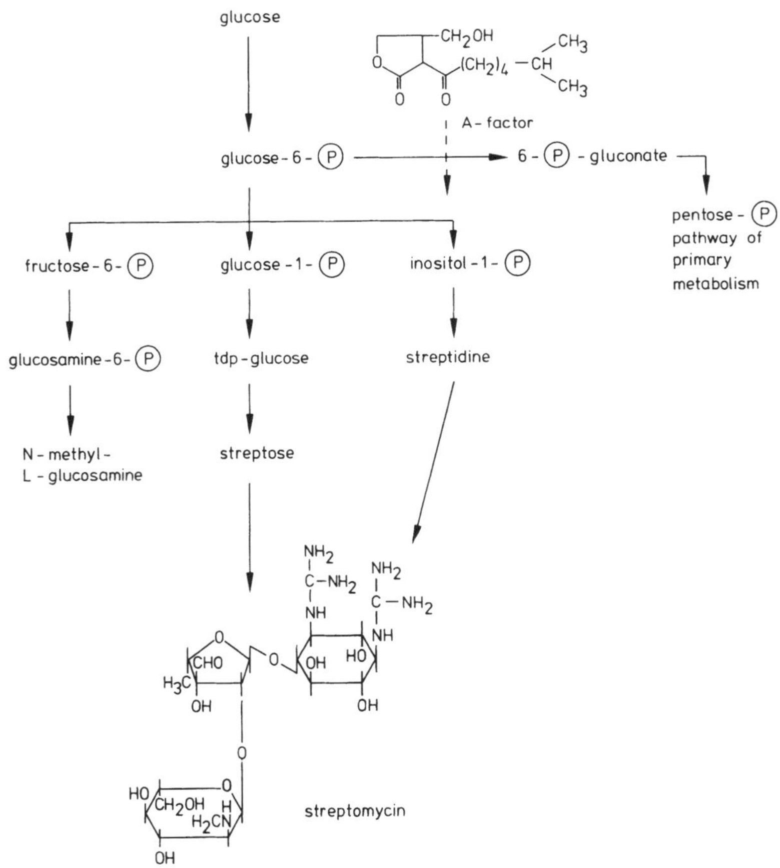


**Fig. 32.** Inhibition of primary metabolism by the A-factor and consequent promotion of streptomycin biosynthesis

Similar induction properties – although possibly not so spectacular as those of a new metabolite – are shown by methionine and norleucine in stimulating cephalosporin C biosynthesis in *Cephalosporium acremonium*[78, 95].

The stimulating effect of 5,5-diethylbarbiturate on the biosynthesis of rifamycin and anthracyclines may also depend on induction of the particular enzyme which hydroxylates aclavin[276)].

Biosynthesis of tetracyclines is induced by the following compounds: benzyl thiocyanate, α-naphthylacetic acid, β-indolylacetic acid, phenylacetic acid and chlorophenoxyacetic acid[524)].

No inducers have been described for antibiotics from the biosynthesis of aromatic amino acids. However, the examples given suggest that such compounds would be found here too by appropriate screening.

## 5.2 Termination of Antibiotic Biosynthesis

Various mechanisms can be envisaged for the termination of antibiotic biosynthesis[346)].

### 5.2.1 Irreversible Decomposition of Enzymes of Antibiotic Biosynthesis

In the biosynthesis of peptide antibiotics, the activity of the corresponding antibiotic synthetase declines only a few hours after the start of synthesis.

Tyrocidin synthetase activity disappears from the cytoplasma of *Bacillus brevis* and is still detectable for a few hours bound to the membrane[291)].

The same holds for bacitracin synthetase in *Bacillus licheniformis* which is membrane-bound shortly after the start of antibiotic production. This membrane binding of antibiotic synthetases may represent a natural form of immobilized enzyme which is then protected against proteolytic enzymes at the end of the growth phase[116, 444)].

Gramicidin S synthetase seems to be subject to oxygen-dependent oxidation[260)] and inactivation of some sulphydryl groups within the enzyme[288)]. If oxygen is replaced by nitrogen, the inactivation of gramicidin S synthetase can be prevented[113, 114)]. This indicates that a lower oxygen transfer in the idiophase of fermentation should increase the production of gramicidin S in vivo.

### 5.2.2 Feedback-Inhibition in Antibiotic Biosynthesis

Studies on feedback regulation assume precise knowledge about the enzyme concerned in the reaction. This assumption is correct in the case of primary metabolism but our knowledge is not always complete for secondary metabolism.

For antibiotics from the primary metabolism of aromatic amino acids, such a regulation mechanism has only been described for chloramphenicol. In its biosynthesis, the first enzyme – arylamine synthetase, which converts chorismic acid into p-aminophenylalanine – is subject to feedback inhibition by the end product of the synthesis, chloramphenicol[223, 224, 315, 316, 332)]. This enzyme only occurs in chloramphenicol producers and is completely repressed at a concentration of 100 $\mu g\ ml^{-1}$ chloramphenicol. This concentration also corresponds approximately to the production maximum in *Streptomyces sp.* 3022 a.

Other examples of feedback regulation are mentioned briefly below. Puromycin inhibits the last enzyme step, O-methyltransferase, in its synthesis[442]. Aurodox at a concentration of 400 μg $ml^{-1}$, which is the maximum amount reached in the fermentation, completely inhibits antibiotic synthesis[309]. Penicillin at a concentration of 200 μg $ml^{-1}$ completely inhibits penicillin synthesis in the producer organism, *Penicillium chrysogenum* NRRL 1951, in which the production maximum is 125 μg $ml^{-1}$ [133]. Feedback regulation has also been observed in the biosynthesis of cycloheximide[272, 273], staphylomycin[567], ristomycin[92], fungicidin[475] and candihexin[351].

# 6 Discussion

Analysis of the origin of secondary from primary metabolism, taking the example of the aromatic amino acids leads to the result that the individual metabolism is favoured by branching within primary metabolism; this is the case with chorismic acid or kynurenine or with the transfer from anabolism to catabolism. Consequently evolution is always possible at points at which the reaction leads to an end product which is not derived from a linear course. Catabolism seems to be favoured for production of secondary metabolites, as with anthranilic aicd. This intermediary metabolite enters into antibiotic biosynthesis from catabolism and not anabolism in the biosynthesis of actinomycin and tryptanthrin.

These branch points in anabolism, as in catabolism, also permit divergent synthetic routes in primary metabolism, whose reaction products are not uniform in their biological activity.

The transition from primary to secondary metabolism is particularly favoured if, as in these antibiotic producers, primary metabolism is not so strongly regulated as in the Enterobacteriaceae. Metabolites then accumulate, like chorismic aicd, at the end of the trophophase when metabolic conversion declines, and are transformed by individual antibiotic synthetases into antibiotic metabolites. On the other hand, if primary metabolism proceeds intact in the trophophase, no antibiotic synthesis is possible because of the inadequate pool of free metabolites from primary metabolism. This theory can confirm that precursors of antibiotic synthesis stimulate the yield if they are added at a given excess to the fermentation medium.

Further, in most cases, antibiotic synthetases are first formed at the end of the trophophase in antibiotic-producing microorganisms. So, in one and the same organism, as in the case of actinomycin synthesis by *Streptomyces parvulus,* one antibiotic synthetase, kynurenine formamidase I, is produced constitutively and the other, kynurenine formamidase II inductively shortly before synthesis of actinomycin[42, 238]. While the function of the constitutive enzyme is not known precisely, the derepressible enzyme seems to be important in the actinomycin synthesizing system.

Addition of appropriate inhibitors of mRNA or protein synthesis demonstrates that many of these antibiotic synthetases represent a de novo synthesis[238]. Such a synthesis, which is initiated by derepession or induction and can lead to up to 60-fold increase in the original enzyme activity, seems to be directly connected with the enrichment of metabolites from primary metabolism or a reduction of essential constituents of the nutrient medium.

At least for the commercial efficiency of metabolites with antibiotic activity, it is an advantage that they are first formed after the trophophase since during this phase, the producers are sensitive to their own antibiotics[9, 124, 179, 332]. Resistance develops in a coordinated manner with antibiotic production by enzymatic modification of the antibiotic, by a change in site of action or reduction in uptake of the antibiotic excreted[76].

Since antibiotic synthetases, at least in complex nutrient media, are often either induced or derepressed in coordination with enrichment of metabolites from primary metabolism, one attempts to see in antibiotics not only products created by chance but to ascribe to them a function in differentiation of bacterial cells[239]. This remains speculation as to how far antibiotics cut off the macromolecular syntheses of DNA, RNA, proteins, cell wall and respiratory chain and so initiate sporulation.

The structural diversification of idiolites arising from the idiophase is illustrated by numerous classes of organic compounds including aminosugars, quinones, coumarins, epoxides, ergot alkaloids, glutarimides, glycosides, indoles, lactones, macrolides, naphthalenes, peptides, phenazines, polyacetylenes, polyenes, pyrroles, quinolines, terpenoids and tetracyclines[24, 343, 424]. Mostly a secondary metabolite base structure is synthesized in numerous modifications by one or more producers. How far this variation in secondary metabolites should be attributed to their specific function in the physiology of the cell or how far these numerous structural types, often with similar modes of action, remain because of the lack of selection pressure, cannot be answered here.

The knowledge of regulation mechanisms[348] for initiation and termination is as important for optimizing production as the consideration of resistance mechanisms. Both may limit the yield. Glucose repression is mostly avoided by slow addition of a carbon source which is fed by control of the dissolved oxygen. Nitrogen repression can be controlled by a slow continuous nitrogen source by addition of proteins or proline. Phosphate repression can be reduced by feeding poorly soluble phosphates.

Appropriate deregulated[98] or highly resistant mutants of producer organisms or consideration of particular constituents of the nutrient medium[354] may have economic significance which justifies studies of biosynthesis and regulation on these grounds.

Although our overall knowledge on the genetic control of antibiotic synthesis in actinomycetes is relatively sparse, there is abundant evidence that plasmids are significant[398]. This can be established either directly, as in the case of methylenomycin A in *Streptomyces coelicolor* A 3, where the structural genes are localized on the plasmid, or the plasmid genes can exert an indirect function. Thus, with chloramphenicol synthesis in *Streptomyces venezuelae,* expression of the chromosomal structural gene is controlled by the plasmid[180]. The methods of gene technology available at present – cloning of DNA fragments to plasmid – or bacteriophage vectors in streptomycetes – offer hopeful means for physical and genetic analysis of structural and regulatory genes for antibiotic synthetases[181]. This opens up the possibility of permitting antibiotic synthesis in bacteria with shorter generation times, which also grow on simpler, more cost-effective nutrient media, in which the antibiotic can be obtained in high yield in a continuous process. This technology, through knowledge of antibiotic biosynthesis, offers the possibility of creating new antibiotics by hybridisation.

## Acknowledgements

I would like to express my sincere thanks to Prof. Dr. R. Engelhorn, Dr. H. Goeth and Prof. Dr. H. Machleidt, Dr. Karl Thomae GmbH, Biberach; Prof. Dr. H. Klupp and Prof. Dr. F. Waldeck, Boehringer Ingelheim, Ingelheim; Prof. Dr. H. Zähner, Universität Tübingen, Lehrstuhl für Mikrobiologie I, Tübingen; Prof. Dr. A. L. Demain, Massachusetts Institute of Technology, Department of Nutrition and Food Science, Cambridge, USA; Prof. Dr. H. G. Floss, Purdue University, Department of Medicinal Chemistry and Pharmacognosy, West Lafayette, USA; Prof. Dr. M. K. Speedie, University of Maryland, College of Pharmacy, Baltimore, USA; and Prof. Dr. J. F. Martin, Universidad de Salamanca, Departmento de Microbiologia, Faculatad de Ciencias, Salamanca, Spain, for their stimulating discussions which inspired the conception of this treatise.

## References

1. Aharonowitz, American Society for Microbiology, Washington, D. C., p. 210–217, 1979
2. Aharonowitz, Y., Demain, A. L.: Can. J. Microbiol. *25,* 61–67 (1979)
3. Ajisaka, M., Kariyone, K., Fomon, K., Yazawa, H., Arima, K.: Agric. Biol. Chem. *33,* 294–295 (1969)
4. Akagawa, H., Okanishi, M., Umezawa, H.: Jpn. J. Genet. *49,* 285 (1974)
5. Akagawa, H., Okanishi, M., Umezawa, H.: J. Gen. Microbiol. *90,* 336–346 (1975)
6. Akagawa, H., Okanishi, M., Umezawa, H.: J. Antibiotics *32,* 610–620 (1979)
7. Akagawa, K., Tsukada, Y.: J. Neurochem. *32,* 269–272 (1979)
8. Albertini, A., Cassani, G., Ciferri, O.: Biochim. Biophys. Acta *80,* 655–664 (1964)
9. Al'Nuri, M. A., Egorov, N. S.: Mikrobiologiya *37,* 413–416 (1968)
10. Arima, K. Imanaka, H., Kousaka, M., Fukuta, A., Tamura, G.: Agr. Biol. Chem. *28,* 575–576 (1964)
11. Arima, K., Imanaka, H., Konsaka, M., Fukuda, A., Tamura, G.: J. Antibiot. *18,* 201–204 (1965)
12. Arima, K., Kohsaka, M., Tamura, G., Imanaka, H., Sakai, H.: J. Antibiotics *25,* 437–444 (1972)
13. Arnstein, H. R. V., Morris, D.: Biochem. J. *76,* 323–327 (1960)
14. Audkya, T. U., Russell, D. W.: J. Gen. Microbiol. *86,* 327–332 (1975)
15. Banerjee, A. B., Bose, S. K.: J. Bact. *87,* 1397–1401 (1964)
16. Baker, T. I.: Biochemistry *5,* 2654–2657 (1966)
17. Baker, T. I.: Genetics *58,* 351–359 (1968)
18. Baker, T. I., Crawford, I. P.: J. Biol. Chem. *241,* 557–584 (1966)
19. Battersby, A. R.: Quarterly Reviews *15,* 259–286 (1961)
20. Bauerle, R. H., Margolin, P.: Cold Spring Harbor Symp. Quat. Biol. *31,* 203–214 (1966)
21. Beer, R. J. S., Broadhurst, T., Robertson, A.: J. Chem. Soc., 1947–1953 (1954)

22. Beer, R. J. S., Jennings, B. E., Robertson, A.: J. Chem. Soc. *885,* 2679–2685 (1954)
23. Belser, W. L., Baron-Murphy, J., Delmer, D. P., Mills, S. E.: Biochem. Biophys. Acta *273,* 1–10 (1971)
24. Berdy, J.: Adv. Appl. Microbiol. *18,* 309–406 (1974)
25. Berl, S., Takagaki, G., Clarke, D. D., Waelsch, H.: J. Biol. Chem. *237,* 2562–2569 (1962)
26. Bernlohr, R. W., Novelli, G. D.: Arch. Biochem. Biophys. *103,* 94–104 (1963)
27. Bertrand, K., Korn, L., Lee, F., Platt, T., Squires, C. L., Squires, C., Yanofsky, C.: Science *189,* 22–26 (1975)
28. Bertrand, K., Squires C., Yanofsky, C.: J. Mol. Biol. *103,* 319–337 (1976)
29. Birch, A. J., Cameron, D. W., Holloway, P. W., Rickards, R. W.: Tetrahedron letters *25,* 26–31 (1960)
30. Birch, A. J., Holloway, P. W., Rickards, R. W.: Biochim. Biophys. Acta *57,* 143–145 (1962)
31. Birch, A. J., Holzapfer, C. W., Rickards, R. W., Djerassi, C., Suzuki, M., Westley, J., Dutcher, J. D., Thomas, R.: Tetrahedron letters 1964, 1485–1490 (1964)
32. Blume, A. J., Balbinder, E.: Genetics *53,* 577–592 (1966)
33. Bondinell, W. E., Vnek, J., Knowles, P. F., Sprecher, M., Sprinson, D. B.: J. Biol. Chem. *246,* 6191–6196 (1971)
34. Borowska, Z. K., Tatum, E. L.: Biochem. Biophys. Acta *114,* 206–209 (1966)
35. Braunshtein, A. E., Goryachenkova, E. V., Paskhina, T. S.: Biokhimiya *14,* 163–179 (1949)
36. Brockmann, H.: Angew. Chem. *76,* 863 (1964)
37. Bronson, M. J., Squires, C., Yankofsky, C.: Proc. Nat. Acad. Sci. USA, *70,* 2335–2339 (1973)
38. Brot, N., Smit, Z., Weissbach, H.: Arch. Biochem. Biophys. *112,* 1–6 (1965)
39. Brown, K. D.: Genetics *60,* 31–48 (1969)
40. Brown, K. D., Doy, C. H.: Biochim. Biophys. Acta *77,* 170–172 (1963)
41. Brown, K. D., Doy, C. H.: Biochim. Biophys. Acta *104,* 377–389 (1965)
42. Brown, K. D., Hitchcock, M. J. M., Katz, E.: Arch. Biochem. Biophys. *202,* 18–22 (1980)
43. Brown, K. D., Somerville, R. L.: J. Bacteriol. *108,* 386–399 (1971)
44. Bu'Lock, J. D., Spencer, B. (Ed.), North Holland Publishing Co., p. 335–346, Vol. 1, Amsterdam 1974
45. Bu'Lock, J. D., Harley-Mason, J.: J. Chem. Soc. *703,* 2248–2252 (1951)
46. Caglioti, L., Cirani, G., Misiti, D., Arcamone, F., Minghetti, A.: J. Chem. Soc. (Perkin I) 1235–1237 (1972)
47. Caglioti, L., Misiti, D., Mondelli, R., Selva, A., Arcamone, F., Cassinelli, G.: Tetrahedron *25,* 2193–2221 (1969)
48. Calvert, R. T., Spring, M. S., Stoker, J. R.: J. Pharm. Pharmac. *24,* 972–978 (1972)
49. Campbell, J. M., Reusser, F., Caskey, C. T.: Biochem. Biophys. Res. Commun. *90,* 1032–1038 (1979)
50. Carlton, B. C., Yanofsky, C.: J. Biol. Chem. *237,* 1531–1534 (1962)
51. Chambers, K., Kenner, G. W., Temple Robinson, M. J., Webster, B. R.: Proc. Chem. Soc. 291–292 (1960)
52. Chang, C. J., Floss, H. G., Hook, D. J., Mabe, J. A., Manni, P. E., Martin, L. L., Schröder, K., Shieh, T. L.: J. Antibiotics *34,* 555–566 (1981)
53. Chirpich, T. P., Zappia, V., Barker, H. A.: Biochim. Biophys. Acta *207,* 505–513 (1970)
54. Chirpich, T. P., Zappia, V., Costilow, R. N., Barker, H. A.: J. Biol. Chem. *245,* 1778–1789 (1970)
55. Cohen, M. M., Simon, G. R., Berry, J. F., Chain, E. B.: Biochem. H. *84,* 43–44 (1962)
56. Cohn, W., Crawford, I. P.: J. Bact. *127,* 267–279 (1976)
57. Cohen, L., Witkop, B.: Angew. Chem. *73,* 253–276 (1961)
58. Cholson, R. K., Ueda, J., Ogasawara, N., Henderson, L. M.: J. Biol. Chem. *239,* 1208–1212 (1964)
59. Corcoran, J. W.: Amsterdam: Elsevier, Vol. 2, p. 339–351, 1973
60. Cotton, R. G. H., Gibson, F.: Biochim. Biophys. Acta *100,* 76–88 (1965)
61. Cotton, R. G. H., Gibson, F.: Biochim. Biophys. Acta *147,* 222–237 (1967)
62. Cotton, R. G. H., Gibson, F.: Biochim. Biophys. Acta *160,* 188–195 (1968)
63. Crawford, I. P., Gunsalus, I. C.: Proc. Nat. Acad. Sci. *56,* 717–724 (1966)
64. Crawford, I. P., Sikes, S., Melhorn, D. K.: Arch. Microbiol. *59,* 72–81 (1967)
65. Creighton, T. E., Yanofsky, C.: J. Biol. Chem. *241,* 980–990 (1966)

66. Creveling, C. R., Daly, J. W., Witkop, B., Udenfriend, S.: Biochim. Biophys. Acta *64*, 125–134 (1962)
67. Dalgliesh, C. E., Knox, W. E., Neuberger, A.: Nature *168*, 20–22 (1951)
68. Dalgliesh, C. E.: Advance Protein Chem. *10*, 31–150 (1955)
69. Daly, J. W., Jerina, D. M., Witkop, B.: Experientia *28*, 1129–1149 (1972)
70. Davidson, B. E., Blackburn, E. H., Dopheide, T. A. A.: J. Biol. Chem. *247*, 4441–4446 (1972)
71. Davis, B. D., Weiss, U.: Arch. exper. Path. u. Pharmakol. *222*, 1–15 (1953)
72. Dayan, J., Sprinson, D. B.: Fed. Prod. *27*, 290 (1968)
73. DeCastro, F. T., Price, J. M., Brown, R. R.: J. Am. Chem. Soc. *78*, 2904–2905 (1956)
74. DeLeo, A. B., Sprinson, D. B.: Biochem. Biophys. Res. Comm. *32*, 873–877 (1968)
75. Demain, A. L.: Clin. Med. *70*, 2045–2051 (1963)
76. Demain, A. L.: Ann. N.Y. Acad. Sci. *235*, 601–612 (1974)
77. Demain, A. L., Matteo, C. L.: Antimicrob. Agents Chemother. *9*, 1000–1003 (1976)
78. Demain, A. L., Newkrik, J. F., Hendlin, D.: J. Bacteriol. *85*, 339–344 (1963)
79. DeMoss, J. A.: Biochim. Biophys. Acta *62*, 279–293 (1962)
80. DiGirolamo, M., Ciferri, O., DiGirolamo, A. B., Albertini, A.: J. Biol. Chem. *239*, 502–507 (1964)
81. Doy, C. H.: Rev. Pure Appl. Chem. *18*, 41–78 (1968)
82. Doy, C. H.: Biochim. Biophys. Acta *198*, 364–375 (1970)
83. Doy, C. H.: Biochim. Biophys. Acta *151*, 293–295 (1968)
84. Doy, C. H.: Biochim. Biophys. Acta *159*, 352–366 (1968)
85. Doy, C. H., Brown, K. D.: Biochim. Biophys. Acta *104*, 377–389 (1965)
86. Doy, C. H., Cooper, J. M.: Biochim. Biophys. Acta *127*, 302–316 (1966)
87. Doy, C. H., Haball, D. M.: Biochim. Biophys. Acta *185*, 432–446 (1969)
88. Doy, C. H., Rivera, Jr., A., Srinivasan, P. R.: Biochem. Biophys. Res. Comm. *4*, 83–88 (1961)
89. Drew, S. W., Demain, A. L.: Antimicrob. Agents Chemother. *8*, 5–10 (1975)
90. Dubois, E., Grenson, M., Wiame, J. M.: Eur. J. Biochem. *48*, 603–616 (1974)
91. Edwards, S. W., Knox, W. E.: J. Biol. Chem. *220*, 79–91 (1956)
92. Egorov, N. S., Toropova, E. G., Suchkova, L. A.: Mikrobiologiya *40*, 475–480 (1971)
93. Ehrlich, J., Bartz, Q. R., Smith, R. M., Joslyn, D. A.: Science *106*, 417–419 (1947)
94. Eikhom, T. S., Jonsen, J., Laland, S., Refsvik, T.: Biochim. Biophys. Acta *76*, 465–468 (1963)
95. Eikhom, T. S., Jonsen, J., Laland, S., Refsvik, T.: Biochem. Biophys. Acta *80*, 648–654 (1964)
96. Eikhom, T. S., Laland, S.: Biochem. J. *92*, 32 (1964)
97. Elander, R. P., Mabe, J. A., Hamill, R. L., Gorman, M.: Appl. Microbiol. *16*, 753–758 (1968)
98. Elander, R. P., Mabe, J. A., Hamill, R. L., Gorman, M.: Folia Microbiol. *16*, 156–165 (1971)
99. Emes, A., Floss, H. G., Lowe, D. A., Westlake, D. W. S., Vining, L. C.: Can. J. Microbiol. *20*, 347–352 (1974)
100. Ensign, S., Kaplan, S., Bonner, D. M.: Biochim. Biophys. Acta *81*, 357–366 (1964)
101. Fantes, P. A., Roberts, L. M., Hütter, R.: Arch. Microbiol. *107*, 207–214 (1976)
102. Fazel, A. M., Jensen, R. A.: J. Bacteriol. *138*, 805–815 (1979)
103. Fickenscher, U., Keller-Schierlein, W., Zähner, H.: Arch. Mikrobiol. *75*, 346–352 (1971)
104. Fickenscher, U., Zähner, H.: Arch. Mikrobiol. *76*, 28–46 (1971)
105. Fiedler, E., Fiedler, H.-P., Gerhard, A., Keller-Schierlein, W., König, W. A., Zähner, H.: Arch. Mikrobiol. *107*, 249–256 (1976)
106. Figenschon, K. J., Froholm, L. O., Laland, S. G.: Biochem. J. *105*, 451–453 (1967)
107. Finkelstein, A., Holz, R.: in: In Membranes, Eisenstein, G. (Ed.), Dekker, pp. 377–408, New York 1973
108. Floss, H. G.: in: "Antibiotics IV Biosynthesis", Corcoran, J. W. (Ed.), Springer-Verlag, pp. 245–253, 258, 1981
109. Floss, H. G., Manni, P. E., Hamill, R. L., Mabe, J. A.: Biochem. Biophys. Res. Comm. *45*, 781–787 (1971)
110. Floss, H. G., Mothes, U., Günther, H.: Z. Naturforschung *19b*, 784–788 (1964)
111. Floss, H. G., Onderka, D. K., Carroll, M.: J. Biol. Chem. *247*, 736–744 (1972)
112. Francis, M. M., Vining, L. C., Westlake, D. W. S.: J. Bacteriol. *134*, 10–16 (1978)

113. Friebel, T. E., Demain, A. L.: FEMS Microbiol. Lett. *1,* 215–218 (1977)
114. Friebel, T. E., Demain, A. L.: J. Bacteriol. *130,* 1010–1016 (1977)
115. Friedman, S., Kaufman, S.: J. Biol. Chem. *240,* 552–554 (1964)
116. Froyshov, O.: FEBS Lett. *81,* 315–318 (1977)
117. Fujikawa, K., Suzuki, T., Kurahashi, K.: Biochim. Biophys. Acta *161,* 232–246 (1968)
118. Gaertner, F. H., DeMoss, J. A.: J. Biol. Chem. *244,* 2716–2725 (1969)
119. Gallo, M., Katz, E.: J. Bacteriol. *109,* 659–667 (1972)
120. Gholson, R. K., Ueda, I., Ogasawara, N. Henderson, L. M.: J. Biol. Chem. *239,* 1208–1214 (1964)
121. Gibson, F., Gibson, M., Cox, G. B.: Biochim. Biophys. Acta *82,* 637–638 (1964)
122. Gibson, M. I., Gibson, F.: Biochem. J. *90,* 248–256 (1964)
123. Gibson, F., Pittard, J.: Bact. Rev. *32,* 465–492 (1968)
124. Ginther, C. O.: Antimicrob. Agents Chemother. *15,* 522–526 (1979)
125. Glaubinger, D., Kohn, K. W., Charney, E.: Biochim. Biophys. Acta *361,* 303–311 (1974)
126. Goldberg, M. E., Creighton, T. E., Baldwin, R. L., Yanofsky, C.: J. Mol. Biol. *21,* 71–82 (1966)
127. Goldstein, M., Contrera, J. F.: J. Biol. Chem. *237,* 1898–1902 (1962)
128. Gollub, E. G., Liu, K. P., Sprinson, D. B.: J. Bact. *115,* 121–128 (1973)
129. Gollub, E. G., Liu, K. P., Sprinson, D. B.: J. Bact. *115,* 1094–1102 (1973)
130. Gollub, E., Sprinson, D. B.: Biochem. Biophys. Res. Comm. *35,* 389–395 (1969)
131. Gollub, E., Zalkin, H., Sprinson, D. B.: J. Biol. Chem. *242,* 5323–5328 (1967)
132. Goodall, McG., Kirshner, N.: J. Biol. Chem. *226,* 213–221 (1957)
133. Gordee, E. Z., Day, L. E.: Antimicrob. Agents Chemother. *1,* 315–322 (1972)
134. Gordee, R. S., Matthews, T. R.: Appl. Microbiol. *17,* 690–694 (1969)
135. Gorman, M., Hamill, R. L., Elander, R. P., Mabe, J.: Biochem. Biophys. Res. Comm. *31,* 294–298 (1968)
136. Gorman, M., Lively, D. H.: Antibiotics *2,* 433–438 (1967)
137. Gottlieb, D., Bhattacharyya, P. K., Anderson, H. W., Carter, H. E.: J. Bacteriol. *55,* 409–417 (1948)
138. Gottlieb, D., Carter, H. E., Legator, M., Gallicchio, V.: J. Bacteriol. *68,* 243–251 (1954)
139. Gottlieb, D., Carter, H. E., Robbins, P. W., Brug, R. W.: J. Bact. *84,* 888–895 (1962)
140. Gottlieb, D., Diamond, L.: Bull. Torrey Botan. Club *78,* 56–60 (1951)
141. Gould, S. J., Chang, C. C.: J. Am. Chem. Soc. *99,* 5496–5497 (1977)
142. Gould, S. J., Chang, C. C.: J. Am. Chem. Soc. *100,* 1624–1626 (1978)
143. Gräfe, U., Bocker, H., Reinhardt, G., Thrum, H.: Z. Allg. Mikrobiol. *14,* 659–673 (1974)
144. Gräfe, U., Bocker, H., Thrum, H.: Z. Allg. Mikrobiol. *17,* 201–209 (1977)
145. Grieshaber, M., Bauerle, R.: Nature New Biol. *236,* 232–235 (1972)
146. Gross, S., Dawson, R. F.: Biochemistry *2,* 186–188 (1963)
147. Greenberg, D. M.: in: "Metabolic Pathways", Greenberg, D. M. (Ed.), Academic Press, vol. 1, p. 153, New York, N.Y. 1969
148. Grunberg, E., Prince, H. N., Titsworth, E., Beskid, G., Tendler, M. D.: Chemotherapia *11,* 249–260 (1966)
149. Guroff, G., Daly, J. W., Jerins, D. M., Renson, J., Witkop, B., Udenfried, S.: Science *157,* 1524–1530 (1967)
150. Guroff, G., Ito, T.: J. Biol. Chem. *240,* 1175–1184 (1965)
151. Guroff, G., Ito, T.: Biochim. Biophys. Acta *77,* 159–161 (1963)
152. Guroff, G., Rhoads, C. A.: J. Biol. Chem. *244,* 142–146 (1969)
153. Haavik, H. J.: J. Gen. Microbiol. *81,* 383–390 (1974)
154. Haavik, H. J.: J. Gen. Microbiol. *84,* 321–326 (1974)
155. Hager, S. E., Gregerman R. I., Knox, W. E.: J. Biol. Chem. *225,* 935–947 (1957)
156. Hagino, H., Nakagawa, K.: Agr. Biol. Chem. *38,* 2125–2134 (1974)
157. Hagino, H., Nakagawa, K.: Agr. Biol. Chem. *38,* 2367–2376 (1974)
158. Hagino, H., Nakagawa, K.: Agr. Biol. Chem. *39,* 323–330 (1975)
159. Hagino, H., Nakagawa, K.: Agr. Biol. Chem. *39,* 331–342 (1975)
160. Hagino, H., Nakagawa, K.: Agr. Biol. Chem. *39,* 351–361 (1975)
161. Hahn, F. E.: Antibiotics *1,* 308–330 (1967)
162. Hahn, F. E.: 5th Int. Congr. Chemother. Proc., Vienna, *4,* 387–390 (1967)

163. Halsall, D. M., Doy, C. H.: Biochim. Biophys. Acta *185,* 432–446 (1969)
164. Hamill, R., Elander, R., Mabe, J., Gorman, M.: Antimicrob. Ag. Chemother. *1967,* 388–396 (1968)
165. Hamill, R. L., Elander, R. P., Mabe, J. A., Gorman, M.: Appl. Microbiol. *19,* 721–725 (1970)
166. Hamilton-Miller, J. M. T.: Adv. Appl. Microbiol. *17,* 109–134 (1974)
167. Harnden, M. R., Wright, N. D.: J. Chem. Soc. Perkin I, 1012–1018 (1977)
168. Hashimoto, M., Hattori, K.: Chem. Pharm. Bull. *16,* 1144 (1968)
169. Hashimoto, M., Hattori, K.: Bull. Chem. Soc. Japan *39,* 410 (1966)
170. Hayaishi, O., Katagiri, M., Rothberg, S.: J. Biol. Chem. *229,* 905–920 (1957)
171. Hayaishi, O., Stanier, R. Y.: J. Bact. *62,* 691–710 (1951)
172. Henderson, E. J., Zalkin, H.: J. Biol. Chem. *246,* 6891–6898 (1971)
173. Hendlin, D.: Arch. Biochem. Biophys. *24,* 435–446 (1949)
174. Henning, U., Helinski, D. R., Chao, F. C., Yanofsky, C.: J. Biol. Chem. *237,* 1523–1550 (1962)
175. Herbert, R. B.: Tetrahedron Letters *51,* 4525–4528 (1974)
176. Hill, R. K., Newkome, G. R.: J. Am. Chem. Soc. *91,* 5893–5894 (1969)
177. Hirota, A., Higashinaka, Y., Sakai, H.: Agric. Biol. Chem. *42,* 147–151 (1978)
178. Hoch, S. O., Anagnostopoulos, C., Crawford, I. P.: Biochem. Biophys. Res. Comm. *35,* 838–844 (1969)
179. Hoeksema, H., Smith, C. G.: Prog. Industr. Microbiol. *3,* 91–139 (1961)
180. Hopwood, D.-P.: Ann. Rev. Microbiol. *32,* 373–392 (1978)
181. Hopwood, D. A., Merrick, M. J.: Bacteriol. Rev. *41,* 595–635 (1977)
182. Hornemann, U., Hurley, L. H., Speedie, M. K., Floss, H. G.: Tetrahedron Lett. *26,* 2255–2258 (1970)
183. Hornemann, U., Hurley, L. H., Speedie, M. K., Floss, H. G.: J. Am. Chem. Soc. *93,* 3028–3035 (1971)
184. Hornemann, U., Speedie, M. K., Kelley, K. M., Hurley, L. A., Floss, H. G.: Arch. Biochem. Biophys. *131,* 430–440 (1969)
185. Hornemann, U., Speedie, M. K., Hurley, L. H., Floss, H. G.: Biochem. Biophys. Res. Comm. *39,* 594–599 (1970)
186. Horwitz, S. B., Chang, S. C., Grollman, A. P., Borkovec, A. B.: Science *174,* 159–161 (1971)
187. Hu, R. C. Y.: Diss. Abstr. Int. B *35,* 5939 (1974)
188. Hurley, L. H., Bialek, D.: J. Antibiot. *27,* 49–56 (1974)
189. Hurley, L. H., Gairola, C.: Antimicrob. Agents Chemother. *15,* 42–45 (1979)
190. Hurley, L. H., Gairola, C., Das, N.: Biochemistry *15,* 3760–3769 (1976)
191. Hurley, L. H., Gairola, C., Zmijewski, Jr., M. J.: J. C. S. Chem. Comm. 120–121 (1975)
192. Hurley, L. H., Lasswell, W. L., Malhotra, R. K., Das, N. V.: Biochemistry *18,* 4225–4229 (1979)
193. Hurley, L. H., Speedie, M. K.: in: Antibiotics, Vol. IV, Biosynthesis, Corcoran, J. W. (ed.), Springer-Verlag, pp. 262–294, 1981
194. Hurley, L. H., Zmijewski, M.: J. C. S. Chem. Comm. 337–338 (1974)
195. Hurley, L. H., Zmijewski, M., Chang, C.-J.: J. Am. Chem. Soc. *97,* 4372–4378 (1975)
196. Hütter, R., DeMoss, J. A.: J. Bact. *94,* 1896–1907 (1967)
197. Hutchinson, M. A., Belser, W. L.: J. Bact. *98,* 109–115 (1969)
198. Hutchinson, C. R., Heckendorf, A. H., Daddona, P. E., Hagaman, E., Wenkert, E.: J. Am. Chem. Soc. *96,* 5609–5616 (1974)
199. Hwang, L. H., Zalkin, H.: J. Biol. Chem. *246,* 6899–6907 (1971)
200. Ichihara, A., Adachi, K., Hosokawa, K., Takeda, Y.: J. Biol. Chem. *237,* 2296–2302 (1962)
201. Ikeda, M., Fahien, L. A., Udenfriend, S.: J. Biol. Chem. *241,* 4452–4456 (1966)
202. Ikeda, M., Tsuji, H., Nakamura, S., Ichiyama, A., Nishizuka, Y., Jayaishi, O. H.: J. Biol. Chem. *240,* 1395–1401 (1965)
203. Im, S. W. K., Davidson, H., Pittard, J.: J. Bact. *108,* 400–409 (1971)
204. Imamoto, F., Ito, J., Yanofsky, C.: Cold Spring Harbor Symp. Quant. Biol. *1,* 235–249 (1966)
205. Imanaka, H., Kousaka, M., Tamura, A., Arima, K.: J. Antibiot. Ser. A *18,* 205–206 (1965)
206. Ito, M., Aida, J., Uemura, T.: Agr. Biol. Chem. *33,* 262–269 (1969)
207. Ito, J., Cox, E. C., Yanofsky, C.: J. Bacteriology *97,* 725–733 (1969)
208. Ito, J., Crawford, I. P.: Genetics *52,* 1303–1316 (1965)

209. Ito, J., Yanofsky, C.: J. Biol. Chem. *241,* 4112–4114 (1966)
210. Jackson, D. A., Yanofsky, C.: J. Biol. Chem. *244,* 4539–4546 (1969)
211. Jackson, D. A., Yanofsky, C.: J. Biol. Chem. *244,* 4526–4538 (1969)
212. Jackson, E. N., Yanofsky, C.: J. Bact. *117,* 502–508 (1974)
213. Jacoby, W. B.: PhD Dissertation, Yale University 1954
214. Jakoby, W. B.: J. Biol. Chem. *207,* 657–664 (1954)
215. Jakoby, W. B., Bronner, D. M.: J. Biol. Chem. *205,* 699–708 (1953)
216. Jensen, R. A., Nasser, D. S.: J. Bact. *95,* 188–196 (1968)
217. Jensen, R. A., Nasser, D. S., Nester, E. W.: J. Bact. *94,* 1582–1593 (1967)
218. Jensen, R. A., Nester, E. W.: J. Mol. Biol. *12,* 468–481 (1965)
219. Jensen, R. A., Nester, E. W.: J. Biol. Chem. *241,* 3365–3372 (1966)
220. Jensen, R. A., Nester, E. W.: J. Biol. Chem. *241,* 3373–3380 (1966)
221. Jensen, R. A., Rebello, J. L.: in: Developments in Indutrial Microbiology, *11*, Corum, C. J. (Ed.), 105–121, 1970
222. Jones, A., Francis, M. M., Vining, L. C., Westlake, D. W. S.: Can. J. Microbiol. *24,* 238–244 (1978)
223. Jones, A., Vining, L. C.: Can. J. Microbiol. *22,* 327–344 (1967)
224. Jones, A., Westlake, D. W. S.: Can. J. Microbiol. *20,* 1599–1611 (1974)
225. Johnson, A. W.: in: Interscience, Waksman, S. A. (Ed.), New York, p. 33, 1968
226. Johnson, A. W., Wolstenholme, G. E. W., O'Connor, C. M.: CIBA Foundation on amino acids and peptides with antimetabolic activity, p. 123, 1958
227. Kagamiyama, H., Matsubara, H., Snell, E. E.: J. Biol. Chem. *247,* 1567–1586 (1972)
228. Kagamiyama, H., Wada, H., Matsubara, H., Snell, E. E.: J. Biol. Chem. *247,* 1571–1572 (1972)
229. Kalan, E. B., Davis, B. D., Srinivasan, P. R., Sprinson, D. B.: J. Biol. Chem. *223,* 907–912 (1956)
230. Kane, J. F., Holmes, W. M., Jensen, R. A.: J. Biol. Chem. *247,* 1587–1596 (1972)
231. Kane, J. F., Jensen, R. A.: J. Biol. Chem. *245,* 2384–2390 (1970)
232. Kannan, L. V., Kozova, J., Rehacek, Z.: Folia Microbiol. *13,* 1–6 (1968)
233. Kariyone, K., Yazawa, H., Kohsaka, M.: Chem. Pharm. Bull. *19,* 2289–2293 (1971)
234. Karlsson, A., Sartori, G., White, R. J.: Eur. J. Biochem. *47,* 251–256 (1974)
235. Karpov, V. L., Romanova, L. G.: Antibiotiki *17,* 419–424 (1972)
236. Katz, E.: in: Antibiotics, Gottlieb, D., Shaw, P. D. (Ed.), Springer-Verlag, p. 276–341, Berlin, Heidelberg and New York 1967
237. Katz, E.: in: Actinomycin, Waksman, S. A. (Ed.), Interscience, p. 45, New York 1968
238. Katz, E.: in: Biochemical and Medical Aspects of Tryptophan Metabolism, Hayaishi, O., Ishimura, Y., Kido, R. (Ed.), pp. 159–177, 1980
239. Katz, E., Demain, A. L.: Bacteriol. Rev. *41,* 449–474 (1977)
240. Katz, E., Weisbach, H.: J. Biol. Chem. *237,* 882 (1962)
241. Katz, E., Weisbach, H.: J. Biol. Chem. *238,* 666–675 (1963)
242. Kaufman, S.: J. Biol. Chem. *226,* 511–524 (1957)
243. Kaufman, S.: J. Biol. Chem. *234,* 2677–2682 (1959)
244. Kaufman, S.: in: The enzymes, Boyer, P. D., Lardy, H., Myrback, K. (Ed.), Second Edition, Academic Press *8,* p. 373, London und New York 1963
245. Kaufman, S., Bridgers, W. F., Eisenberg, F., Friedman, S.: Biochem. Biophys. Res. Comm. *9,* 497–502 (1962)
246. Kaufman, S., Friedman, S.: J. Biol. Chem. *240,* 552–554 (1965)
247. Kenner, G. W., Chambers, K., Temple Robinson, M. J., Webster, B. R.: Proc. Chem. Soc. *1960,* 291–292 (1960)
248. Kenner, G. W., Bunton, C. A., Temple Robinson, M. J., Webster, B. R.: Tetrahedron *19,* 1001–1010 (1963)
249. Khokhlov, A. S., Anisova, L. N., Tovarova, J. J., Kleiner, E. M., Kovalenko, J. V., Krasilnikova, O. J., Kornitskaja, E. Y., Pliner, S. A.: Z. Allg. Mikrobiol. *13,* 647–655 (1973)
250. Khokhlov, A. S., Tovarova, J. J.: Luckner, M., Schreiber, K. (Ed.) in: Regulation of secondary product and plant hormone metabolism, , Pergamon Press, p. 133–145, New York 1979
251. Kim, K.-H., Tchen, T. T.: Biochim. Biophys. Acta *59,* 569–576 (1962)
252. King, T. P., Craig, L. C.: J. Am. Chem. Soc. *77,* 6627–6631 (1955)

253. Kinsky, S. C.: Ann. Re. Pharmacol. *10,* 119–142 (1970)
254. Kirby, G. W., Ogunkoya, L.: Chem. Comm. 546–547 (1965)
255. Kirshner, N., Goodall, McC.: Biochim. Biophys. Acta *24,* 658–659 (1957)
256. Kleiner, E. M., Pliner, S. A., Soifer, V. S., Onoprienko, V. V., Balashova, T. A., Rozynev, B. V., Khokhlov, A. S.: Bioorg. Khim. *7,* 1142–1147 (1976)
257. Kleinkauf, H.: Planta medica *35,* 1–18 (1979)
258. Kleinkauf, H.: Chimia *34,* 344 (1980)
259. Kleinkauf, H.: Chemie in unserer Zeit *14,* 105–114 (1981)
260. Kleinkauf, H., Koischwith, H.: Lipmann Symposium: Energy, biosynthesis and regulation in molecular biology, Walter de Gruyter, p. 336–344, New York 1974
261. Knox, W. E.: Biochem. J. *49,* 686–693 (1951)
262. Knox, W. E.: Biochem. J. *53,* 379–385 (1953)
263. Knox, W. E., Edwards, S. W.: J. Biol. Chem. *216,* 479–487 (1955)
264. Knox, W. E., Edwards, S. W.: J. Biol. Chem. *216,* 489–498 (1955)
265. Knox, W. E., Goswami, M. N. D.: J. Biol. Chem. *235,* 2662–2666 (1960)
266. Knox, W. E., Mehler, A. H.: J. Biol. Chem. *187,* 419–430 (1950)
267. Koch, G. L. E., Shaw, D. C., Gibson, F.: Biochim. Biophys. Acta *212,* 375–386 (1970)
268. Koch, G. L. E., Shaw, D. C., Gibson, F.: Biochim. Biophys. Acta *212,* 387–395 (1970)
269. Kohn, K. W., Glaubinger, D., Spears, C. L.: Biochim. Biophys. Acta *361,* 288–302 (1974)
270. Kominek, L. A.: in: Antibiotics, Gottlieb, D., Shaw, D. P. (Ed.), Springer-Verlag *II,* p. 231, 1967
271. Kominek, L. A.: Antimicrob. Agents Chemother. *1,* 123–134 (1972)
272. Kominek, L. A.: Antimicrob. Agents Chemother. *7,* 861–863 (1975)
273. Kominek, L. A.: Antimicrob. Agents Chemother. *7,* 856–860 (1975)
274. Kotake, Y.: J. Chem. Soc. Japan *60,* 632–640 (1939)
275. Kotake, Y., Masayama, I.: Z. physiol. Chem. *243,* 237–244 (1936)
276. Kralovcova, E., Glumanerova, M., Vanek, Z.: Folia Microbiol. *22,* 182–188 (1977)
277. Kruinski, V. M., Robbers, J. E., Floss, H. G.: J. Bacteriol. *125,* 158–165 (1976)
278. Kumagai, H., Kashima, N., Yamada, H.: Biochem. Biophys. Res. Comm. *39,* 796–801 (1970)
279. Kumagai, H., Matsui, H., Ohgishi, H., Ogata, K.: Biochem. Biophys. Res. Comm. *34,* 266–270 (1969)
280. Kurahashi, K., Yamada, M., Mori, K., Fujikawa, K., Kambe, M., Imae, E., Takahashi, H., Sakamoto, Y.: Cold Spring Habour Symp. Quant. Biol. *34,* 815–826 (1969)
281. Kurylo-Berowska, Z., Abramsky, T.: Biochim. Biophys. Acta *264,* 1–10 (1972)
282. Labowitz, A. M., Slayman, C. W.: J. Bacteriol. *112,* 1020–1022 (1972)
283. Lack, L.: J. Biol. Chem. *236,* 2835–2840 (1961)
284. Lack, L.: Biochim. Biophys. Acta *34,* 117–123 (1959)
285. La Du, B. N., Jr., Greenberg, D. M.: Science *117,* 111–112 (1953)
286. La Du, B. N., Jr., Greenberg, D. M.: J. Biol. Chem. *190,* 245–255 (1951)
287. La Du, B. N., Zannoni, V. G.: J. Biol. Chem. *217,* 777–788 (1955)
288. Laland, S. G., Zimmer, T. L.: Essays Biochem. *9,* 31–57 (1973)
289. Lampen, J. O.: Am. J. Clin. Pathol. *52,* 138–146 (1969)
290. Lechevalier, H., Acker, R. F., Corke, C. T., Haensler, C. M., Waksman, S. A.: Mycologia *45,* 155–171 (1953)
291. Lee, S. G., Littau, V., Lipmann, F.: J. Cell Biol. *66,* 238–242 (1975)
292. Lee, F., Squires, C. L., Yanofsky, C.: J. Mol. Biol. *103,* 383–393 (1976)
293. Leeper, L. C., Udenfriend, S.: Federation Proceedings *15,* 298 (1956)
294. Leimgruber, W., Batcho, A. D., Schenker, F.: J. Am. Chem. Soc. *87,* 5793–5795 (1965)
295. Leimgruber, W., Stefanovic, V., Schenker, F., Karr, A., Berger, J.: J. Am. Chem. Soc. *87,* 5791–5793 (1965)
296. Le May-Knox, M., Knox, W. E.: Biochem. J. *49,* 686–693 (1951)
297. Lester, G.: J. Bact. *107,* 193–202 (1971)
298. Levin, E. Y., Kaufman, S.: J. Biol. Chem. *236,* 2043–2049 (1961)
299. Levin, J. G., Sprinson, D. B.: Biochem. Biophys. Res. Comm. *3,* 157–163 (1960)
300. Levin, J. G., Sprinson, D. B.: J. Biol. Chem. *239,* 1141–1150 (1964)
301. Lingens, F.: Angew. Chem. Int. Ed. Engl. *7,* 350–360 (1968)
302. Lingens, F., Goebel, W., Uesseler, H.: Biochem. Z. *346,* 357–367 (1966)

303. Lingens, F., Goebel, W., Uesseler, H.: European J. Biochem. *2,* 442–447 (1967)
304. Lingens, F., Goebel, W., Uesseler, H.: European J. Biochem. *1,* 363–374 (1967)
305. Lingens, F., Sprössler, B., Goebel, W.: Biochim. Biophys. Acta *121,* 164–166 (1966)
306. Liras, P., Lampen, J. O.: Biochim. Biophys. Acta *372,* 141–153 (1974)
307. Liras, P., Lampen, J. O.: Biochim. Biophys. Acta *374,* 159–163 (1974)
308. Liras, P., Villanueva, J. R., Martin, J. F.: J. Gen. Microbiol. *102,* 269–278 (1977)
309. Liu, C. M., Hermann, T., Miller, P. A.: J. Antibiotics *30,* 244–251 (1977)
310. Liu, C. M., McDaniel, L. E., Schaffner, C. P.: J. Antibiotics *25,* 116–121 (1972)
311. Liu, C. M., McDaniel, L. E., Schaffner, P.: Antimicrob. Agents Chemother. *7,* 196–202 (1975)
312. Lively, D. H., Gorman, M., Haney, M. E., Mabe, J. A.: Antimicrob. Agents Chemother. 462–469 (1966)
313. London, J., Goldberg, M. E.: J. Biol. Chem. *247,* 1566–1570 (1972)
314. Lorence, J. H., Nester, E. W.: Biochemistry *6,* 1541–1552 (1967)
315. Lowe, D. A., Westlake, D. W. S.: Can. J. Biochem. *49,* 448–455 (1971)
316. Lowe, D. A., Westlake, D. W. S.: Can. J. Biochem. *50,* 1064–1069 (1972)
317. Luckner, M.: The Secondary Metabolism of Plants and Animals. Academic Press, pp. 176–181, 305–313, New York, N.Y. 1972
318. McBride, T. J., Oleson, J. J., Woolf, D.: Cancer Research *26,* 727–732 (1966)
319. McDonald, J. C.: Can. J. Microbiol. *6,* 27–34 (1960)
320. McGeoch, D., McGeoch, J., Morse, D.: Nat. New Biol. *245,* 137–140 (1973)
321. McGrath, R., Siddiquellah, M., Vining, L. C., Sala, F., Westlake, D. W. S.: Biochem. Biophys. Res. Comm. *29,* 567–581 (1967)
322. McGrath, R., Vining, L. C., Sala, F., Westlake, D. W. S.: Can. J. Biochem. *46,* 587–594 (1968)
323. Mach, B., Reich, E., Tatum, E. L.: Biochemistry *50,* 175–181 (1963)
324. Mach, B., Tatum, E. L.: Proc. Nat. Acad. Sci. *52,* 876–884 (1964)
325. Mach, B., Tatum, E. L.: 6. Internat. Congress of Biochemistry, New York, 1964
326. Magasnaik, B., Prival, M., Brenchley, J., Tyler, P., Deleo, A., Streicher, S., Bender, R., Paris, C.: Curr. Top. Cell Regul. *8,* 118–138 (1974)
327. Majumdar, M. K., Majumdar, S. K.: Biochem. J. *120,* 271–278 (1970)
328. Malik, V. S.: Adv. Appl. Microbiol. *15,* p. 297–336 (1972)
329. Malik, V. S., Vedpal, S.: Adv. Appl. Microbiol. *25,* 75–93 (1979)
330. Malik, V. S., Vining, L. C.: Can. J. Microbiol. *16,* 173–179 (1970)
331. Malik, V. S., Vining, L. C.: Can. J. Microbiol. *17,* 1287–1290 (1971)
332. Malik, V. S., Vining, L. C.: Can. J. Microbiol. *18,* 137–143 (1972)
333. Malik, V. S., Vining, L. C.: Can. J. Microbiol. *18,* 583–590 (1972)
334. Mann, D. F., Floss, H. G.: Lloydia *40,* 136–145 (1977)
335. Manney, T. R., Duntze, W., Janosko, N., Salazar, J.: J. Bact. *99,* 590–596 (1969)
336. Manson, M. D., Yanofsky, C.: J. Bact. *126,* 679–689 (1976)
337. Marinus, M. G., Loutit, J. S.: Genetics *63,* 547–556 (1970)
338. Marsh, W. S., Garretson, A. L., Rao, K. V.: Brit. 1,012,684 (Cl. 007 g), Dec. 8 (1965)
339. Marsh, W. S., Garretson, A. L., Wesel, E. M.: Antibiot. Chemother. *10,* 316–320 (1960)
340. Martin, J. F.: Develop. Industr. Microbiol. *17,* 223–231 (1976)
341. Martin, J. F.: in: Schlessinger, D. (Ed.), American Society for Microbiology, Washington, D.C., pp. 548–552, 1976
342. Martin, J. F.: Ann. Rev. Microbiol. *31,* 13–38 (1977)
343. Martin, J. F.: Academic Press Inc. New York, p. 19–37 (1978)
344. Martin, A. R.: in: Wilson, C. O., Gisvold, O., Doerge, R. F. (Ed.), Textbook of organic medicinal and pharmaceutical chemistry, 7th Ed., J.B. Lippincott company, p. 269–347, Philadelphia, PA., USA 1977
345. Martin, J. F., Demain, A. L.: Biochem. Biophys. Res. Comm. *71,* 1103–1109 (1976)
346. Martin, J. F., Demain, A. L.: Microbiol. Rev. *44,* 230–251 (1980)
347. Martin, J. F., Demain, A. L.: Edward Arnold, London, *3,* p. 426–450 (1978)
348. Martin, J. F., Gil, J. A., Naharro, G., Gras, P., Villanneva, J. R.: American Society of Microbiology, Washington, D.C., p. 205–209, 1979
349. Martin, J. F., Liras, P.: Antibiotics *29,* 1306–1309 (1976)

350. Martin, J. F., Liras, P., Demain, A. L.: FEMS Microbiol. Letters *2,* 173–176 (1977)
351. Martin, J. F., McDaniel, L. E.: Dev. Ind. Microbiol. *15,* 324–337 (1974)
352. Martin, J. F., McDaniel, L. E.: Biochim. Biophys. Acta *411,* 186–194 (1975)
353. Martin, J. F., McDaniel, L. E.: Eur. J. Appl. Microbiol. *3,* 135–144 (1976)
354. Martin, J. F., Naharro, G., Liras, P., Villanneva, J. R.: J. Antibiotics *32,* 600–606 (1979)
355. Mason, H. S.: J. Biol. Chem. *168,* 433–438 (1947)
356. Mason, H. S.: J. Biol. Chem. *172,* 83–99 (1949)
357. Mason, H. S.: Pigment Cell Biol., Proc. Conf. Biol. Normal and Atypical Pigment Cell Growth, 4th, p. 563–582, Houston, Tex. 1959
358. Mason, H. S.: Advan. Biol. Skin *8,* 293–312 (1967)
359. Matteo, C. C., Glade, M., Tanaka, A., Piret, J., Demain, A. L.: Biotechnol. Bioeng. *17,* 129–142 (1975)
360. Mattern, J. E., Pittchard, J.: J. Bacteriol. *107,* 8–15 (1971)
361. Mehler, A. H. in: "Amino acid metabolism", McElroy, W. D., Glass, B. (Ed.), Johns Hopkins Press, p. 882, Baltimore, Maryland 1955
362. Mehler, A. H., Knox, W. E.: J. Biol. Chem. *187,* 431–438 (1950)
363. Mehler, A. H., McDaniel, E. G., Hundley, J. M.: J. Biol. Chem. *232,* 323–330 (1958)
364. Mesentsev, A. S., Kuljaeva, V. V., Rubasheva, L. M.: J. Antibiotics *27,* 866–873 (1974)
365. Methler, D. E., Ikawa, M., Snell, E. E.: J. Am. Chem. Soc. *76,* 648–653 (1954)
366. Michalik, J., Emilianowicz-Czerska, W., Switalski, L., Raczynska-Bojanowska, K.: Antimicrob. Agents Chemother. *8,* 526–531 (1975)
367. Miller, A. L., Walker, J. B.: J. Bacteriol. *104,* 8–12 (1970)
368. Miozzari, G., Yanofsky, C.: Proc. Natl. Acad. Sci. *75,* 5580–5584 (1978)
369. Mitoma, C., Leeper, L. C.: Federation Proc. *13,* 266 (1954)
370. Mitoma, C.,Weissbach, H., Udenfriend, S.: Nature *175,* 994–995 (1955)
371. Mitsuhashi, S., Davis, B. D.: Biochim. Biophys. Acta *15,* 54–61 (1954)
372. Mitsuhashi, S., Davis, B. D.: Biochim. Biophys. Acta *15,* 268–280 (1954)
373. Mohler, W. C., Suskind, S. R.: Biochim. Biophys. Acta *43,* 288–299 (1960)
374. Molinoff, P. B., Axelrod, J.: Ann. Rev. Biochem. *40,* 465–487 (1971)
375. Moore, K., Subba, P. V., Towers, G. H. N.: Biochem. J. *106,* 507–514 (1968)
376. Morino, Y., Snell, E. E.: J. Biol. Chem. *242,* 2793–2799 (1967)
377. Morino, Y., Snell, E. E.: J. Biol. Chem. *242,* 5591–5601 (1967)
378. Morino, Y., Snell, E. E.: J. Biol. Chem. *242,* 5602–5610 (1967)
379. Mosteller, R. D., Yanofsky, C.: J. Bact. *105,* 268–275 (1971)
380. Mothes, K., Schuette, H. R.: Angew. Chem. *75,* 265–281 (1963)
381. Nagano, H., Zalkin, H., Henderson, E. J.: J. Biol. Chem. *245,* 3810–3820 (1970)
382. Nagatsu, T., Levitt, M., Udenfriend, S.: Anal. Biochem. *9,* 122–126 (1964)
383. Nagatsu, T., Levitt, M., Udenfriend, S.: J. Biol. Chem. *239,* 2910–2917 (1964)
384. Nakamura, S., Ikeda, M., Tsuji, H., Nishizuka, Y., Hayaishi, O.: Biochem. Biophys. Res. Comm. *13,* 285–290 (1963)
385. Nakatsukasa, W. M., Nester, E. W.: J. Biol. Chem. *247,* 5972–5979 (1972)
386. Nasser, D., Henderson, G., Nester, E. W.: J. Bact. *98,* 44–50 (1969)
387. Nazario, M., Kinsey, J. A., Ahmad, M.: J. Bact. *105,* 121–126 (1971)
388. Neish, A. C.: Phytochemistry *1,* 1–24 (1961)
389. Nester, E. W., Jensen, R. A.: J. Bact. *91,* 1594–1598 (1966)
390. Nester, E. W., Lorence, J. H., Nasser, D. S.: Biochemistry *6,* 1553–1562 (1967)
391. Newton, W. A., Snell, E. E.: Proc. Nat. Science *54,* 382–389 (1964)
392. Newton, W. A., Morino, Y., Snell, E. E.: J. Biol. Chem. *240,* 1211–1218 (1965)
393. Nishida, M., Matsubara, T., Watanabe, N.: J. Antibiotics *18,* 211–219 (1965)
394. Nishizuka, Y., Hayaishi, O.: J. Biol. Chem. *238,* 3369–3377 (1963)
395. Nose, M., Arima, K.: J. Antibiotics *22,* 135–143 (1969)
396. Novick, A., Szilard, L.: in: "Dynamics of growth processes", Princeton University Press, p. 21–32, Princeton, N.J. 1954
397. Oleson, J. J., Calderella, L. A., Mjos, K. J., Reith, A. R., Thie, R. S., Toplin, I.: Antibiot. Chemother. *11,* 158–164 (1961)
398. Okanishi, M., Manome, T., Umezawa, H.: J. Antibiotics *33,* 88–91 (1980)
399. Onderka, D. K., Floss, H. G.: J. Am. Chem. Soc. *91,* 5864–5866 (1969)

400. O'Neill, W. P., Nystrom, R. F., Rinehart, K. L., Jr., Gottlieb, D.: Biochemistry *12,* 4775–4784 (1973)
401. Ortega, M. V., Brown, G. M.: J. Biol. Chem. *235,* 2939–2945 (1960)
402. Ostrander, J. M., Hurley, L. H.: J. Natural Products – Lloydia *42,* 693 (1979)
403. Ostrander, J. M., Hurley, L. H., McInnes, A. G., Smith, D. G., Walter, J. A., Wright, J. L. C.: J. Antibiotics *33,* 1167–1171 (1980)
404. Padeiskaya, E., Pershin, G. N., Kutchak, S. N., Gerchikov, L. N., Egorova, E. F., Preobrazhenskaya, M. N., Suvorov, N. N.: Tetrahedron *24,* 6131–6143 (1968)
405. Pass, L., Raczynska, Bojanowska, K.: Acta Biochim. Pol. *15,* 355–367 (1968)
406. Pastan, J., Adhya, S.: Bacteriol. Rev. *40,* 527–551 (1976)
407. Paulus, H., Gray, E.: J. Biol. Chem. *239,* 865–871 (1964)
408. Perlman, D., Otani, S., Perlman, K. L., Walker, J. E.: J. Antibiotics *26,* 289–296 (1973)
409. Pestka, S.: Biochem. Biophys. Res. Comm. *36,* 589–595 (1969)
410. Pestka, S.: Proc. Nat. Acad. Sci. *64,* 709–714 (1969)
411. Pirt, S. J., Righelato, R. C.: Appl. Microbiol. *15,* 1284–1290 (1967)
412. Pitt, B. M.: Nature *196,* 272–273 (1962)
413. Pittard, J., Camakaris, J., Wallace, B. J.: J. Bacteriol. *97,* 1242–1247 (1969)
414. Platt, T.: in: "The operon", Miller, J. H., Reznikoff, W. S. (Ed.), Cold Spring Harbor Lab., pp. 263–302, New York 1918
415. Proctor, A. R., Kloos, W. E.: J. Bact. *114,* 169–177 (1973)
416. Queener, S. F., Gunsalus, I. C.: Proc. Nat. Acad. Sci. *67,* 1225–1232 (1970)
417. Rao, K. V., Cullen, W. P.: Antibiot. Ann. *1959–60,* 950–953 (1960)
418. Rao, K. V.: Antibiot. Chemother. *10,* 312–315 (1960)
419. Rao, K. V., Biemann, K., Woodward, R. B.: J. Am. Chem. Soc. *85,* 2532–2533 (1963)
420. Raper, H. S.: Biochem. J. *21,* 89–96 (1927)
421. Raper, H. S.: J. Chem. Soc. 125–130 (1938)
422. Ratledge, C.: Nature *203,* 428–429 (1964)
423. Ravdin, R. G., Crandall, D. L.: J. Biol. Chem. *189,* 137–149 (1951)
424. Read, G., Westlake, D. W. S., Vining, L. C.: Can. J. Biochem. *47,* 1071–1079 (1969)
425. Rebello, J. L., Jensen, R. A.: J. Biol. Chem. *245,* 3738–3744 (1970)
426. Rehacek, Z.: Antimicrob. Agents Chemother. *1963,* 530–540 (1964)
427. Rehacek, Z., Kennan, L. V., Ramankutty, M., Puza, M.: Hindustan Antibiot. Bull. *10,* 280–286 (1968)
428. Robb, F., Hutchinson, M. A., Belser, W. L.: J. Biol. Chem. *246,* 6908–6912 (1971)
429. Rogers, H. J., Lomakina, N., Abraham, E. P.: Biochem. J. *97,* 579–586 (1965)
430. Rogers, H. J., Newton, G. G. F., Abraham, E. P.: Biochem. J. *97,* 573–578 (1965)
431. Roscoe, J., Abraham, E. P.: Biochem. J. *99,* 793–800 (1966)
432. Rose, J. K., Squires, C. L., Yanofsky, C., Yang, H.-L., Zubay, G.: Nat. New. Biol. *245,* 133–137 (1973)
433. Rose, J. K., Yanofsky, C.: J. Mol. Biol. *69,* 103–118 (1972)
434. Rossi, A., Corcoran, J. W.: Biochem. Biophys. Res. Comm. *50,* 597–602 (1973)
435. Rudd, B. A. M., Hopwood, D. A.: J. Gen. Microbiol. *114,* 35–43 (1979)
436. Rudd, B. A. M., Floss, H. G., Hopwood, D. A.: J. Natural Products-Lloydia *42,* 691 (1979)
437. Ruttenberg, M. A., Mach, B.: Biochemistry *5,* 2864–2869 (1966)
438. Saito, Y., Hayaishi, O., Rothberg, S.: J. Biol. Chem. *229,* 921–934 (1937)
439. Salcher, O., Lingens, F.: Tetrahedron Lett. 3101–3102 (1978)
440. Salzman, L. A., Katz, E., Weissbach, H.: J. Biol. Chem. *239,* 1864–1866 (1964)
441. Salzman, L. A., Weissbach, H., Katz, E.: Arch. Biochem. Biophys. *130,* 536–546 (1969)
442. Sankaran, L., Poged, B. M.: Antimicrob. Agents Chemother. *8,* 721–732 (1975)
443. Scannell, J. P., Pruess, D. L., Demny, R. C., Williams, T., Stempel, A.: J. Antibiotics *23,* 618–619 (1970)
444. Schaeffer, P.: Bacteriol. Rev. *33,* 48–71 (1969)
445. Schepartz, B.: J. Biol. Chem. *193,* 293–298 (1951)
446. Schindler, F., Zähner, H.: Arch. Mikrobiol. *79,* 187–203 (1971)
447. Schmit, J. C., Zalkin, H.: Biochemistry *8,* 174–181 (1969)
448. Schmit, J. C., Zalkin, H.: J. Biol. Chem. *246,* 6002–6010 (1971)
449. Schmit, J. C., Artz, S. W., Zalkin, H.: J. Biol. Chem. *245,* 4019–4027 (1970)

450. Schroeder, W., Bannister, B., Hoeksema, H.: J. Am. Chem. Soc. *89,* 2448–2453 (1967)
451. Sealock, R. R., Goodland, R. L.: Science *114,* 645–646 (1951)
452. Seidenberg, M., Martinez, R. J., Guthrie, R.: Arch. Biochem. Biophys. *97,* 470–473 (1962)
453. Shiio, I., Miyajima, R., Nakagawa, M.: J. Biochem. *72,* 1447–1455 (1972)
454. Shiio, I., Sugimoto, S., Nakagawa, M.: Agr. Biol. Chem. *39,* 627–635 (1975)
455. Shiio, I., Sugimoto, S.: J. Biochem. *79,* 173–183 (1976)
456. Shiio, I., Sugimoto, S.: J. Biochem. *83,* 879–886 (1978)
457. Shimada, K., Hook, D. J., Warner, G. F., Floss, A. G.: Biochemistry *17,* 3054–3058 (1978)
458. Shimizu, Y., Shimizu, N., Hayashi, M.: Proc. Nat. Acad. Sci. USA *70,* 1990–1994 (1973)
459. Shirahata, K., Hayashi, T., Deguchi, T., Suzuki, T., Matsubara, I.: Agr. Biol. Chem. *36,* 2229–2232 (1972)
460. Siddiquellah, M., McGrath, R., Vining, L. C.: Can. J. Biochem. *45,* 1881–1889 (1967)
461. Sivak, A., Melconi, M. L., Nobili, F., Katz, E.: Biochim. Biophys. Acta *57,* 283–289 (1962)
462. Smith, C. G.: Bacteriol. *75,* 577–583 (1958)
463. Smith, L. C., Ravel, J. M., Lax, S. R., Shive, W.: J. Biol. Chem. *237,* 3566–3570 (1962)
464. Smith, L. G., Hinmann, J. W.: Progr. Industr. Microbiol. *4,* 137–163 (1963)
465. Smith, R. L., Bungay, H. R., Pittenger, R. C.: Appl. Microbiol. *10,* 293–296 (1962)
466. Smithers, L. M., Engel, P. P.: Genetics *78,* 799–808 (1974)
467. Snoke, J. E., Cornell, M.: J. Bacteriol. *89,* 415–420 (1965)
468. Snow, M. L., Lauinger, C., Ressler, C.: J. Org. Chem. *33,* 1774–1780 (1968)
469. Snyder, H. R., Matteson, D. S.: J. Am. Chem. Soc. *79,* 2217–2221 (1957)
470. Soltero, F. V., Johnson, J.: Appl. Microbiol. *1,* 52–57 (1953)
471. Somerville, R. L., Yanofsky, C.: J. Mol. Biol. *11,* 747–759 (1965)
472. Speedie, M. K., Hornemann, U., Floss, H. G.: J. Biol. Chem. *250,* 7819–7825 (1975)
473. Speedie, M. K., Park, M. O.: J. Antibiotics *33,* 579–584 (1980)
474. Spencer, I. D.: in: "Comprehensive Biochemistry", Florkin, M., Stotz, E. H. (Ed.), Elsevier Publishing Co., *20,* p. 270–286, Amsterdam 1968
475. Spizek, J., Malek, J., Suchy, J., Vondracek, M., Vanek, Z.: Folia Microbiol. *10,* 263–266 (1965)
476. Spoerl, E., Carleton, R.: J. Biol. Chem. *210,* 521–529 (1954)
477. Sprinson, D. B.: Advanc. Carbohydr. Chem. *15,* 235–270 (1960)
478. Sprinson, D. B., Gollub, E. G., Hu, R. C., Liu, K. P.: Acta Microbiol. Acad. Sci. Hung. *23,* 167–170 (1976)
479. Squires, C., Lee, F., Bertrand, K., Squires, C. L., Bronson, M. J., Yanofsky, C.: J. Mol. Biol. *103,* 351–381 (1976)
480. Squires, C. L., Lee, F., Yanofsky, C.: J. Mol. Biol. *92,* 93–111 (1975)
481. Squires, C. L., Rose, J. K., Yanofsky, C.: Nature New Biol. *245,* 131–133 (1973)
482. Srinivasan, P. R., Katagiri, M., Sprinson, D. B.: J. Am. Chem. Soc. *77,* 4943–4944 (1955)
483. Srinivasan, P. R., Katagiri, M., Sprinson, D. B.: J. Biol. Chem. *234,* 713–715 (1959)
484. Srinivasan, P. R., Rivera, Jr., A.: Biochemistry *2,* 1059–1062 (1963)
485. Srinivasan, P. R., Rothschild, J., Sprinson, D. B.: J. Biol. Chem. *238,* 3176–3182 (1963)
486. Srinivasan, P. R., Shigeura, H. T., Sprecher, M., Sprinson, D. B., Davis, B. D.: J. Biol. Chem. *220,* 477–497 (1956)
487. Srinivasan, P. R., Sprinson, D. B., Kalan, E. B., Davis, B. D.: J. Biol. Chem. *223,* 913–920 (1956)
488. Stanier, R. Y., Hayaishi, O.: Science *114,* 326–330 (1951)
489. Staub, M., Dénes, G.: Biochim. Biophys. Acta *178,* 588–598 (1969)
490. Staub, M., Dénes, G.: Biochim. Biophys. Acta *178,* 599–608 (1969)
491. Stebbing, N.: J. Cell Sci. *9,* 701–717 (1971)
492. Stefanovic, V.: Biochem. Pharm. *17,* 315–323 (1968)
493. Stein, W. H., Paladini, A. C., Hirs, C. H. W., Moore, S.: J. Am. Chem. Soc. *76,* 2848–2849 (1954)
494. Stenmark-Cox, S., Jensen, R. A.: Arch. Biochem. Biophys. *167,* 540–546 (1975)
495. Stevens, C. O., Henderson, L. M.: J. Biol. Chem. *234,* 1188–1190 (1959)
496. Stratton, C. D., Rebstock, M. C.: Arch. Biochem. *103,* 159–163 (1963)
497. Subba Rao, P. V., Sreeleela, N. S., Premakumar, R., Vaidyanathan, C. S.: J. Bact. *107,* 100–105 (1971)

498. Sugimoto, S., Nakagawa, M., Tsuchida, T., Shiio, J.: Agr. Biol. Chem. *37,* 2327–2336 (1973)
499. Sugimoto, S., Shiio, I.: J. Biochem. *81,* 823–833 (1977)
500. Sugiyama, S., Yano, K., Tanaka, H., Komagata, K., Arima, K.: J. Gen. Appl. Microbiol. *4,* 223–240 (1958)
501. Suzuki, T., Honda, H., Katsumata, R.: Agr. Biol. Chem. *36,* 2223–2228 (1972)
502. Takashima, K., Fujimoto, D., Tamiya, N.: J. Biochem. *55,* 122–125 (1964)
503. Tamir, H., Srinivasan, P. R.: Proc. Nat. Acad. Sci. *66,* 547–551 (1970)
504. Tamura, A., Takeda, J., Naruto, S., Yoshimura, Y.: J. Antibiotics *24,* 270 (1971)
505. Tanabe, M.: in: "Biochemistry" (Specialist Periodical Reports), The Chemical Society, London *2,* 243–246, 1973
506. Tanaka, H., Sugiyama, S., Yano, K., Arima, K.: Bull. Agr. Chem. Soc. Japan *21,* 67–68 (1957)
507. Taniuchi, H., Hatanaka, M., Kuno, S., Hayaishi, O., Nakajima, M., Kurihara, N.: J. Biol. Chem. *239,* 2204–2211 (1964)
508. Tendler, M. D., Korman, S.: Nature *199,* 501 (1963)
509. Tomino, S., Yamada, M., Itoh, H., Kuraraski, K.: Biochemistry *6,* 2552–2560 (1967)
510. Toropova, E. G., Egorov, N. S., Suchkova, L. A.: Antibiotiki (Moscow) *18,* 587–590 (1973)
511. Tripathi, R. K., Gottlieb, D.: J. Bacteriol. *100,* 310–318 (1969)
512. Tristam, H.: Sci. Progr. *56,* 449–477 (1968)
513. Tsai, H., Suskind, S. R.: Biochim. Biophys. Acta *284,* 324–340 (1972)
514. Tsai, M. D., Floss, H. G., Rosenfeld, H. J., Roberts, J.: J. Biol. Chem. *254,* 6437–6443 (1979)
515. Uchida, M., Suzuki, S., Ichihara, K.: J. Biochem. *41,* 41–65 (1954)
516. Udenfriend, S., Clark, C. T., Titus, E.: J. Am. Chem. Soc. *75,* 501–502 (1953)
517. Udenfriend, S., Cooper, J. R.: J. Biol. Chem. *194,* 503–511 (1952)
518. Udenfriend, S., Cooper, J. R., Clark, C. T., Baer, J. E.: Science *117,* 663–665 (1953)
519. Udenfriend, S., Wyngaarden, J. B.: Biochim. Biophys. Acta *20,* 48–52 (1956)
520. Umbarger, H. E.: Ann. Rev. Biochem. *38,* 323–370 (1969)
521. Umezawa, H.: Cancer Chemother., Proc. Takeda Int. Conf., Osaka 1966, 197–203, 1967
522. Umezawa, H.: University Park Press, State College, p. 210, Pennsylvania 1967
523. Van Tamelen, E. E., Dickie, J. P., Loomans, M. E., Dewey, R. S., Strong, F. M.: J. Am. Chem. Soc. *83,* 1639–1646 (1961)
524. Vanek, Z.: Folia Biol. *4,* 100–105 (1958)
525. Vining, L. C., Westlake, D. W.: Can. J. Microbiol. *10,* 705–716 (1964)
526. Volcani, B. E., Snell, E. E.: Proc. Soc. Exp. Biol. Med. *67,* 511–513 (1954)
527. Vypijack, A. M., Egorov, N. S., Zharkova, G. G.: Antibiotiki *15,* 392–395 (1970)
528. Waksman, S. A., Lechevalier, A. A., Schaffner, C. P.: Review Bull. Wld. Hlth. Org. *33,* 219–226 (1965)
529. Wallace, B. J., Pittard, J.: J. Bact. *93,* 237–244 (1967)
530. Wallace, B. J., Pittard, J.: J. Bact. *94,* 1279–1280 (1967)
531. Wallace, B. J., Pittard, J.: J. Bact. *97,* 1234–1241 (1969)
532. Waltho, J. A.: Biochim. Biophys. Acta *320,* 232–241 (1973)
533. Wat, C.-K., Malik, V. X., Vining, L. C.: Can. J. Chem. *49,* 3655–3656 (1971)
534. Wegman, J., DeMoss, J. A.: J. Biol. Chem. *240,* 3781–3788 (1965)
535. Wegman, J., Carwford, I. P.: J. Bacteriol. *95,* 2325–2335 (1968)
536. Weinberg, E. D.: Dev. Ind. Microbiol. *15,* 70–81 (1974)
537. Weiss, U., Davis, B. D., Mingioli, E. S.: J. Am. Chem. Soc. *75,* 5572–5576 (1958)
538. Weissbach, H., Katz, E.: J. Biol. Chem. *236,* PC 16 (1961)
539. Werner, R. G.: Antimicrob. Agents Chemother. *18,* 858–862 (1980)
540. Werner, R. G., Demain, A. L.: Appl. Envir. Microbiol. *40,* 675–677 (1980)
541. Werner, R. G., Demain, A. L.: J. Antibiotics *34,* 551–554 (1981)
542. Werner, R. G., Reuter, W.: Arzneim.-Forsch./Drug Res. *29,* 59–63 (1979)
543. Werner, R. G., Thorpe, L. F., Reuter, W., Nierhaus, K. H.: Eur. J. Biochem. *68,* 1–3 (1976)
544. Westlake, D. W. S., Vining, L. C.: Biotechnol. Bioeng. *11,* 1125–1134 (1969)
545. White, R. J., Martinelli, E.: FEBS Lett. *49,* 233–236 (1974)
546. White, R. J., Martinelli, E., Gallo, G. G., Lancini, G., Beynon, P.: Nature *243,* 273–277 (1973)
547. White, R. J., Martinelli, E., Lancini, G.: Proc. Natl. Acad. Sci. *71,* 3260–3264 (1974)

548. Williams, J. N., Jr., Sreenivasan, A.: J. Biol. Chem. *203,* 109–116 (1953)
549. Williams, J. N., Jr., Sreenivasan, A.: J. Biol. Chem. *203,* 606–512 (1953)
550. Williams, J. N., Jr., Sreenivasan, A.: J. Biol. Chem. *203,* 613–623 (1953)
551. Wilson, D. A., Crawford, I. P.: J. Biol. Chem. *240,* 4801–4808 (1965)
552. Wilson, W. L., Labra, C., Barrist, E.: Antibiotics and Chemother. II, *11,* 147–150 (1961)
553. Winnick, R. E., Lis, H., Winnick, T.: Biochim. Biophys. Acta *49,* 451–462 (1961)
554. Wiss, O.: Helv. Chim. Acta *32,* 1694–1698 (1949)
555. Wiss, O., Fuchs, H.: Helv. Chim. Acta *32,* 2553–2556 (1949)
556. Wiss, O., Weber, F.: Z. Physiol. Chem. *304,* 232–240 (1956)
557. Witkop, B., Goodwin, S.: Experimentia *8,* 377–379 (1952)
558. Witz, D. F., Hessler, E. J., Miller, T. L.: Biochemistry *10,* 1128–1132 (1971)
559. Wong, D. T., Airall, J. M.: J. Antibiotics *23,* 55–62 (1970)
560. Wong, D. T., Horug, J.-S., Gordee, R. S.: J. Bacteriol. *106,* 168–173 (1971)
561. Wood, W. A., Gunsalus, I. C., Umbreit, W. W.: J. Biol. Chem. *170,* 313–321 (1947)
562. Woolf, L. I.: Biochem. J. *49* ix (1951)
563. Wyler, H., Chiovini, J.: Helv. Chim. Acta *51,* 1476–1494 (1968)
564. Yamada, H., Kumagai, H., Kashima, N., Torii, H.: Biochem. Biophys. Res. Comm. *46,* 370–374 (1972)
565. Yamamoto, A., Omori, T., Yasui, H.: Nippon Nogei Kagaku Kaishi *40,* 152–160 (1966)
566. Yamashita, T., Miyairi, N., Kunugita, K., Shimizu, K., Sakai, H.: J. Antibiotics *23,* 537–541 (1970)
567. Yanagimoto, M., Terni, G.: J. Ferment. Technol. *49,* 604–610 (1971)
568. Yaniv, H., Gilvarg, C.: J. Biol. Chem. *213,* 787–795 (1955)
569. Yanofsky, C.: J. Bact. *68,* 577–584 (1954)
570. Yanofsky, C.: in: "Amino acid metabolism", McElroy, W. D., Glass, B. (Ed.), Johns Hopkins Press, p. 930, Baltimore, Maryland 1955
571. Yanofsky, C.: Bact. Rev. *24,* 221–245 (1960)
572. Yanofsky, C.: J. Am. Med. Assoc. *218,* 1026–1035 (1971)
573. Yanofsky, C.: in: "Alfred Benton symposium IX: Control of ribosome synthesis", Kjeldgaard, N. C., Maaloe, O. (Ed.), Copenhagen: Munksgaard, pp. 149–163, 466, 1975
574. Yasunobu, K., Tanaka, T., Knox, W. E., Mason, H. S.: Fed. Prod. *17,* 340 (1958)
575. Zähner, H.: Folia Microbiol. *24,* 435–443 (1979)
576. Zähner, H.: in: "Antibiotics and Other Secondary Metabolites", Hütter, R., Leisinger, T., Nuesch, J., Wehrli, W. (Ed.), Academic Press, London 1978
577. Zähner, H.: in: "Handbuch der Biotechnologie", Präve, P., Faust, U., Sittig, W., Sukatsch, D. A. (Hrsg.), Akademische Verlagsgesellschaft, Wiesbaden 1982
578. Zalkin, H., Chen, S. H.: J. Biol. Chem. *247,* 5996–6003 (1972)
579. Zalkin, H., Hwang, L. H.: J. Biol. Chem. *246,* 6899–6707 (1971)
580. Zannoni, V. G., La Du, B. N.: J. Biol. Chem. *235,* 2667–2671 (1960)
581. Zee, L., Hornemann, U., Floss, H. G.: Biochem. Physiol. Planz. *168,* 19–25 (1975)
582. Zharikova, G. G., Silaev, A. B., Sushkova, J. V.: Antibiotiki *8,* 425–430 (1963)
583. Zharikova, G. G.: Prikl. Biokhim. Mikrobiol. *1,* 83–89 (1965)
584. Zubay, G., Morse, D. E., Schrenk, W. J., Miller, J. H. M.: Proc. Nat. Acad. Sci. *69,* 1100–1103 (1972)

# Clinical Pharmacology of Benzodiazepines*

**Prof. Dr. Ulrich Klotz**

Dr. Margarete Fischer-Bosch Institute of Clinical Pharmacology, Auerbachstraße 112, D-7000 Stuttgart 50, West-Germany

Benzodiazepines are extensively used throughout the world. An increasing number of very similar substances has emerged on the market. The continuing scientific interest in the clinical and pharmacological properties of these drugs has greatly expanded our knowledge about them. The numerous substances share a similar pharmacodynamic profile (anxiolytic, sedative/hypnotic, anticonvulsive, muscle relaxant properties), which is mediated by interacting with neuronally localized receptors within the brain. Significant differences exist primarily in the pharmacokinetics of the benzodiazepines. Thus, onset and duration of action can be explained by specific features of their pharmacokinetic characteristics, which can be modified by many factors. Most of the side effects (excessive depression of the central nervous system is predominant) are an individual exacerbation of the desired therapeutic effect. The safety of the benzodiazepines is remarkable. However, abuse and their potential of dependence should be kept in mind.



* Supported by the Robert Bosch Foundation Stuttgart

Progress in Clinical Biochemistry
and Medicine, Vol. 1

# 1 Chemistry

In the development of psychoactive drugs, the discovery of chlordiazepoxide by the chemist Sternbach and its pharmacological characterization by Randall in 1957 represents a milestone[1]. The new class of benzodiazepines (see Fig. 1) has become almost synomymous with minor tranquilizers, although the large number of the series all show to a similar degree other properties such as sleep-inducing, muscle-relaxant and anticonvulsant activity.

Within a few years over 3000 compounds were synthesized and pharmacologically tested. It was soon realized that the substitution pattern played an important role with the electro-negative substituent at the 7 position (ring A) being of paramount importance. The biological potency is increased by halogens (F, Cl) in position 2′ (ring C) but strongly reduced by a substituent in 4′. Substituents larger than a hydroxyl or methyl group in position 3 (ring B) result in the loss of activity. Many different substituents have been tried in position 1 – while the methyl group seems to enhance the biological activity the 1-tert-butyl homologue of diazepam is almost completely inactive[2].

As can be seen in Fig. 1, the original 5-phenyl-1,4-benzodiazepine ring can be replaced by isosteric ring systems without loss of biological activity: e.g. ring A by a thiophene, pyrazole or pyridine ring; the nitrogens of ring B can be shifted to the 1,5 or 2,4 position; ring C can be replaced by a 2-pyridyl or cyclohexenyl group. More recently the fusion of additional rings especially on the diazepine ring has been introduced, resulting in the oxazinobenzodiazepine ketazolam and the oxazolobenzodiazepines oxazolam and cloxazolam. Since the additional ring is rapidly removed by metabolism without loss of pharmacological potency, the three compounds can be considered as pro-drugs. A different type of annelated benzodiazepines represent the triazolobenzodiazepines estazolam and triazolam (see Fig. 1). The additional ring is not removed by biotransformation, and they bind strongly and directly to the benzodiazepine receptors. The injectable imidazol-benzodiazepine midazolam differs from the classic benzodiazepines by its basicity, stability in aqueous solutions and rapid metabolism[3]. Some parts of the molecule resemble the specific benzodiazepine receptor antagonist Ro 15-1788 (see Fig. 1).

Surprisingly, an opioid benzodiazepine (tifluadom) was recently characterized (Fig. 1), which has lost all affinity for benzodiazepine binding sites but which has moderate to high affinity for opiate receptors[4].

Although the safety of the benzodiazepines is remarkable, even in overdose, the continuous proliferation of new compounds suggests that we have not yet found the ideal drug and the question may arise of what might follow[5].

# 2 Metabolism and Clearance

Before benzodiazepines can be excreted into the urine, they have to be metabolized. Thus, hepatic biotransformation accounts for the elimination of all benzodiazepines

5–Phenyl–1,3–dihydro–2H–
1,4–benzodiazepin–2–ones

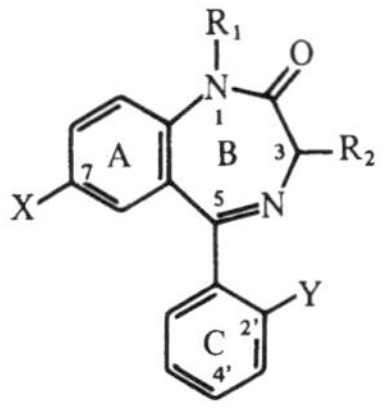

Examples of isosters

Ring A

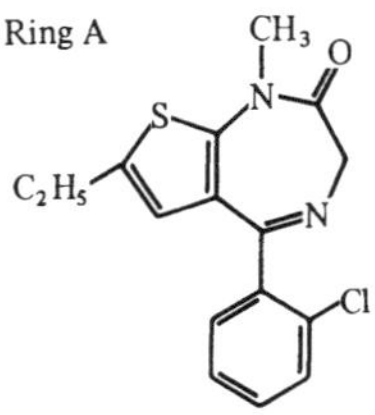

Clotiazepam

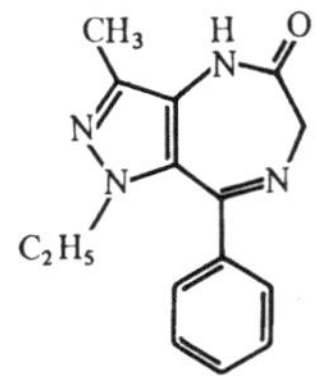

Ripazepam

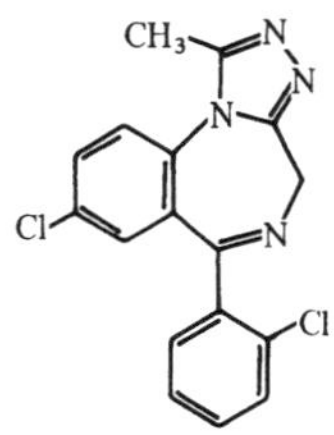

Triazolam

Ring B

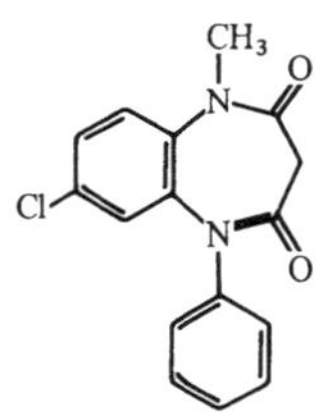

Clobazam

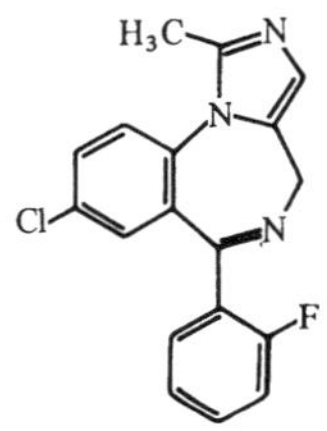

Midazolam

Ring C

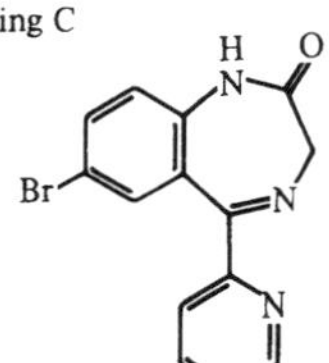

Bromazepam

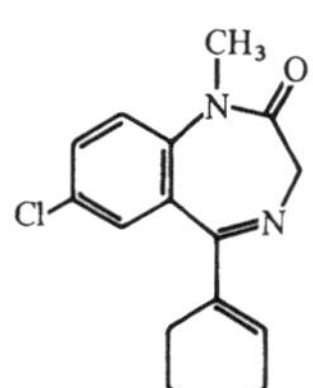

Tetrazepam

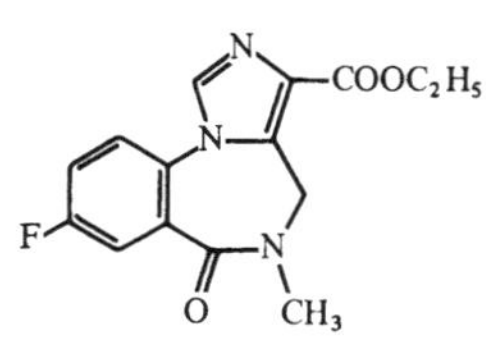

Ro 15–1788

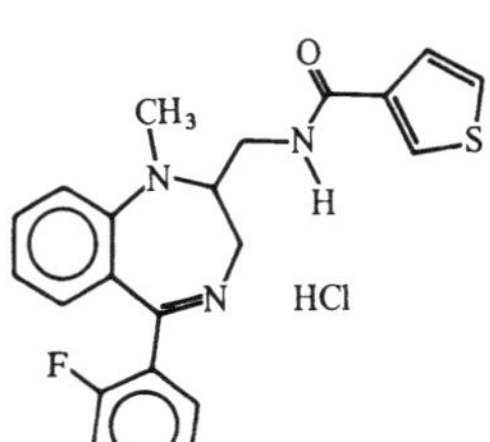

Tifludom

**Fig. 1.** Chemical structures of different benzodiazepine derivates

in man. Two principal pathways are involved: hepatic microsomal oxidation (e.g. N-dealkylation and hydroxylation), which is accomplished by the cytochrome P 450 dependent mixed function oxidase system (MFOS) and glucuronide conjugation (phase II reaction), which sometimes has to be preceded by hydroxylation (phase I reaction). In Fig. 2 some examples for newer benzodiazepines are given. Both types of metabolic pathways are differently controlled and differently influenced by various endogenous and exogenous factors (see Sect. 5.3). If a benzodiazepine possesses a nitro group in position 7 (e.g. clonazepam, nitrazepam, flunitrazepam), this has to be reduced to an amino group before subsequent acetylation (see part of Fig. 3). In Fig. 4 the major metabolic pathways of the two classical benzodiazepines diazepam and chlordiazepoxide are outlined. From the simplified scheme of Fig. 5 it is obvious that many benzodiazepines share common, still active metabolites (e.g. oxazepam) and some compounds, e.g. prazepam, halazepam, chlorazepate (see also Fig. 2), pinazepam, oxazolam, ketazolam have to be considered as precursors of desmethyldiazepam (nordiazepam). These pro-drugs undergo nearly complete presystemic (gastrointestinal, hepatic) metabolism and act predominantly just like desmethyldiazepam. In Table 1 the most common benzodiazepines and their metabolites are listed (reviews in[6–8]).

Indirect conclusions about drug metabolism can be derived from pharmacokinetic considerations applying the so-called clearance concept. The elimination of benzodiazepines by the liver may depend on a number of factors. These include the total blood flow to the organ, binding of the drug to blood constituents, uptake and transport across the hepatocyte plasma membrane, the distribution of flow (shunting) within the organ, intracellular transport and the metabolism, excretion or secretion of the drug. The rate-limiting step for a given compound may vary, and many physiological and pathological factors may interfere with the process. In an attempt to simplify these complex biological events, hepatic drug removal (i.e. clearance, $CL_H$) has been conceptualized as a function of both perfusion (blood flow, $Q_H$), and the capacity of the liver to remove the drug in the absence of flow limitations (intrinsic clearance, $CL_{intr.}$). The mathematical model of this relationship is given by the equation:

$$CL_H = Q_H \frac{CL_{intr.}}{Q_H + CL_{intr.}}$$

This model further relates the extraction ratio E (the fraction of drug removed in a single pass through the liver) to intrinsic clearance and blood flow as:

$$E = \frac{CL_{intr.}}{Q_H + CL_{intr.}}$$

The elimination of drugs with a very high intrinsic clearance (and thus a large $E > 0.7$) would be predominantly dependent on blood flow (flow-limited). In contrast, the elimination of most benzodiazepines with a low intrinsic clearance (small $E < 0.3$) would be dependent on intrinsic clearance and be largely independent of flow (capacity-limited). There are benzodiazepines with values of E intermediate

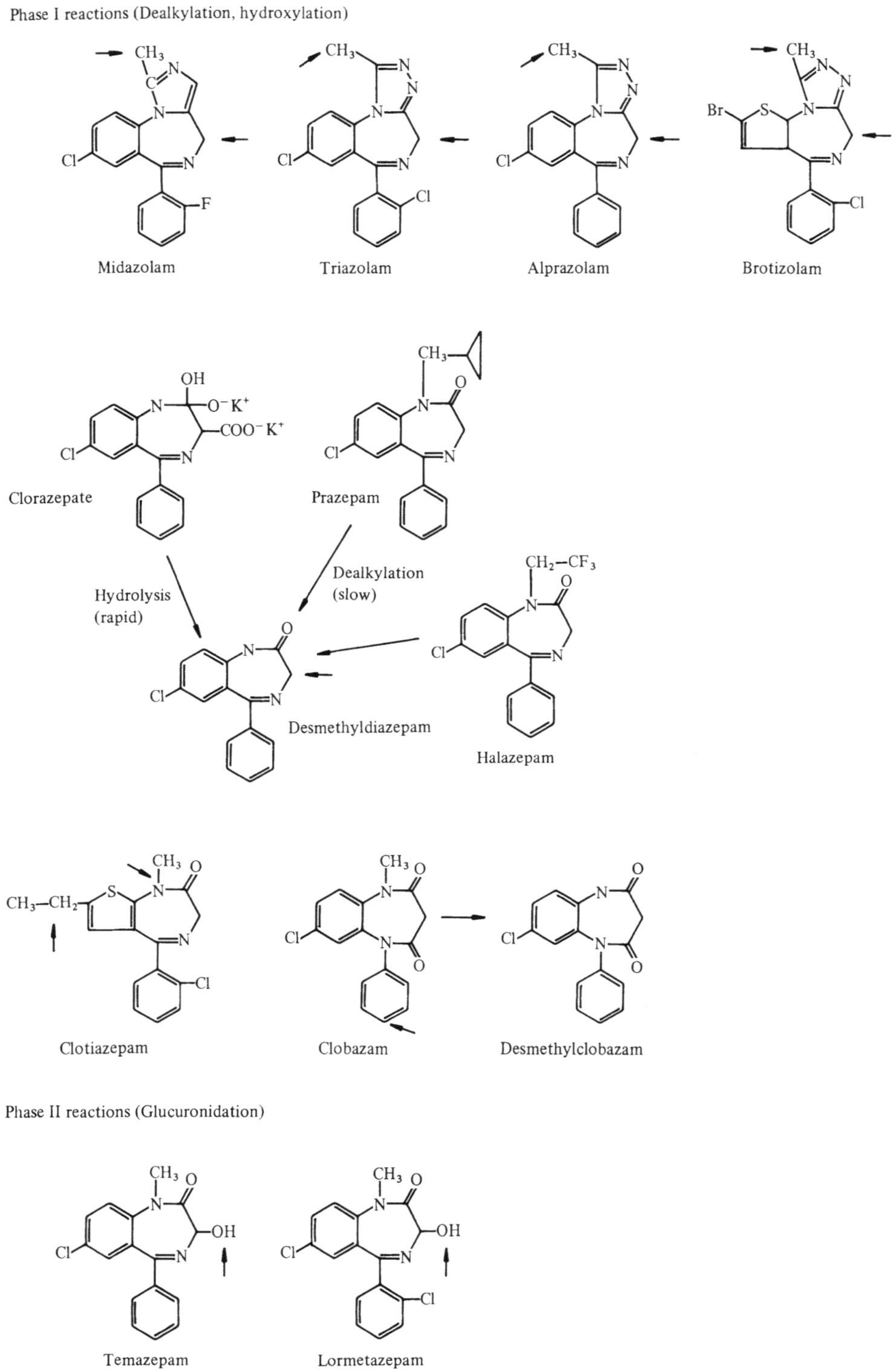


**Fig. 2.** Structures of some newer benzodiazepines and the sites (↑) for metabolic reactions

**Table 1.** Benzodiazepines and their metabolites

| drug | major metabolic reactions/ metabolites | active metabolites | minor pathways/ metabolites | active metabolites |
|---|---|---|---|---|
| alprazolam | hydroxylation | + | | |
| bromazepam | oxidation | (+) | | |
| brotizolam | oxidation | ++ | | |
| camazepam | oxidation | + | | |
| chlordia-zepoxide | demethylation, hydroxylation | ++ | oxazepam | ++ |
| clazepam | desmethyldiazepam, hydroxylation | ++ | | |
| clobazam | demethylation | ++ | | |
| clonazepam | nitroreduction, conjugation | ? | glucuronidation, acetylation | – |
| clorazepate | desmethyldiazepam | ++ | oxazepam | ++ |
| clotiazepam | demethylation, hydroxylation | + | | |
| diazepam | demethylation | ++ | oxazepam, temazepam | ++ |
| estazolam | oxidation | ? | | |
| flunitrazepam | nitroreduction, demethylation | ++ | glucuronidation | – |
| flurazepam | dealkylation, hydroxylation | ++ | | |
| halazepam | desmethyldiazepam | ++ | oxazepam | ++ |
| ketazolam | diazepam, desmethyldiazepam | ++ | oxazepam | ++ |
| lorazepam | glucuronidation | – | | |
| lormetazepam | glucuronidation | – | oxazepam | ++ |
| medazepam | demethylation, hydroxylation | ++ | oxazepam | ++ |
| midazolam | hydroxylation | + | | |
| nitrazepam | nitroreduction, acetylation | – | 3-hydroxynitrazepam | ++ |
| oxazepam | glucuronidation | – | | |
| oxazolam | desmethyldiazepam | ++ | oxazepam | ++ |
| pinazepam | desmethyldiazepam | ++ | oxazepam | ++ |
| prazepam | desmethyldiazepam | ++ | oxazepam | ++ |
| temazepam | glucuronidation | – | oxazepam | ++ |
| tetrazepam | oxidation | (+) | | |
| triazolam | hydroxylation | + | | |

++ metabolites biologically active as parent compound
\+ some biological activity of metabolites
(+) some biological activity likely
? no data available

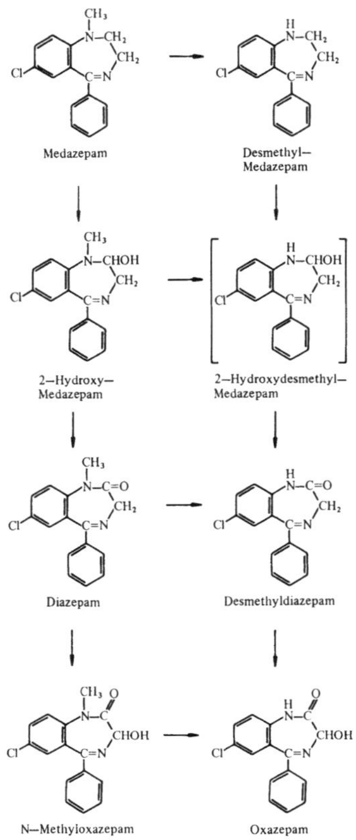


Metabolic pathways of medazepam in humans.

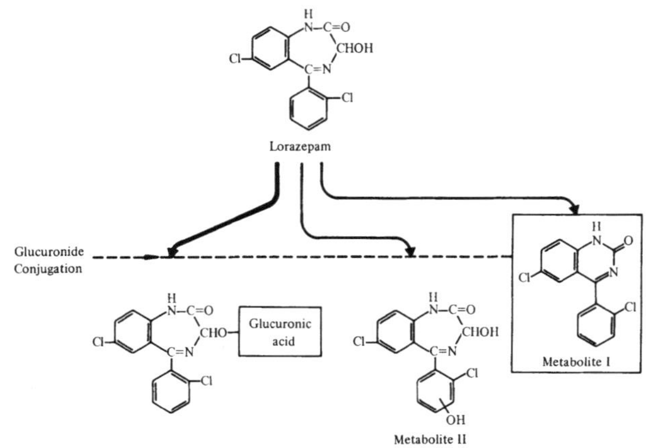


Urinary metabolites of lorazepam in humans. Lorazepam glucuronide is the only clinically important metabolite.

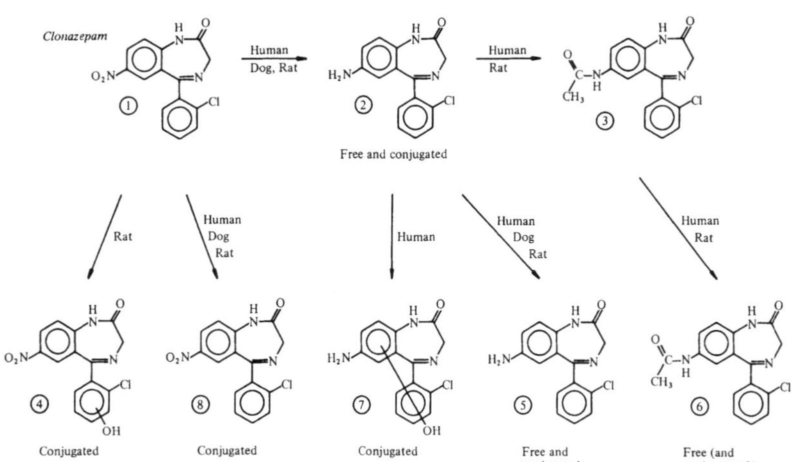


Major metabolic pathways of clonazepam.

**Fig. 3.** Metabolic pathways of some older benzodiazepines

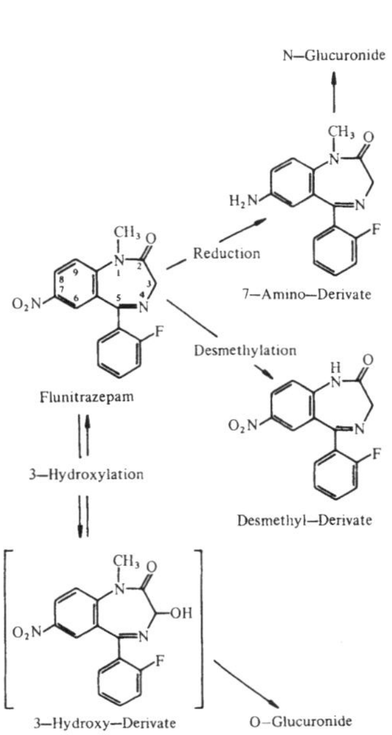


Metabolic pattern of flunitrazepam in humans.

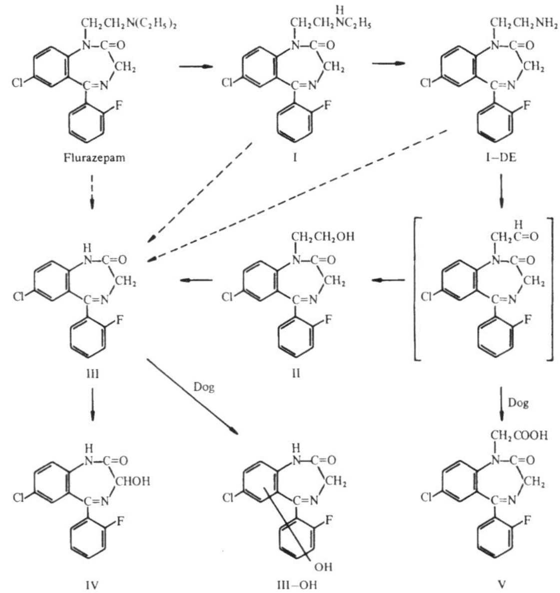


Metabolic pathways of flurazepam (FRZ) in humans (and dog).
I = monodesethylFRZ, I–DE = didesethylFRZ, II = hydroxyethylFRZ, III = desalkylFRZ (main metabolite in plasma), IV = desalkyl–3–hydroxyFRZ, V = FRZ–$N_1$–acetic acid

**Fig. 3.** (continued)

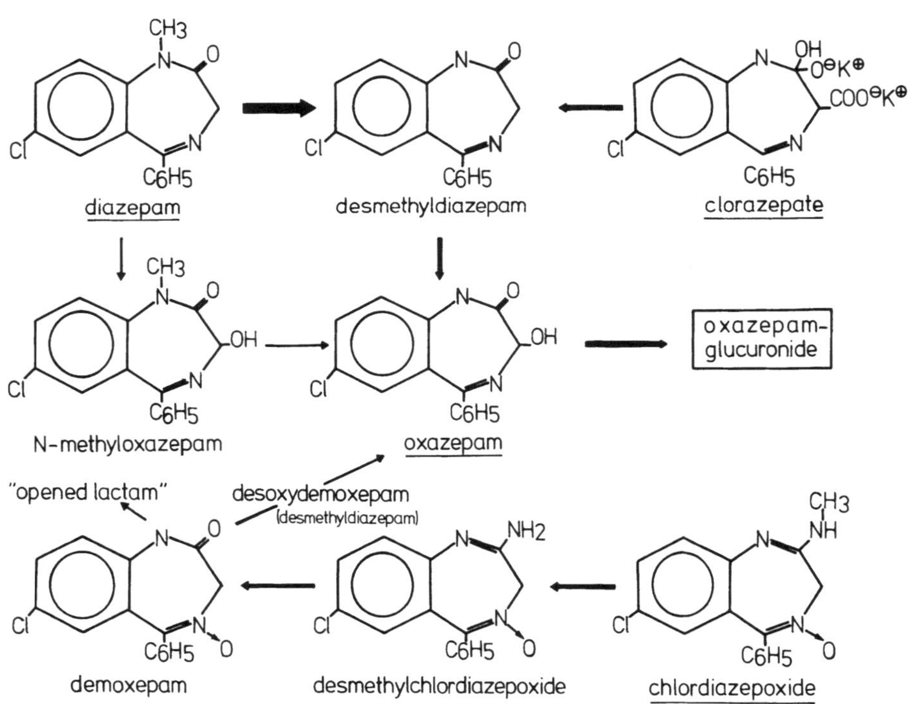


**Fig. 4.** Metabolic pathways of the two "classical" benzodiazepines diazepam and chlordiazepoxide

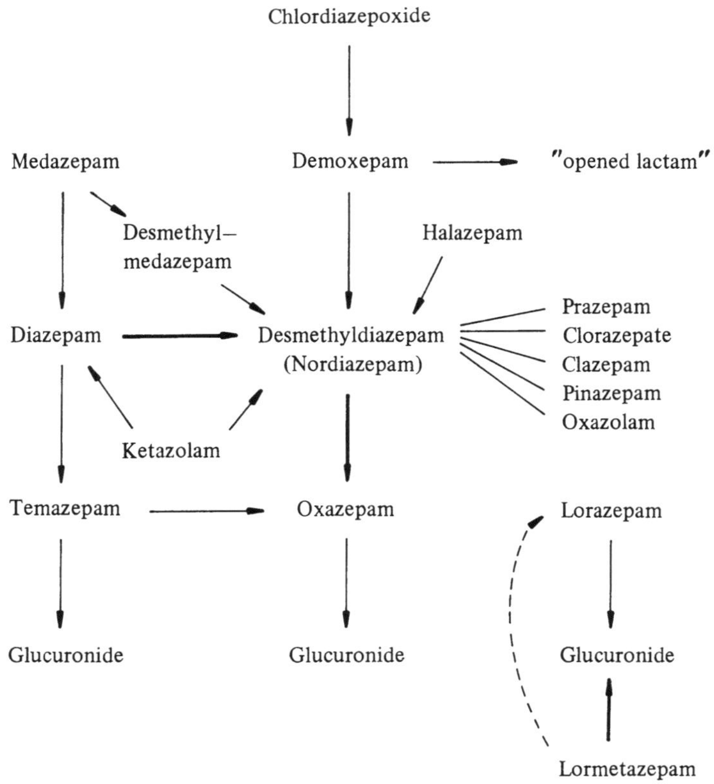


**Fig. 5.** Simplified metabolic scheme for some important benzodiazepines

between these extremes, where both flow and intrinsic clearance exert a separate influence. The extraction ratio also determines the fraction of an orally administered drug which reaches the systemic circulation (bioavailability, F) such that $F = 1 - E$.

# 3 Pharmacology

One reason for the popularity of benzodiazepines is their rather broad spectrum of pharmacologic activity. Historically, the first benzodiazepine, chlordiazepoxide, was found to have potent sedative, muscle-relaxant, taming and anticonvulsant activity[9]. Initial clinical trials indicated that it had antianxiety effects in human subjects[10] and that this effect might be separated from its sedative action, which induced drowsiness and improvement of the sleep pattern[11]. The anticonvulsant properties could be demonstrated first by the increase in threshold to pentylenetetrazol convulsions in human epileptics[12]. The next benzodiazepines that followed chlordiazepoxide (introduced in 1960), e.g. diazepam (1963), oxazepam (1965), medazepam (1968), clorazepate (1969), lorazepam (1972), prazepam (1973), flurazepam (1974), nitrazepam (1974) and clonazepam (1976) exhibit an almost identical pharmacological profile, which was characterized by various tests (e.g. electrical shocks, pentylenetetrazol, rotating rod, continuous avoidance, behaviour check) in different animal models (e.g. mice, rats, cats, monkeys). With these screening tests also the biological activity of identified metabolites was evaluated (review in[13]). The antianxiety, antiagression, anticonvulsant, muscle relaxation and hypnotic effects of the "first generation" of benzodiazepines have been comprehensively reviewed in 1973 and 1974[14, 15]. From more recent overviews it is obvious that the newer benzodiazepines share very similar pharmacological properties and that there are few differences between the various benzodiazepine derivatives. The main distinguishing features are differences in pharmacokinetics and the presence or absence of pharmacologically active metabolites[6, 7, 16–22]. All benzodiazepines are effective in the management of anxiety and insomnia, and their classification into "anxiolytics" and "hypnotics" does not appear to be justified.

## 4 Mode of Action

In the last years scientific interest was focused on the mechanism of action of benzodiazepines. Electrophysiological studies indicated that the inhibitory neurotransmitter γ-aminobutyric acid (GABA) was involved (review in[23]). In 1977, specific benzodiazepine binding sites in the central nervous system were postulated[24, 25]. In vitro binding studies with tritium labelled diazepam and flunitrazepam revealed highly specific, stereoselective, reversible and saturable binding sites localized on the nerve cell membranes. The binding affinity of active benzodiazepine derivatives correlate well with their relative potencies of anxiolytic, anticonvulsant and muscle relaxant activity in animal models[24, 25]. An intriguing question is how the primary event of binding at the recognition site of the proposed receptor initiates a chain of events (see Fig. 6) leading finally to the pharmacological response, e.g. tranquility, sedation and prevention of seizures. The bimolecular association of the ligand with the receptor subsequently leads to conformational changes of the ligand-receptor complex. Several structurally differing ligands with high affinity for benzodiazepine receptors have been found to act as benzodiazepine antagonists. The imidazobenzodiazepine Ro 15–1788 (see Fig. 1) selectively prevents or blocks the actions of benzodiazepines without possessing intrinsic effects in most neuropharmacological procedures[26, 27]. In contrast, so-called β-carbolines antagonize the effects of benzodiazepines at doses which produce behavioural and electrophysiological effects that are opposite to those of benzodiazepines[28–30]. The effects of β-carbolines (and diazepam) also can be blocked by Ro 15–1788[31]. It is suggested that these three compounds interact with the benzodiazepine receptor to stabilize one of two possible functional states[32].

In the numerous binding studies certain correlations between the influence of GABA on binding and the intrinsic activity of the benzodiazepines were noticed. Thus, the benzodiazepine receptor has been rationalized as a heterotropic site on the GABA-receptor-chloride-ionophore-complex. The complex nature of the bind-

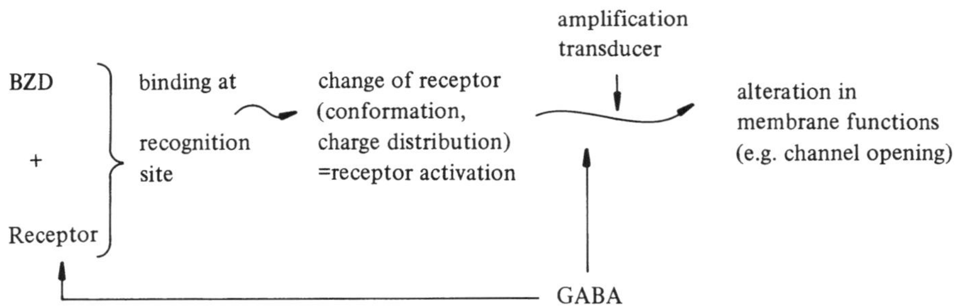


**Fig. 6.** Schematic diagram of the events following the binding of benzodiazepines to their recognition site

ing of structurally different compounds suggested the presence of two major classes of benzodiazepine receptors with high (type I) and low (type II) affinity. The relative proportions of both benzodiazepine receptors vary in different regions of the brain, e.g. the cerebellum contains predominantly type I (review in[33]). The functional significance of the subtypes is unknown. For the different pharmacological effects only 25–70% of the heterogeneous benzodiazepine binding sites have to be occupied[34]. Since tolerance can develop to the sedative and anticonvulsant effect of benzodiazepines, it is discussed if prolonged administration of these drugs induce changes in receptor number and/or affinity (review in[35]).

The idea that benzodiazepines interact with GABA is derived mainly from electrophysiological studies. Benzodiazepines potentiate the inhibitory effects of GABA in a variety of neuronal systems and GABA enhances the affinity of benzodiazepines (review in[36]). This strongly suggests a functional link between both binding sites. In addition, this receptor complex is associated with chloride channels. In the allosteric model for benzodiazepine receptor function, GABA exerts its inhibitory effect by opening two-state (open/closed) channels that are permeable to chloride ions. The flow of chloride ions into the cell moves the membrane's electrical potential away from the threshold for triggering an action potential, thus excitation is inhibited. Benzodiazepines do not alter the elementary ionic conductance of the chloride channels opened by GABA. Potentiation of GABA responses occurs by an increase in opening frequency and open lifetime at which GABA molecules successfully open chloride ion channels (review in[37]). The receptor model is further complicated by a protein called GABA-modulin, which is present in brain and competitively inhibits both GABA- and benzodiazepine binding[38]. By analogy to the enkephalins, attempts are made to identify endogenous ligands for benzodiazepine receptors. Several substances have been proposed including hypoxanthine, inosine, nicotinamide and β-carbolines. However, so far no definite answer on the existence of such ligands can be given. Based on their pharmacological properties β-carbolines could be characterized as anxiogenic benzodiazepine antagonists, since they can induce severe anxiety in humans[39] and animals[40] and reverse the hypnotic action of flurazepam[41].

While the neuronal existence of the benzodiazepine-GABA-receptor-chloride channel (ionophore)-complex is unquestionable[42], different models have been proposed in terms of a three-state[32] or allosteric[33] model (Fig. 7). The intimate associa-

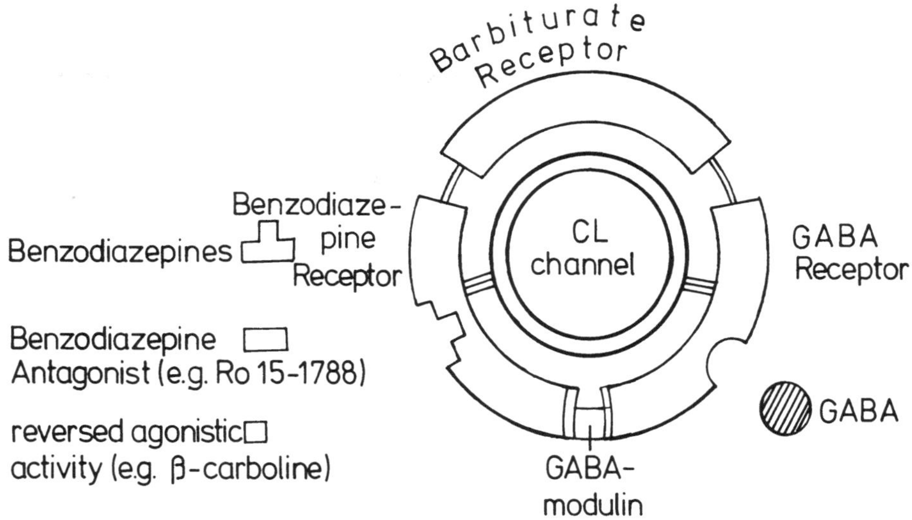


**Fig. 7.** Simplified model of the benzodiazepine-GABA-chloride-ionophore-receptor-complex

tion of benzodiazepine and GABA receptor and chloride channel suggest that benzodiazepine act physiologically as modulators of GABA. Therefore, the benzodiazepine receptor might be viewed as a coupling or regulatory unit. Whether this coupling function is modulated by endogenous ligands remains questionable. When such an endogenous agonist actually exists, then benzodiazepines would antagonize those effects[43)].

Based on the present evidence, ligands for the benzodiazepine receptor may belong to either of the following four categories:

a) full agonists (e.g. benzodiazepine)
b) antagonists (e.g. Ro 15–1788)
c) antagonists with partial agonistic activity
d) antagonists with partial inverse agonistic activity (e.g. β-carbolines)

In first controlled clinical trials, Ro 15–1788 did not demonstrate central effects following a single oral dose of up to 600 mg[44)] or 100 mg intravenously[45)]. However, the hypnogenic effect of flunitrazepam[46)] and the central effects of 3-methylclonazepam[47)] and diazepam[48)] were antagonized.

In humans somewhat contraversial results have been reported with physostigmine. Reversal of diazepam-induced hypnosis was observed[49–51)] but it failed to reverse clinical, psychomotor or EEG effects of lorazepam[52)] or diazepam[53)]. Similarly surprising was the clinical observation that aminophylline antagonizes diazepam sedation[54, 55)]. From these findings one could speculate whether additional mechanisms might be involved in the central action of benzodiazepines.

# 5 Clinical Pharmacokinetics

For practicing physicians decision-making is complicated by an increasing number of marketed benzodiazepines. Since qualitative differences in the pharmacological potency are at best subtle, only a very limited number of benzodiazepines are really needed. Advances in understanding the pharmacokinetics of the benzodiazepines can provide, at least partly, support in this decision process. Therefore, one should be familiar with the absorption (following oral, intramuscular and rectal application), distribution and hepatic elimination of these drugs and their determinants, which might modify drug disposition. The time of onset, intensity and duration of clinical action depends on the physico-chemical and pharmacokinetic properties of the particular derivative. Thereby some differences exist whether the benzodiazepine is used for short-term (single dosing) or long-term (sub/-chronic dosing) indications.

## *5.1 Single versus Multiple Dosing*

Onset of benzodiazepine action depends on the rate of absorption and penetration in specific brain regions. Both processes are accomplished by passive diffusion and depend mainly on the lipid solubility at physiologic pH. Increasing lipophilicity is associated with more rapid diffusion across plasma and/or cell membranes. Since benzodiazepines are highly lipophilic molecules, they enter rapidly into brain tissues. Following a single intravenous injection (excluding the slower and rate-determining absorption), the sedative/hypnotic effects can be observed within 20 to 30 seconds (one single blood circulation time) and no more than after a few minutes. Highly lipophilic derivatives (e.g. diazepam, midazolam) act faster than less lipophilic compounds (e.g. chlordiazepoxide, lorazepam). Thus, there appears to be some association between the onset of drug effect and the rapidity of passage of benzodiazepines into the central nervous system (review in[56)]). However, contrary to studies with diazepam and flunitrazepam the rapid clinical effect of the highly lipophilic midazolam cannot be explained by rapid passage into human cerebrospinal fluid, where only in some cases very low concentrations (maximal 5 ng/ml) could be detected[57)].

Following a single oral dose, the same principle holds true for the rate-limiting process of absorption: e.g. the more polar oxazepam is absorbed slower than diazepam. In addition, the characteristics of the pharmaceutical formulation (e.g. particle size, soft or hard gelatine capsules) is important. Another variable has to be taken into account when the precursors of the active moiety of desmethyldiazepam are administered: while the pro-drug clorazepate is very rapidly hydrolyzed to desmethyldiazepam, prazepam is slowly transformed into the active substance. A rapid onset of action may be desirable if patients benefit therapeutically. However, if patients experience an unwelcome feeling of drowsiness or excessive relaxation, a slower rate of absorption with less pronounced peak concentrations might be preferable.

For intramuscular administration, the site of injection and the injecting person (physician or nurse) are of critical importance[58, 59]. More rapid and complete absorption is achieved by injection into the area of the deltoid muscle. With chlordiazepoxide[60–62] and diazepam[58, 63] a slow, incomplete and erratic absorption has been described, which is due to the likelihood of local precipitation. In contrast, lorazepam is rapidly, completely and reliably absorbed[64]. Diazepam is also available through a rectal tube for acute therapy (e.g. febrile convulsions in children). Absorption is rapid (peak plasma levels after about 17 min) and complete[65].

Following a single dose, the duration of action differs among benzodiazepines. The major determinant seems to be the rate of redistribution from the brain to peripheral tissue sites (e.g. adipose tissue). Apparently, lipophilicity and blood perfusion are the driving forces for the shift of active drug from the central nervous system to some storage sites. Thus, less lipophilic derivatives (e.g. lorazepam) may persist longer in effective brain concentrations than more lipophilic agents (e.g. diazepam). As illustrated in Fig. 8 the time- and concentration-dependent sedative effect of a single intravenous dose of diazepam (0.1 mg/kg) persists for about two h. During that time period, plasma levels decline rapidly, and this initial distribution phase is responsible for the relatively short duration of action[66].

Systemic studies with eight benzodiazepines in anesthetized rats recently suggested that benzodiazepine distribution in vivo is determined largely by lipid solubility and that duration of acute EEG effects are longest for the least lipophilic drugs having the smallest apparent volume of distribution[67].

During multiple dosing, benzodiazepines, and in many cases also their biological active metabolites, accumulate according to their pharmacokinetic properties. Within 4 to 5 elimination half-lives ($t_{1/2}$) the corresponding steady-state concentrations ($C^{ss}$) are achieved. When $t_{1/2}$ is long, accumulation is slow (e.g. diazepam, desmethyldiazepam). Conversely, benzodiazepines with a short $t_{1/2}$ (e.g. midazolam) rapidly attain the steady state, which is characterized by the balance of drug intake and drug removal. The daily maintenance dose substitutes for the loss of drug from the body, and as long as effective plasma levels are maintained, therapeutic/toxic effects can be anticipated. According to the equation:

$$C^{ss}_{av} = \underbrace{\frac{F \cdot D}{\tau}}_{\text{input rate}} \times \underbrace{\frac{1}{CL}}_{\text{elimination rate}} \qquad \tau = \text{dosing interval}$$

the average $C^{ss}$ is proportional to the input rate of the bioavailable dose ($F \cdot D$) and reversibly proportional to the total body clearance (CL) of the drug/metabolite. In the case of benzodiazepines, hepatic metabolism is the determinant for the CL. Physiological or disease-induced changes in $t_{1/2}$ and/or CL differently modify the accumulation pattern.

The day-by-day buildup of active substances in the blood and brain during initiation of (sub)-chronic therapy influences the clinical effects of benzodiazepines. Depending on the extent of drug accumulation, drowsiness, mental confusion, motor incoordination, and impairment of intellectual and psychomotor performance will occur. These side effects are nothing else than a pronounced (exaggerated) pharmacological action of the therapeutic agents, and they are more frequent with

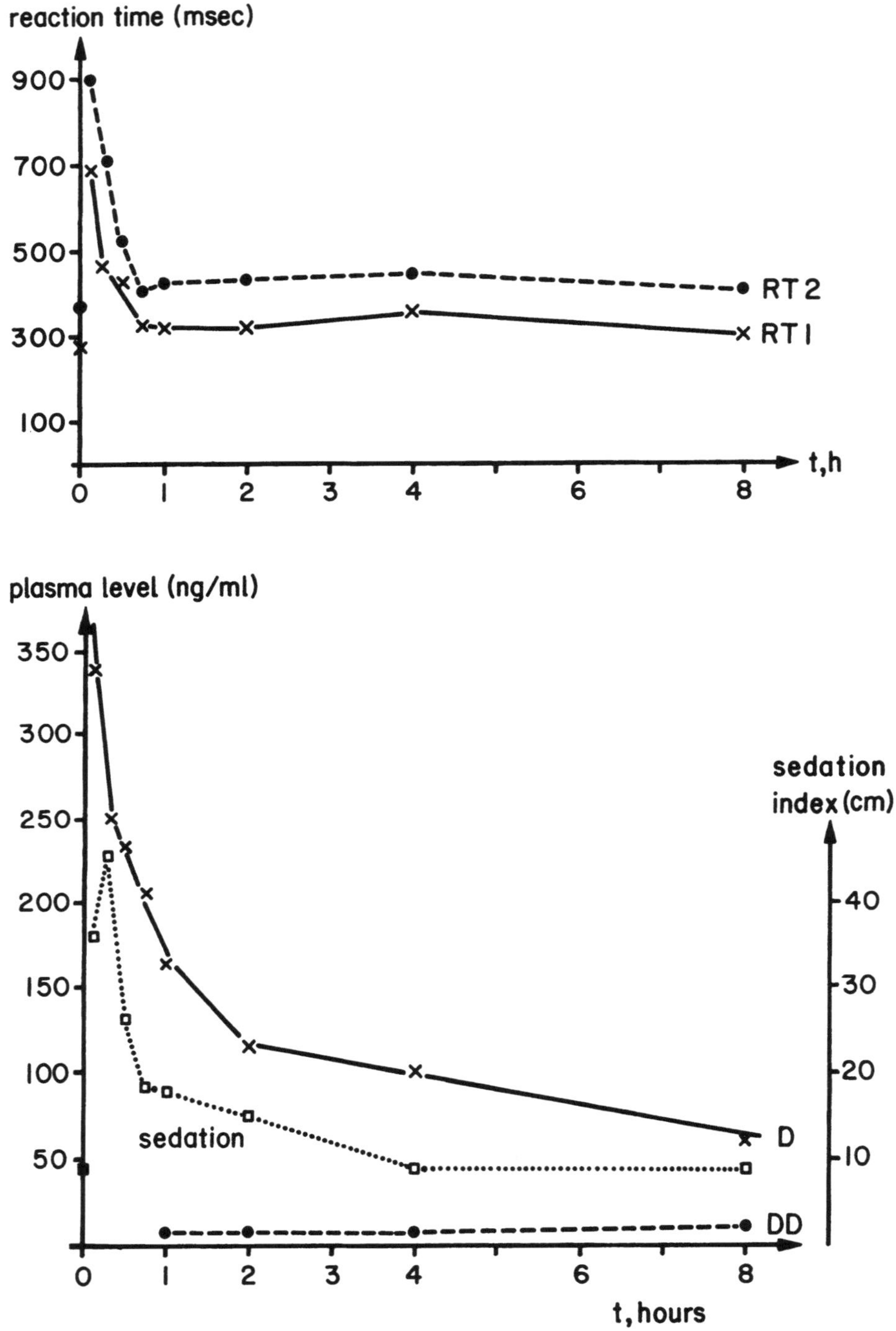


**Fig. 8.** Plasma level time profile of diazepam (D) and its major metabolite desmethyldiazepam (DD) during the distribution phase following a single intravenous dose of 0.1 mg/kg in one healthy subject (bottom). In addition, the time-dependent pharmacodynamic response (sedation and choice reaction time RT 1/RT 2) is given (see also Reference[66])

benzodiazepines possessing a long $t_{1/2}$[68–70]. There are some indications that adaptation ("tolerance") to the sedative effects might develop, especially during the early course of therapy.

The duration of therapeutic effects (e.g. treatment of symptoms such as anxiety or insomnia), unwanted side effects and the potential for drug interactions after discontinuation of drug intake are dependent on the rate of drug disappearance from the blood and the receptor site. Thus, recurrence of symptoms and waning of side effects will correspond to the drug's and/or active metabolite(s)' elimination half-life. Therefore, attempts were made to classify, similar to the barbiturates, the numerous benzodiazepines according to their elimination half-life (see Table 2).

During multiple dosing, these characteristics should be kept in mind for any particular benzodiazepine in considering the risks of day-time sedation (hang-over drowsiness) and interactions with other central active drugs, especially alcohol.

**Table 2.** Elimination half-lives ($t_{1/2}$) of benzodiazepines and their biological active metabolites

| benzodiazepine | $t_{1/2}$, h | active metabolite (s) | overall rate of elimination of active substances |
|---|---|---|---|
| chlordiazepoxide: | 10–18 | 20–80 | slow |
| clobazam: | 10–30 | desmethylclobazam: 36–46 | slow |
| clonazepam: | 24–56 | ? | slow |
| clorazepate: | 1.5– 2.5 | desmethyldiazepam: 50–80 | slow |
| clazepam: | ? | desmethyldiazepam: 50–80 | slow |
| diazepam: | 30–45 | desmethyldiazepam: 50–80 | slow |
| flurazepam: | 2 | 47–100 | slow |
| halazepam: | 1– 2 | desmethyldiazepam: 50–80 | slow |
| ketazolam: | 1.5 | diazepam: 30–45<br>desmethyldiazepam: 50–80 | slow |
| medazepam: | 2 | 20–80 | slow |
| nitrazepam: | 20–50 | ? | slow |
| oxazolam: | – | desmethyldiazepam: 50–80 | slow |
| pinazepam: | 16 | desmethyldiazepam: 50–80 | slow |
| prazepam: | 1–3 | desmethyldiazepam: 50–80 | slow |
| alprazolam: | 10–18 | ? | intermediate |
| bromazepam: | 12–24 | ? | intermediate |
| camazepam: | 20–24 | ? | intermediate |
| estazolam: | 8–31 | ? | intermediate |
| flunitrazepam: | 10–25 | 20–30 | intermediate |
| tetrazepam: | 12 | ? | intermediate |
| loprazolam: | 7– 9 | ? | intermediate |
| clotiazepam: | 3–15 | ? | intermediate/rapid |
| oxazepam: | 5–18 | – | intermediate/rapid |
| lorazepam: | 10–18 | – | intermediate/rapid |
| lormetazepam: | 9–15 | – | intermediate/rapid |
| temazepam: | 6–16 | – | intermediate/rapid |
| brotizolam: | 4– 8 | 9.5 | rapid |
| triazolam: | 2– 4 | 3– 8 | rapid |
| midazolam: | 1– 3 | ? | very rapid |

## *5.2 Pharmacokinetic Properties of Specific Compounds*

In this paragraph the basic pharmacokinetic parameters will be summarized. It should be emphasized that so-called normal values sometimes vary widely in the literature. This variability is due to differences in analytical measurements, pharmacokinetic evaluation of the data and patient selection. In addition, the disposition of benzodiazepines is affected by a variety of different physiological factors (see the following sections).

*Alprazolam* is a new triazolo-benzodiazepine derivative with anxiolytic-antidepressant properties. The major metabolic pathway in humans involves hydroxylation in at least 2 positions[71)]. Absorption appears to be rapid since peak plasma concentrations are reached within 2 h. In young subjects $t_{1/2}$ and CL averaged 11 h and 1.3 ml/min/kg, respectively. The free fraction in plasma ($f_u$) is approximately 30%[72)].

*Bromazepam's* elimination can be characterized by a $t_{1/2}$ of 14.4 ± 4.9 h and an oral CL of 59 ± 19.6 ml/min. Its absorption rate seems to be somewhat variable since maximal plasma concentrations were achieved within 1 to 6 h in 10 healthy volunteers[73)]. The mean absolute bioavailability (F) of an oral solution, a conventional tablet and an enteric-coated tablet is 52, 62 and 99%, respectively. These differences can be explained by a pH-dependent ring opening[74)]. Predictable steady-state blood levels are maintained at multiple daily-dosing[75)].

*Brotizolam* is a new triazolothieno-diazepine. It is extensively metabolized (hydroxylation at the methyl group and at the diazepine ring). Both metabolites are pharmacologically almost as active as the parent compound. Brotizolam is absorbed quickly ($t_{1/2a}$ = 0.2 to 1.4 h), and oral bioavailability averaged 70%. The protein unbound fraction varied from 5 to 10%. Its rapid elimination can be characterized by a $t_{1/2}$ of 4 to 8 h and a CL of 113 ml/min[76)].

*Chlordiazepoxide* was the first clinically available benzodiazepine. It is extensively metabolized to active compounds (e.g. desmethylchlordiazepoxide, demoxepam, desmethyldiazepam, oxazepam). Absorption appears to be relatively slow (peak levels within 2 to 6 h; $t_{1/2}$ of absorption 0.7 h) and complete (F 80–100%) following oral administration. After intramuscular injection, lower peak concentrations are reached later (7.6 h post-dosing; $t_{1/2}$ of absorption 3 to 5 h) and often a biphasic ("erratic") blood concentration profile can be observed caused by precipitation at the injection site (F = 86%). Plasma protein binding is about 95%. In young healthy volunteers $t_{1/2}$ ranges between 10 and 18 h and CL averages 30 ml/min (reviews in[6, 77)]).

*Clazepam* is rapidly biotransformed to a hydroxylated metabolite and to desmethyldiazepam. It is suggested that it acts mainly through the formation of *desmethyldiazepam*[78)].

*Clobazam* represents a 1,5-benzodiazepine, which, like diazepam, is primarily demethylated to the active desmethylclobazam. Both substances cumulate during multiple dosage. It is reasonably rapidly absorbed (peak levels within 2 h) and highly bound to plasma proteins ($f_u$ = 10–15%). Elimination of clobazam is characterized by a $t_{1/2}$ between 10 and 30 h and a CL of about 40 ml/min in young subjects (review in[79]).

*Clonazepam* is a drug of first choice in the treatment of status epilepticus. It is extensively metabolized, including nitroreduction, hydroxylation, acetylation and glucuronidation. Absorption is almost complete (F = 95%) but its rate is variable (peak levels between 1–8 h). Plasma protein binding has been found to be about 82%. For $t_{1/2}$ and CL ranges of 24 to 56 h and 60 to 80 ml/min, respectively, have been reported (review in[6]).

*Clorazepate* can be regarded as a pro-drug of *desmethyldiazepam.* Following oral intake the presystemic conversion already takes place in the acidic stomach as well as during the first passage through the liver. After i.v. injection, unchanged drug can be detected during the first two hours[80, 81]. The elimination of the highly bound ($f_u$ = 2–4%) desmethyldiazepam is very slow ($t_{1/2}$ = 50–80 h, CL = 8–15 ml/min) and seems to be faster in treated epileptics (review in[6]).

*Clotiazepam,* a thienodiazepine, is metabolized both by demethylation and by hydroxylation. Both metabolites appear to be pharmacologically active. Protein binding of clotiazepam is extensive ($f_u$ about 1%) and elimination with a $t_{1/2}$ between 3 and 15 h relatively rapid[82, 83].

*Diazepam* represents the most extensively studied benzodiazepine. Its major blood-metabolite, desmethyldiazepam, contributes to its action, especially during multiple dosing when plasma concentrations of the active metabolite exceed those of the parent compound. After oral administration, absorption of diazepam is rapid (peak levels within 90 min) and complete (F = 100%). Following intramuscular injection, a poor (F = 60–80%) and irregular absorption, dependent on the site and depth of injection, has been observed.

Single dose kinetics have revealed a $t_{1/2}$ between 30 and 45 h and a CL of about 20 to 35 ml/min. During multiple treatment elimination might be slightly slower ($t_{1/2}$ = 30–60 h, CL = 20 ml/min). Normal protein binding ranges between 96 and 98% (review in[6]).

*Estazolam,* a new triazolo-benzodiazepine hypnotic agent, is relatively slowly absorbed (peak concentration attained within 6 h) and $t_{1/2}$ ranged from 8 to 31 h[84].

*Flunitrazepam* is mainly used in anaesthesiology (premedicant, induction agent).The complex metabolic pattern includes nitroreduction, demethylation, hydroxylation and conjugation. At least two metabolites are still biological active. Following oral administration F averaged 80% and absorption is rapid (effective plasma concentrations after 10 to 20 min). The plasma protein binding is about 80%. The normal range of $t_{1/2}$ and CL is between 10–25 h and 210–465 ml/min, respectively (review in[6]). A recent publication reports a mean CL-value of only 94 ml/min[85].

*Flurazepam* was mainly introduced as a hypnotic. Its major blood metabolite, desalkylflurazepam (still active), possesses a long $t_{1/2}$ (47–100 h) and therefore cumulates after repeated doses of flurazepam. Free and conjugated hydroxyethylflurazepam are mainly found in urine. Flurazepam is absorbed rapidly from the gastrointestinal tract; however, unchanged drug is only occasionally found in the blood shortly after administration (review in[6]). This is due to first-pass metabolism of flurazepam in the small bowel mucosa as well as in the liver[86].

*Halazepam* represents another pro-drug of *desmethyldiazepam*. The precursor is rapidly converted ($t_{1/2}$ = 1–2 h) to its active substance[7, 87].

*Ketazolam* is rapidly biotransformed ($t_{1/2}$ = 1.5 h) directly and indirectly via *diazepam* to *desmethyldiazepam*. Both metabolites seem to be the active moieties[88].

*Lorazepam's* structure is very similar to oxazepam. The addition of only one CL atom in the phenyl ring enforces the sedative-hypnotic properties. It is relatively rapidly eliminated ($t_{1/2}$ = 10–18 h, CL = 50–90 ml/min) in form of an inactive glucuronide. This major metabolite exhibits a similar $t_{1/2}$ (12–21 h) and a renal CL of about 37 ml/min. Following intramuscular injection, sublingual or oral ingestion absorption of lorazepam is rapid (peak concentration within 1 to 4 h) and complete (F = 95–100%). Normally, 85–94% of lorazepam and approximately 65% of the conjugate are bound to plasma proteins (reviews in[6, 89, 90]).

*Lormetazepam* differs from lorazepam by an additional methyl group in $N_1$-position. It is mainly conjugated with glucuronic acid. A minor pathway includes demethylation to lorazepam with subsequent rapid glucuronidation. Usually $t_{1/2}$ ranges from 9–15 h with a CL between 180 and 300 ml/min. Oral doses are 70 to 80% bioavailable[91, 92].

*Medazepam* is extensively metabolized to active substances, such as desmethylmedazepam, diazepam, desmethyldiazepam and oxazepam. After rapid oral absorption (peak levels within 1–2 h) F varies from 49 to 76%. Unchanged medazepam is almost completely bound to plasma proteins ($f_u$ = 0.2%). The parent compound has a short $t_{1/2}$ (2 h); however, the cumulating active metabolites contribute to a prolonged action, especially during multiple dosing (review in[6]).

*Midazolam* is a new imidazo-benzodiazepine derivate mainly used as an induction agent in anaesthetic practice. Metabolism involves hydroxylation at two different positions on the molecule. At physiological pH it is extremely lipid-soluble and, therefore, an oral dose is very rapidly absorbed (peak effects and concentrations within 20 to 60 min). Protein binding of midazolam is extensive ($f_u$ = 2–6%). Since its blood clearance is relatively high (500–1100 ml/min) in relation to liver blood flow (approximately 1500 ml/min), the hepatic elimination of midazolam is partly perfusion dependent and a considerable hepatic first pass effect occurs, which accounts for a bioavailability of only 40 to 60%. Midazolam is the most rapidly eliminated benzodiazepine with $t_{1/2}$ of 1–3 h[93, 94].

*Nitrazepam's* metabolic pathways are similar to those of clonazepam. The major metabolites, 7-aminonitrazepam and 7-acetamidonitrazepam, have no benzodiazepine-like effects in clinically realistic concentrations. The acetylation reaction is genetically polymorphic (e.g. slow/rapid acetylators). A variable oral bioavailability was observed (range 54–93%) and 85 to 88% of nitrazepam are bound to plasma proteins. Elimination is relatively slow ($t_{1/2}$ = 20–50; oral CL = 50–120 ml/min). Free and conjugated 7-aminonitrazepam demonstrate a similar $t_{1/2}$ of about 45 h (reviews in[6, 95]).

*Oxazepam,* like diazepam, can be considered as a well-studied standard benzodiazepine. It is predominantly metabolized by conjugation to form a glucuronide that is pharmacologically inactive and which is subsequently excreted by the renal route. Oral absorption is incomplete (F = 70–86%) and peak levels are achieved within 2 to 4 h. Plasma protein binding ranges from 87 to 95%. Elimination is relatively rapid ($t_{1/2}$ = 5–18 h; CL = 76–136 ml/min) and there is almost no accumulation during multiple dosing (review in[6]).

*Oxazolam* is another pro-drug of *desmethyldiazepam.* After oral administration, unchanged drug was absent in plasma and urine, as it is metabolized prior to reaching the systemic circulation. The appearance $t_{1/2}$ of desmethyldiazepam was 1.5 h and its elimination $t_{1/2}$ averaged 56 h[96].

*Pinazepam* acts also as a pro-drug of *desmethyldiazepam.* Unchanged drug reaches maximal plasma concentrations within 2 h and is eliminated with a $t_{1/2}$ of 16 h. Thirty minutes after oral administration the plasma level of the active metabolite becomes higher than that of the parent compound which is probably due to an extensive hepatic first pass effect[97].

*Prazepam's* absorption is relatively slow. However, it is almost completely converted to the active *desmethyldiazepam* during its first passage through the liver. Thus, only the metabolite, reaching its maximal levels between 3 and 5 h, is available for systemic action[98, 99].

*Temazepam* can be regarded as $N_1$-methyloxazepam and it is also demethylated to oxazepam in a minor metabolic pathway. However, temazepam is primarily conjugated to the major metabolite temazepam glucuronide[100]. Absorption depends on whether a soft gelatin (used in Europe) or a hard gelatin (used in the USA) capsule is utilized[101]. From the soft gelatin capsule, peak levels of temazepam are generated more rapidly (0.8 h) than from the hard gelatin capsule (1.4 h). The unbound fraction of temazepam is about 2.5 to 3%. Elimination proceeds with a $t_{1/2}$ of 8 to 16 h and a CL of approximately 90 ml/min[101–103].

*Tetrazepam's* structure (see Fig. 1) and metabolism (demethylation, hydroxylation) both closely resemble diazepam. The manufacturer reports a $t_{1/2}$ of 12 ± 2 h and plasma protein binding of 70%.

*Triazolam's* metabolic pathways involve hydroxylation in two different positions. The metabolites may have some pharmacological activity; however, they appear in plasma in negligible amounts[104]. Following oral administration, peak concentrations are observed between 1 and 2 h. Elimination is rapid with a $t_{1/2}$ of 2 to 4 h and CL of about 350 ml/min[105, 106].

*Camazepam*[107], *loprazolam*[108] and *quazepam*[109] are new effective benzodiazepines whose clinical pharmacokinetics have not been yet fully investigated.

As already mentioned, benzodiazepines are lipophilic and highly protein-bound drugs, which are widely distributed into the various tissues (apparent distribution volume larger than actual body spaces) and extensively metabolized in the liver before the metabolites are finally excreted into the urine. Thus, any physiological and/or disease-induced change in body composition, protein content, liver and other organ function can modify the pharmacokinetics of the benzodiazepines. Today, this class of drugs can be already regarded as model compounds for evaluating the effects of different factors and pathophysiological conditions on drug disposition.

## *5.3 Determinants of Clinical Pharmacokinetics*

In this section some examples will be given which can explain the clinical variability in the pharmacokinetics of benzodiazepines and their action.

### *5.3.1 Variability in Absorption*

As a group the benzodiazepines are rapidly and nearly completely absorbed after oral administration. There are some differences of the rate of absorption among the drugs, with prazepam being the slowest[110]. As has been demonstrated for temazepam, the galenic formulation will be one major determinant for the rate of absorption[101]. Since medications are often taken at meal time, food intake may be of some clinical importance[111, 112]. The rate of absorption was reduced but the extent of absorption slightly increased when diazepam was given with food[113]. Meals had no effect on the extent of absorption of clobazam[114, 115] and oxazepam[116], but lower peak levels of clobazam occurred later. Most pharmacokinetic studies with oral application are performed after an overnight fast and food is allowed at 2 to 3 h post-dosing. Thus, from the limited data one could generalize that absorption of benzodiazepine might be somewhat delayed by food.

### *5.3.2 Plasma Protein Binding*

One important determinant in drug disposition is the binding to plasma proteins, and this is especially true for drugs where more than 90–95% are bound (reviews in[117, 118]. Most of the benzodiazepines commonly used exhibit such an extensive binding, and only the low plasma protein binding of flurazepam (15%) can be ignored (96.5% active metabolite desalkylflurazepam is bound). Over a wide concentration range (10–10 000 ng/ml) binding of benzodiazepines is concentration-independent[119)]. Among the many derivatives, especially the pharmacokinetics and plasma binding of diazepam has been thoroughly investigated. Diazepam can be considered as a model drug of this group for the evaluation of a variety of factors influencing protein binding. Even in a relatively homogenous population of 27 healthy and drug-free male volunteers an approximately five-fold variation in the free fraction of diazepam or desmethyldiazepam was seen[120, 121)]. In twin-studies the genetic contribution to the observed variability in the plasma binding of diazepam was examined. Since the within-pair variances were not greater in 11 dizygotic than in 18 monozygotic twin pairs, a greater contribution of environmental than of genetic factors will explain the intersubject variation[121)]. However, no significant linear relationships were observed between serum albumin or total protein and the binding of chlordiazepoxide, diazepam[121)], lorazepam and oxazepam[122)].

In pharmacokinetic studies a postprandial increase in plasma levels of diazepam has sometimes been observed. Therefore, it was investigated whether food intake might influence these plasma levels via changes in plasma binding. There was no significant change in diazepam bound before and after meals, suggesting that normal food intake does not alter plasma binding of diazepam[123–125)]. In addition, no influence in the binding of chlordiazepoxide, oxazepam and lorazepam between the fed and fasting state has been reported[125)]. Surprisingly, a diurnal variation of ± 20% in the free fraction of diazepam was described[126)], which could not be confirmed with midazolam[127)].

### *5.3.3 Age*

In view of the increased incidence of chronic illness and disability, the elderly consume more than a proportionate share of benzodiazepines. In addition to the increased exposure to drugs, the elderly incurs a number of physiological changes that are likely to alter the response to drugs. As people grow older there is a wasting of muscle tissue, accumulation of fat and a decrease in body water, liver mass and brain weight. Progressive decreases with age are found for cardiac output (hepatic perfusion), lung and kidney function. While there are many ways in which older patients respond differently from younger people, it is difficult in many instances to separate the overlapping effects of "normal" aging from the effects of degenerative diseases. Age-related changes in benzodiazepine disposition and/or pharmacodynamics are considered as causative factors for an altered drug response.

There is little definitive evidence that demonstrates an age-related decline in intestinal absorption. Reduced intestinal perfusion secondary to the decreased cardiac output might decrease the absorption of certain drugs. However, high-clearance drugs (e.g. midazolam) that exhibit first pass elimination could be cleared more slowly, resulting in a higher bioavailability. A number of factors influence the apparent volume of distribution (V) of benzodiazepines, such as changes in body composition (accumulation of fat, loss of lean body mass and body water) and plasma protein binding with age. The slight decrease in albumin, which might be larger in poorly nourished, ill or severely debilitated elderly, causes a small increase in the free and biological active fraction of some drugs, which consequently will be subject to more intensive distribution and a faster elimination.

It has been demonstrated by two independent studies that elderly eliminate chlordiazepoxide more slowly than young subjects[128, 129]. With diazepam somewhat different results have been reported: $t_{1/2}$ exhibited a significant age dependency ($r = 0.83$, $p < 0.001$) ranging from about 20 h (20 years) to about 100 h (80 years), which was associated with a parallel increase in V ($r = 0.74$, $p < 0.001$). Plasma binding was not affected by age. In the 33 healthy adults (of whom 27 were men) CL did not significantly ($r = 0.39$, $p = 0.09$) decline with age[130]. Confirmatory results have been published recently[131]: $t_{1/2}$ of diazepam among males increased with age ($r = 0.53$, $p < 0.005$) as did V ($r = 0.67$, $p < 0.001$). There was only a weak tendency of a decrease in CL ($r = 0.32$) and an increase of the free fraction of diazepam ($r = 0.14$). For desmethyldiazepam also an age-dependent elimination has been found in men but not in women[132–134]. The prolonged $t_{1/2}$ seen in the elderly was mainly due to a decrease in CL.

Studies with other oxidized benzodiazepines yielded similar results. Desalkylflurazepam, the principle and active metabolite of flurazepam, had a longer $t_{1/2}$ in elderly compared to young men. Steady-state plasma levels of this metabolite were also significantly higher in elderly than in young men[135]. Similarly, $t_{1/2}$ of midazolam was longer in elderly (mean 4.33 h) than in young patients (2.77 h), which was caused, as in the case of diazepam, by a significant increase in V[136]. Clobazam's $t_{1/2}$ is significantly longer in the elderly, which is mainly due to an increase in V. Additionally, CL declined with age in men[137]. With triazolam a reduced CL and an unaffected $t_{1/2}$ has been found in old age[138]. Alprazolam's CL also seems to be age-dependent[139]. The increase in $t_{1/2}$ of brotizolam with age was attributable to a decrease in CL[164].

No significant change with age in the disposition of oxazepam could be observed[140, 141]. Similarly, the aging process was associated only with minor and clinically irrelevant changes in the disposition of lorazepam[142, 143], temazepam[144], nitrazepam[145–147] and flunitrazepam[148].

These kinetic studies were carried out with single doses of the different benzodiazepines. Thus, the question arises whether this represents the clinical situation, where such drugs are administered for longer periods of time. Therefore, we performed clinically more relevant steady-state cross-over studies in the elderly with diazepam, bromazepam and oxazepam. Compared to healthy subjects between 20 to 34 years of age, hospitalized patients with a mean age of 81 years demonstrated no change in the disposition of oxazepam (50 mg given daily for one week). However, the same six elderly patients showed a prolonged $t_{1/2}$ of bromazepam (3 mg/day for one week) and diazepam (5 mg/day for two weeks). In both cases the longer $t_{1/2}$ was associated with an increase in V, but oral CL was not significantly decreased[149]. Since the results of the multiple-dose studies with diazepam and oxazepam are similar to what has been found after single dosing of both drugs, one could assume that comparative single-dose studies will give representative answers to the question on whether the disposition of a particular drug will be affected by the age of the patients.

If one summarizes and attempts to draw any conclusions from the numerous age studies with the various benzodiazepines, one fact seems very obvious: drugs which are solely eliminated by glucuronidation (so-called phase II reaction), such as oxazepam, lorazepam and temazepam, are not significantly affected by age. Similarly, disposition of nitroreduced benzodiazepines (e.g. flunitrazepam, nitrazepam) seems to be age-independent. The picture is not as simple for derivates where the enzymatic mixed function oxidase system is involved, e.g. the so-called phase I reactions dealkylation and hydroxylation. For some drugs CL, as a measure of hepatic elimination and metabolism, is significantly decreased in the elderly, for other derivatives CL is only marginally reduced or only affected in men. Since benzodiazepines are very lipophilic drugs, their V increases with age. Thus, the disposition of a particular benzodiazepine can be altered in several different ways in the elderly.

### *5.3.4 Gender*

Sex-related changes in drug disposition might alter the action of drugs[150], and prescription/consumption of benzodiazepines in females is twice as high as in males. Since body composition (e.g. percentage of adipose tissue) is different in both sexes, changes in the distribution of the lipophilic benzodiazepines can be anticipated. Thus, it is not too surprising that chlordiazepoxide[151], diazepam[152, 153] and midazolam[154] show a significantly larger V among females as compared to age-matched males, suggesting more extensive drug distribution. However, V of alprazolam[139], triazolam and temazepam[150] seems to be gender-independent. Values of $t_{1/2}$ and CL for chlordiazepoxide[151], midazolam[154] and alprazolam[139] did not differ between sexes. Only in elderly women $t_{1/2}$ of triazolam was shorter and $t_{1/2}$ of temazepam longer than in elderly men[150]. Correspondingly, CL for triazolam was higher and for temazepam lower only in aged female subjects as compared to age-matched male subjects. For diazepam, only in young females a longer $t_{1/2}$ and a reduced CL compared to young males has been observed[152, 153].

Currently, no clear picture has been derived from these complex data; this might be due to confounding cofactors, such as obesity or intake of oral contraceptives in young females.

### *5.3.5 Obesity*

About 20% of our population is affected by obesity, and this can influence the disposition of benzodiazepines, e.g. $t_{1/2}$ of alprazolam[156], diazepam, desmethyldiazepam[157, 158] and midazolam[154] in obese subjects were more than two-fold prolonged in comparison to controls, which was due entirely to a large increase in V. In contrast, the prolonged $t_{1/2}$ of triazolam seen in obese subjects was caused by a decrease in CL[156]. When V and CL of lorazepam and oxazepam were normalized to body weight, both increased values in obese subjects approached control values. Since V and CL increased with body weight, $t_{1/2}$ being dependent on both parameters was not significantly different in this population for both drugs[159].

### 5.3.6 *Liver Function*

All benzodiazepines are metabolized in the liver, and, therefore, the function of this drug-eliminating organ plays a key role for their pharmacokinetics. Liver disease in humans is not a single entity, but consists of an assortment of pathophysiological disturbances, such as damage of hepatic cells, reduction or shunting of liver blood perfusion and alteration in plasma protein synthesis. All these factors have separate and potentially additive or competitive effects on the pharmacokinetics of benzodiazepines. In addition, the degree of functional abnormality varies widely within any diagnosed disease group, and the clinical and laboratory criteria for rating the severity of the particular hepatic dysfunction are relatively crude. Difficulties also occur in the selection of the right and comparable control individuals, which becomes obvious, if one keeps in mind all the factors influencing the distribution and elimination of drugs. Based on these criteria, diazepam is more slowly eliminated ($t_{1/2}$ increased, CL decreased) in patients with acute hepatitis and alcoholic cirrhosis. In addition, an increased V also has been observed in such patients, which is due to a decrease in plasma protein binding[130, 161]. Similar results in respect to a prolonged $t_{1/2}$, decreased CL and protein binding and a larger V in patients with hepatic dysfunction have been found for chlordiazepoxide[160, 162], desmethyldiazepam[161] and brotizolam[165]. With nitrazepam no significant differences in $t_{1/2}$, CL and V between patients with alcoholic cirrhosis and age-matched controls have been described; however, if the decrease in plasma protein binding is taken into consideration, the CL of unbound nitrazepam was significantly lower[147].

In contrast, oxazepam[140, 162] and lorazepam[142], two benzodiazepines which are eliminated by hepatic conjugation with glucuronic acid, did not show significant changes in their elimination in patients with liver disease, compared to age-matched controls. From these data it can be assumed that conjugation of drugs is less effected by liver disease than hydroxylation and dealkylation.

### 5.3.7 *Alcohol Withdrawal*

Chronic alcoholics in moderate-severe withdrawal and with apparently normal liver function exhibited no significant differences in the protein binding, distribution and elimination of diazepam[163].

### 5.3.8 *Kidney Function*

It is generally proposed that drugs (e.g. benzodiazepines) which are extensively metabolized can be given in normal doses to patients with renal insufficiency. However, water soluble metabolites will accumulate. If they are still pharmacologically active, they can contribute to clinical effects. In uremic patients the apparent terminal $t_{1/2}$ of oxazepam was 24–91 h, which was due to secondary plasma concentration peaks about 24 h after the dose. In contrast, CL appeared to be normal[166]. Conjugated metabolites accumulated and their renal CL was significantly correlated to creatinine CL[166, 167]. Plasma protein binding of oxazepam[167] and diazepam[121, 168] was reduced in patients with renal disease, which might be due to hypoproteinaemia or displacement by accumulated endogenous ligands. After a single oral dose of lorazepam, $t_{1/2}$ of unchanged drug was not different in patients with chronic renal failure. High concentrations of the glucuronide accumulate in plasma in the presence of severe renal function impairment[169)]. This might be one reason why in two patients with renal failure $t_{1/2}$, following subchronic administration of lorazepam, was about 50% prolonged[170)]. In patients with different degrees of renal failure the mean range of $t_{1/2}$ for brotizolam (6.9–8.2 h) was similar to control values (3.6–7.9 h). Thus, no dose adjustment seems to be necessary in renal failure[171)]. In patients with end-stage renal insufficiency on maintenance haemodialysis $t_{1/2}$ of diazepam (37 h) was greatly reduced to controls (92 h) and CL of total drug increased. These changes were largely related to disease-induced changes in plasma binding (7% free diazepam in renal patients vs. 1.4% in controls), which resulted in a reduced V and no change in CL for unbound (active) diazepam[172)].

### 5.3.9 *Thyroid Function*

Several studies have suggested that hyperthyroidism may be associated with altered drug disposition[173, 174)]. However, none of the kinetic variables for total and unbound diazepam in patients with newly diagnosed hyperthyroidism differed significantly from those in controls matched for age and sex[172)]. The pharmacokinetics of oral oxazepam was investigated in hyperthyroid patients before and after treatment with radio-iodine or carbimazole. Prior to be treatment, elimination was much faster ($t_{1/2}$ = 3 h, CL = 248 ml/min) than 3 months later ($t_{1/2}$ = 9.9 h, CL = 86 ml/min). Thus, higher doses or more frequent administration may be required when treating hyperthyroid patients with oxazepam. There was no significant change in oxazepam elimination in hypothyroid subjects[175)].

### 5.3.10 *Severe Burns*

Scanty information is available on the disposition of benzodiazepines following thermal injury. Recently, in patients with burns covering an average of 65% of body surface, a significant increase in free fraction of diazepam together with a significant reduction in the clearance of free drug has been observed. Consequently, also $t_{1/2}$ was greater in burn patients (72 h) than in control patients (36 h). The reduced elimination may be attributable to burn injury and/or concomitant administration of other drugs, especially because the known drug-metabolizing inhibitor cimetidine was given[176)].

### 5.3.11 *Genetics*

Ethnic differences may be a cause of interindividual variation in response to drugs[177)]. While the free fraction and $t_{1/2}$ of diazepam were almost identical in young Caucasians and Orientals (both living in the United States), V was slightly larger in the former race (1.1 vs. 0.9 ml/kg). CL was higher in Caucasians (0.4 ml/min/kg) than in Orientals (0.3 ml/min/kg). It was concluded that the differences in kinetics between the two races were due to genetic and not to environmental factors[177)].

### 5.3.12 Smoking

Pharmacokinetic changes also can be induced by smoking (review in[179]). While clearances of desmethyldiazepam[180] and oxazepam[181] were increased in smokers, the disposition of diazepam[130] and chlordiazepoxide[182] did not demonstrate significant differences between smokers and non-smokers.

### 5.3.13 Pregnancy and Newborns

Normal physiological changes in pregnancy can modify the pharmacokinetics of some drugs[183]. The obstetrical use of benzodiazepines, especially diazepam, has increased steadily. They should be avoided in early pregnancy because the absolute safety of these products cannot be assessed and because there are seldomly clear indications. However, the acute use at parturition is indicated to induce relaxation of the pelvic musculature, to prevent and to inhibit eclamptic convulsions, or merely for decreasing the mother's anxiety. During labour the maternal dose should be as low as possible to avoid the so-called floppy infant syndrome (review in[6]).

Decreased serum protein binding of diazepam during pregnancy has been observed[6, 184, 185] and there was a linear correlation with gestational age[185]. Whether this can be attributed to the lower serum albumin concentrations in pregnant women and/or to other factors (e.g. endogenous substances and inhibitors) is not clearly understood[184, 185]. However, in cord plasma there was significantly reduced binding capacity for diazepam with albumin levels of less than 40 g/l[6, 185].

The disposition of diazepam[186] and its major metabolite desmethyldiazepam[187] are changed in pregnant women. At parturition, $t_{1/2}$ of diazepam (mean 65 h) was significantly longer than in age-matched nonpregnant controls (29 h). Since CL was comparable in both groups, the prolonged $t_{1/2}$ should be related to changes in the distribution of diazepam[6]. Qualitatively identical results were seen with desmethyldiazepam after intramuscularly injection of its precursor clorazepate: $t_{1/2}$ was prolonged (180 h vs. 60 h), CL unchanged and V increased (3.1 l/kg vs. 1.8 l/kg) during pregnancy. Clorazepate itself exhibits lower peak plasma concentrations and a higher clearance in pregnant (37 to 42 weeks) women[187].

The lipid soluble benzodiazepines penetrate easily into the placenta, and generally this transfer is faster in late than in early pregnancy. After a single intramuscular dose of diazepam the feto-maternal ratio during the first 6 h ranged between 1.2–2.0. The continued maternal oral use of diazepam caused a feto-maternal ratio of 0.4 for diazepam and its metabolite desmethyldiazepam (review in[6]).

The transplacental passage of diazepam is rapid and the distribution equilibrium between mother and fetus is approached within 5 to 10 min after intravenous injection of the drug[188]. Following a single dose of clorazepate (20 mg i.m.) a balance was established between the fetal and maternal circulation after 4 h. The maximum concentrations in the umbilical artery and vein were found between 2 and 4 h after injection[189]. About 12 h after single oral doses of oxazepam[190] and nitrazepam[191] the feto-maternal ratio for total oxazepam (free and conjugated) and nitrazepam was about 0.6. In a steady-state situation the feto-maternal ratio of oxazepam and nitrazepam approached unity[6, 190, 191].

The distribution of benzodiazepines between mother's blood and breast milk is complex, lipid solubility and protein binding being the major determinants. Plasma levels of diazepam[192] and desmethyldiazepam[189, 192] were about 2 to 10 times higher than in breast milk; similar relations have been found for nitrazepam and flunitrazepam (review in[6]). The milk to maternal plasma ratio of the more polar lormetazepam (0.06) seems to be somewhat lower[193].

Diazepam is often used in newborns and infants as an anticonvulsant or sedative drug. In the premature and full-term newborn it is efficiently and rapidly absorbed after oral and rectal (but not intramuscular) administration. The plasma protein binding of diazepam is considerably reduced at birth (about 86% bound). Apparently related to the degree of maturation of metabolic activity, $t_{1/2}$ of diazepam in premature infants ranged from 40 to 400 h, while in full-term neonates $t_{1/2}$ varied from 20 to 50 h. Infants demonstrated an accelerated elimination with a $t_{1/2}$ between 8 to 14 h (reviews in[6, 194]). In four neonates a $t_{1/2}$ between 73 and 138 h was calculated for desmethyldiazepam[189]. Glucuronide-conjugation also gradually increases with maturation. Therefore, $t_{1/2}$ of oxazepam was 2- to 4-fold longer in the newborns (12–27 h) than in the mothers (5–8 h)[195].

**Table 3.** Factors which can modify the pharmacokinetics of benzodiazepines

| | |
|---|---|
| a) | age, obesity, gender |
| b) | liver-, kidney-, thyroid function |
| c) | severe burns (?) |
| d) | pregnancy |
| e) | co-medication, food, smoking |
| f) | genetics |

## *5.4 Drug Interactions*

The world-wide consumed benzodiazepines are often concomitantly given with a variety of drugs. Thus, during multi-drug therapy, interactions can occur at the levels of absorption, distribution, elimination and at receptor sites.

### *5.4.1 Absorption Interferences*

Most of the benzodiazepines are ingested orally. Rate and extent of absorption can determine onset and duration of action. Patients with gastrointestinal symptoms often receive antacids and benzodiazepines. Clorazepate is converted to its active form by pH-dependent hydrolysis and decarboxylation in the stomach[196]. If clorazepate was taken concurrently with Mg-Al-hydroxide followed by a second dose one hour later, time and peak of the maximal concentration of desmethyldiazepam were significantly affected. Absorption of the single dose was delayed by a factor of two, and the bioavailability reduced by approximately 10%. Subjective effects of generalized sedation occurred earlier and were more profound when clorazepate was taken with water as opposed to Mg-Al-hydroxide[197]. Similarly, when gastric juices were maintained above pH 6 for 2 h with sodium bicarbonate, the bioavailability of clorazepate was reduced between 5–75% compared to the same four subjects with gastric acidity levels lower than pH 3[196].

However, in two other studies a magnesia/alumina antacid suspension only tended to slow absorption of desmethyldiazepam, but did not alter the extent[198]; and steady-state plasma levels of desmethyldiazepam were not changed, if 30 ml Maalox were given four times daily[199]. In addition, it is very unlikely that this interaction is of general importance, since Mg-Al-hydroxide had no influence on the completeness of absorption of chlordiazepoxide[200], and diazepam's absorption was not affected by magnesium trisilicate plus aluminum hydroxide or Mg-Al-hydroxide (cited in[197]).

### *5.4.2 Protein Binding Displacement Interactions*

In general, benzodiazepines are strongly bound to plasma albumin ("silent receptors"). Free fatty acids are known to displace diazepam *in vitro*[201]. The antiepileptic drug valproic acid is a fatty acid, which at increasing (but therapeutic) concentrations competitively reduced the *in vitro* binding of diazepam from 98.1 ± 0.08 to 96.1 ± 0.21%[202]. Sex differences in disposition studies might be obscured by endogenous and/or exogenous steroids. Females taking oral contraceptives had a significantly higher free fraction of diazepam (1.99 ± 0.34) than age-matched controls (1.67 ± 0.20)[203]. Similar results have been observed with chlordiazepoxide[204]. Women taking oral contraceptive steroids had a lower plasma binding (from 93.6 ± 1.5% to 95.5 ± 1.5%) and a higher volume of distribution (from 0.62 ± 0.23 l/kg to 0.40 ± 14 l/kg) than women not incorporating such "drugs". Ethanol (or its metabolites and/or metabolic consequences) also lowered plasma binding of chlordiazepoxide from 94.7 ± 0.6% to 93.4 ± 1.3%[205].

It is very common in pharmacokinetic studies to use repeated small doses of heparin to maintain the patency of an intravenous cannula for collecting blood. Heparin can increase the circulating levels of free fatty acids (FFA), which consequently might displace benzodiazepines from its plasma binding sites. Several studies noticed an increase in the free fraction of diazepam, desmethyldiazepam, chlordiazepoxide and oxazepam when different doses of heparin were administered to healthy subjects or patients. However, the observed concomitant increases in FFA and free drug concentrations are probably caused by the *in vitro* formation of FFA due to continued lipase activity on endogenous substrate[206]. Thus, the heparin-induced protein binding changes are to a large extent *in vitro* artifacts, as could be recently shown for diazepam[207].

### 5.4.3 *Interaction in Elimination*

Hepatic elimination of benzodiazepines can be altered by induction or inhibition. In the treatment of epileptic patients, clonazepam is often co-administered with other anticonvulsants. In eight subjects plasma levels of clonazepam were lowered significantly after pre-treatment for 19 days with phenytoin, which was due to an increased clearance by 46–58%. After phenobarbital pre-treatment, the mean plasma clonazepam concentration was slightly lowered (11%), and clearance increased by 19–24%. The overall effect of phenytoin was greater than the effect of phenobarbital[208)]. Decreased levels of clonazepam were also observed in patients when phenytoin was given[209)].

Controlled steady-state levels of clonazepam declined to a lower steady state over 5 to 15 days after addition of carbamazepine. The decline ranged between 19 and 37%. Induced half-lives of clonazepam (22.5 ± 11.5 h) were shorter than control values (32.1 ± 16.6 h; [210)]). It appears that in treated epileptics also the apparent oral clearance of clorazepate is greater than in normals, which might reflect differences in hepatic metabolism[211)]. The elimination of diazepam in epileptics on different anticonvulsants was also faster ($t_{1/2}$ = 13.1 h, CL = 51.7 ml/min) than in age-matched controls ($t_{1/2}$ = 34 h, CL = 20 ml/min)[212)].

In a group of seven tuberculous patients on triple therapy with isoniazid, ethambutol and rifampin the disposition of a single intravenous dose of diazepam was compared with that in healthy drug-free controls matched for age and sex. Mean half-lifes of the drug among patients (14 h) was significantly shorter than in controls (58 h) and total clearance correspondingly increased from 0.37 to 1.50 ml/min/kg. This effect is probably due to the enzyme-inducing rifampin[213)].

While induction phenomena can diminish the effects of benzodiazepines, metabolic inhibition might result in drug overdose. As already mentioned, oral contraceptives (OC) may be a complicating factor in pharmacokinetic analysis. Women taking such drugs exhibited a prolonged $t_{1/2}$ of chlordiazepoxide (from 14.8 ± 5.9 h to 24.3 ± 12 h) and clearance of unbound drug was lower (from 8.7 ± 5.0 to 5.1 ± 3.0 ml/min/kg) than in those not using them, but these differences did not reach statistical significance[204)]. However, in a more recent study, intake of OC resulted in a significant prolonged $t_{1/2}$ (20.6 vs. 11.6 h) and a reduced CL (13.4 ml/min vs. 33.2 ml/min), while plasma binding was not significantly decreased[214)].

Plasma clearance of diazepam in five women taking oral contraceptives (median 14.0 ml/min) is significantly less than in men (median 23.4 ml/min) or in ten young women not taking oral contraceptives (median 26.8 ml/min)[215)]. In another study, $t_{1/2}$ of diazepam was significantly longer (69 vs. 47 h) and CL significantly less (0.27 vs. 0.45 ml/min/kg) in OC users than in control subjects[216)]. While both $t_{1/2}$ and CL of total nitrazepam were not affected by intake of OC, CL of unbound drug was significantly decreased[217)]. Surprisingly, $t_{1/2}$ for lorazepam (6.0 vs. 14.0 h) and oxazepam (7.7 vs. 12.1 h) were significantly reduced in women taking OC, which was due to a significant increase in CL of lorazepam (289 vs. 78 ml/min) and oxazepam (251 vs. 107 ml/min), respectively[214)]. Thus, OC exert a differential effect on the elimination of benzodiazepines, whereby phase I reactions (e.g. oxidation, dealkylation) are impaired and phase II reactions (e.g. glucuronidation) are even enhanced.

During treatment with isoniazid (INH), the half-life of diazepam increased from 34 to 45 h, and total clearance declined from 0.54 to 0.40 ml/min/kg in nine healthy subjects, indicating impairment of drug metabolism[213]. INH co-administration also prolonged triazolam's $t_{1/2}$ (3.3 vs. 2.5 h) and reduced the apparent oral CL (3.9 vs. 6.8 ml/min/kg) but had no influence on the pharmacokinetics of oxazepam[218]. These interactions are the second example for the differential control of drug oxidation and conjugation.

Disulfiram impairs the elimination of diazepam and chlordiazepoxide but not that of oxazepam and lorazepam[219, 220]. Pre-treatment during 14 days with 0.5 g disulfiram reduced the elimination of chlordiazepoxide by a factor of two and that of diazepam by about 40%.

Cimetidine is very often co-administered with benzodiazepines. Systematic studies with various benzodiazepines eliminated by different metabolic steps revealed that cytochrome P 450-dependent phase I reactions are impaired, while conjugations with glucuronic acid (phase II reaction) were not affected. If healthy volunteers were pretreated (even for one day) and/or given concomitantly therapeutic doses of cimetidine, elimination of single doses of diazepam[221, 222], desmethyldiazepam[223] and chlordiazepoxide[224, 225] were significantly impaired. This inhibition is reversible, since disposition of chlordiazepoxide was normalized 2 days after termination of cimetidine intake[224]. In contrast, pharmacokinetics of oxazepam[223, 226] and lorazepam[226] were not altered by cimetidine. With respect to diazepam, the same situation applies to steady-state conditions. Co-administration of cimetidine (but not ranitidine) increased steady-state plasma levels of diazepam significantly (about 70%), which was due to a reduction of CL from 20.9 ± 9.5 to 14.0 ± 3.0 ml/min. The half-life was also prolonged from 39.5 ± 16.6 h to 101 ± 58.1 h by cimetidine[227].

In other single-dose studies the elimination of alprazolam[228], clobazam[229], nitrazepam[230] and triazolam[228] was impaired by cimetidine, while temazepam was not affected[231]. The differential effect of cimetidine is due to its binding to cytochrome P 450, the system needed for drug oxidation/dealkylation but not for glucuronidation (reviews in[232, 233]).

There is clear clinical evidence that alcohol does enhance the central effects of various benzodiazepines. The interaction is probably caused by different mechanisms. There is some controversy about the effect of alcohol on the absorption of diazepam. In some studies initially higher plasma levels of diazepam have been observed, when alcohol or alcoholic drinks[234–236] were concomitantly ingested, which could suggest an accelerated rate of absorption. In contrast, other studies indicated no significant absorption changes[237, 238]. Since all these studies varied in their design, no definite conclusion can be drawn.

The effect of alcohol on the serum levels following a single dose of chlordiazepoxide was also examined. Concentrations were only significantly increased at 120 and 150 min after alcohol ingestion as compared to those after a placebo drink. However, the bioavailability of chlordiazepoxide (and diazepam) were not significantly modified after alcohol[239]. The kinetics of a single oral dose of chlordiazepoxide was compared in male volunteers (mean age 31.2 years) and in male patients (mean age 51.6 years) with diagnosis of acute alcohol intoxication. The habitual drinkers showed a prolonged $t_{1/2}$ and a reduced CL. This impaired elimination was due to external factors such as alcohol, other drugs, nutrition and possibly age and liver disease[240].

Healthy male volunteers ingested ethanol (0.8 g/kg) as a 25% solution in orange juice 1 h before chlordiazepoxide (0.6 mg/kg) was injected intravenously. To maintain plasma ethanol concentrations of 50–150 mg/100 ml for 32 h, additional ethanol (0.5 g/kg) was given orally every 5 h. CL of chlordiazepoxide fell from 26.6 ± 2.6 ml/min without ethanol, to 16.6 ± 3.1 ml/min with ethanol (unbound drug clearance 468 ± 51 ml/min vs. 264 ± 38 ml/min); $t_{1/2}$ was prolonged from 7.1 ± 1.9 h to 11.8 ± 6 h with ethanol. Thus, short-term ethanol ingestion in moderate doses impairs the elimination of chlordiazepoxide[205]. In a similar cross-over study, healthy male subjects received 10 mg diazepam intravenously preceded by 60 min with oral ethanol (0.7 g/kg) and followed for 8 h by ethanol (0.15 g/kg/h) to maintain the blood concentrations between 800 to 1000 mg/l. The area under the curve for total and free diazepam increased in all subjects by 30% and 26.5%, respectively, after ethanol, indicating inhibition of hepatic intrinsic clearance of free drug[241].

An interaction was also demonstrated with a combination of clobazam and alcohol, where clobazam serum levels were more than 50% elevated for 340 min after administration of alcohol. The enhanced absorption was associated with a stronger impairment in different pharmacodynamic performance tests[242]. From these studies one can generalize that the combined effects of benzodiazepines and alcohol are more hazardous than the effects of alcohol alone.

Besides these established pharmacokinetic interactions, there is additional experimental evidence that ethanol interacts with benzodiazepines also at the receptor site in the central nervous system (CNS). It was found that ethanol-pretreated rats demonstrated a six-fold higher brain concentration of diazepam than controls. Ethanol, like benzodiazepines, also selectively enhanced the γ-aminobutyric acid (GABA)-mediated inhibitory transmission in the CNS. Binding of diazepam to solubilized brain fraction (benzodiazepine-GABA-ionophore complex) was dose-dependently enhanced by ethanol. This interaction will result in facilitation of GABAergic transmission, and will add to the observed enhanced central effects[243].

### *5.4.4 Interaction at Receptor Site*

In addition to the "social drug" alcohol, other compounds might interact with benzodiazepines at the brain level. Sedation induced by intravenous diazepam was partly reserved for about one hour by a bolus injection of 100 mg of doxapram[244]. Similarly, physostigmine, a centrally and peripherally acting anticholinesterase, has been reported to reverse the sedation/hypnosis induced by diazepam[49–51, 245] and lormetazepam[246]. However, in two other studies physostigmine failed to antagonize the clinical, psychomotor or EEG-effects of lorazepam[52] and diazepam[53]. Thus, whether the action of benzodiazepines in humans is also mediated by central cholinergic synapses and/or receptors remains to be clarified.

Methylxanthines, such as caffeine[247] and theophylline[54, 55, 248] also seem to antagonize the diazepam-induced sedation and psychomotor impairment. These pharmacodynamic interaction studies provide some evidence that purinergic (adenosine-mediated) mechanisms might be involved in the action of diazepam.

Several in vitro studies with synaptosomal brain fractions indicated that the new imidazodiazepine Ro 15–1788 (see Fig. 1) inhibits the specific binding of benzodiazepines at their receptor site, thus antagonizing the action of this class of drugs[26, 27]. Subsequently, in first clinical trials it was found that the selective benzodiazepine antagonist Ro 15–1788 can reverse the sedative-hypnotic effects of 3-methylclonazepam[47], diazepam[48] and flunitrazepam[46]. In addition, Ro 15–1788 was well tolerated without significant pharmacological effects by healthy volunteers when administered as a single oral dose up to 600 mg[44] and as intravenous bolus up to 100 mg[45]. Since the antagonistic efficacy of Ro 15–1788 was of relatively short duration[46–48], it can be assumed that this compound is rapidly eliminated. We found recently that the hepatic elimination of Ro 15–1788 is actually very rapid ($t_{1/2}$ = 1 h; CL = 691 ml/min)[249].

The search for endogenous ligands of the benzodiazepine receptor led to the discovery of ethyl-β-carboline-3-carboxylate (β-CCE), a potent agent in antagonizing the pharmacological and electrophysiological effects of benzodiazepines[40, 41, 250]. In healthy volunteers severe anxiety could be induced by β-CCE[39, 251]. Since β-CCE inhibits specific binding of benzodiazepines, its interaction with brain benzodiazepine receptors is probably responsible for its pharmacological potency.

It can be assumed that in the near future partial and/or more pure agonist/antagonist of the benzodiazepine-receptor will be developed to differentiate the various actions of this important class of drugs and that the existence of endogenous ligand(s) will be clarified.

## 6 Plasma Concentrations and Clinical Response

Several studies attempting to correlate benzodiazepine plasma levels with their clinical effects have been discouraging. The difficulties in designing appropriately controlled studies are due to various problems: 1. many of the benzodiazepines have (multiple) active metabolites with varying biological potencies and pharmacokinetics; 2. the nature and severity of the clinical symptoms (e.g. anxiety) are hardly to define and to measure quantitatively in an objective manner; 3. the patient populations are not homogenous and frequently spontaneous remissions will occur, which necessitate the use of placebo control groups; 4. the probable development of tolerance.

### *6.1 Diazepam/Desmethyldiazepam*

In a randomized, double-blind, placebo-controlled study with diazepam (20 mg/day) acute clinical anxiety was rated by the Hamilton scale (HS) and a symptom scale. A significant correlation was found between the steady-state plasma levels of diazepam

($r = 0.51$; $p < 0.05$) or desmethyldiazepam ($r = 0.66$; $p < 0.05$) and the improvement rating of some patient's complaints. Minimal effective concentrations for diazepam ranged from 341–472 ng/ml[252)]. In a study in severely anxious outpatients with clorazepate (7.5 mg t.i.d.) and diazepam (5 mg t.i.d.) it was speculated that the slightly greater improvement observed with clorazepate compared to diazepam is related to higher plasma levels of desmethyldiazepam in the former group (570 ng/ml vs. 350 ng/ml)[253)]. A significant ($r = 0.35$; $p < 0.05$) but modest relationship between desmethyldiazepam concentrations (derived from clorazepate) and symptom relief in nine patients with chronic anxiety was reported. However, with diazepam no significant relationships were found[254)]. Hospitalized chronic alcoholic patients undergoing alcohol withdrawal were treated orally with varying dosages of diazepam. There was a significant negative correlation between total hostility outward scores and diazepam ($r = 0.64$; $p < 0.05$) and desmethyldiazepam plasma concentrations ($r = 0.71$; $p < 0.02$)[255)]. A study of 23 psychiatric outpatients with moderately severe anxiety suggested an effective diazepam threshold around 300 ng/ml. There was a trend that total diazepam (+ desmethyldiazepam) levels were related to efficacy[256)]. In 13 healthy male volunteers temporal changes in EEG after administration of diazepam (10 mg p.o.) paralleled for the recorded first 2 h with the development of blood levels. Blood levels of 100 ng/ml were associated with significant changes in EEG beta activity[257)]. Another objective and convenient measure of brainstem reticular formation function is the determination of the peak velocity of saccadic eye movement (PVSEM). Following a single oral dose of 10 mg of diazepam, a significant ($r = 0.49$; $p < 0.05$) negative log-linear correlation between PVSEM and plasma levels, both measured until 12 h after administration, was found. Surprisingly, the relationship was absent after administering 10 mg desmethyldiazepam[258)].

In a study with normal volunteers, anxious outpatients and hospitalized psychiatric patients the effects of a single dose of diazepam were measured by several scales and check-lists. An almost flat dose-response curve (change in mood) was observed in those subjects who received different dosages, whereas a linear log-dose response curve described the increasing sedation. In individual subjects no relationships between blood levels or pharmacokinetic parameters and effects of diazepam were noted[259)].

In balanced order, but in a flexible dosage schedule, 20 patients with anxiety states were treated double-blind with medazepam (10–50 mg/day), diazepam (2–20 mg/day), chlordiazepoxide (10–50 mg/day) and placebo. No significant correlations were observed between plasma levels of any of the benzodiazepines and the clinical ratings or behavioural measures[260)]. In similar studies with diazepam[261–263)] or diazepam (+ amylobarbitone)[264)] and desmethyldiazepam (+ amylobarbitone)[265)], no relationships could be established between kinetic and clinical measurements.

In summary, there is some controversy on whether relationships between the clinical effects and plasma levels of diazepam/desmethyldiazepam do exist. This is mainly due to methodological difficulties (e.g. patient characterisation and response quantification).

## 6.2 Chlordiazepoxide

In a subgroup of anxious volunteers a significant decrease in anxiety scores was observed when plasma levels of chlordiazepoxide were maintained above 700 ng/ml, which might indicate a certain borderline for clinical effectivity[266)]. In a well-designed study with 15 anxious volunteers there was no correlation between anti-anxiety effect and plasma levels of chlordiazepoxide (see also[260)]), whereas the active metabolites (desmethylchlordiazepoxide, demoxepam) correlated significantly ($p < 0.02$)[267)]. Additional studies are necessary to test these interesting findings.

## 6.3 Clonazepam

In epileptic patients with successful clinical response to clonazepam, the steady-state plasma levels are usually between 30 and 60 ng/ml. Whether this range (or 20 to 70 ng/ml) can be regarded as a therapeutic guide remains to be established (reviews in[6, 268)]).

## 6.4 Nitrazepam/Flunitrazepam

The sedative effects of nitrazepam (single oral doses of 5 and 10 mg) correlated significantly only as the plasma concentrations were increasing (up to 4 h)[269)]. The neurophysiological measurement of PVSEM correlated significantly ($r = -0.49$; $p < 0.05$) with plasma concentration following a single oral dose of 5 mg[258)].

The number of errors in a pencil tracking test and self-ratings of sedation in healthy volunteers after a single i.v. administration of flunitrazepam correlated significantly with its plasma levels. The sleep-inducing effect is postulated with plasma concentrations exceeding 8–10 ng/ml[270)].

## 6.5 Midazolam

Following single intravenous and oral dosing as well as steady-state infusions, significant linear correlations (r-values between 0.54 and 0.97) between plasma levels of midazolam and dynamic effects, as assessed by the d2 letter cancellation test and a sedation index formed from visual analogue scales, were observed in healthy subjects[93)].

## 6.6 Other Benzodiazepines

While there was a significant ($r = -0.63$) log-linear correlation between PVSEM and serum temazepam concentrations (200 mg po), such a relationship could not be observed with flurazepam[258]. Similarly, the relationship between measurable bromazepam blood levels and the degree of EEG changes was not significant[257].

After 21 days of oral ketazolam administration, no significant correlations occurred between drug blood concentrations and anxiety or hostility levels[255].

## *6.7 Conclusions*

Whether relationships between plasma levels of benzodiazepines and their clinical response can be seen probably depends on the sensitivity and specificity of the pharmacodynamic measurements as well as on the stratification of the populations examined. It is also conceivable that the rate of change in the concentrations of benzodiazepines is more important than the absolute levels in determining the time course and intensity of their pharmacological action.

There also remains the open question of whether the pharmacological measurements correspond to the clinical symptoms.

In addition, tolerance and pharmacokinetic pecularities might complicate the situation. Obviously, there is a need for further well-controlled studies to resolve this interesting and practical relevant problem.

# 7 Clinical Use of Benzodiazepines

Benzodiazepines as a group possess more pharmacological and clinical similarities than dissimilarities. However, the therapeutic merit of this class of drugs is unquestionable and will be discussed briefly in the following sections.

## *7.1 Anxiety and Related Conditions*

We know very little about the origins of anxiety or the determinants of its appearance in humans, but we have realized that anxiety is all around us. It occurs in 2 to 5% of the general population and might be differentiated into situational, performance, anticipatory and separation anxieties. Anxiety can be also secondary to somatic dysfunctions, such as angina pectoris, cardiac disturbancies, medications and drug withdrawal or postmenopausal symptoms. Depression might be involved with anxiety and vice versa. The efficacy of benzodiazepines in the symptomatic treatment of nonpsychotic anxiety has been well established and they are being increasingly prescribed for general stress responses. In hundreds of well-controlled studies it has been shown that benzodiazepines are superior to placebo, barbiturates and meprobamate (reviews in[271–274]. In general, moderate to marked improvement can be obtained in about 65 to 75% of benzodiazepine-treated patients. Anxious patients respond best if they suffer from high levels of emotional and somatic symptoms of anxiety and from low levels of depression and interpersonal problems[275].

Anxiety present with or triggered by physical illnesses is also a disorder where benzodiazepines might be used profitably. A significant predictor for clinical efficacy seems to be the initial response to a benzodiazepine, e.g. of patients who report considerable improvement after one week of treatment with diazepam, 90% were markedly improved at the end of 6 weeks on diazepam[276)]. In another study, no additional improvement occurred by increase of daily dosage as well as prolonged drug treatment in patients resistant to treatment initially. Tolerance did not develop over the total 6-month treatment period[277)]. Similar findings were reported by other groups[278, 279)]. Nevertheless, the American FDA mandates caution against anxiolytic treatment of more than four months' duration, especially because the risk of dependence will increase. Since the newer benzodiazepine anxiolytics appear clinically similar or equivalent in both efficacy and toxicity compared to the "gold standard" diazepam, the choice of drug should be guided by the temporal pattern of the anxiety state. Diazepam with its long $t_{1/2}$ is appropriate if anxiety levels are high and sustained. For episodic anxiety, shorter-acting compounds (e.g. oxazepam) might offer some advantages.

## *7.2 Depressive Disorders*

Highly variable effects of benzodiazepines have been observed in depression, which may reflect difficulties in correct diagnosis or heterogeneity of patients. A similar picture emerges if one critically reviews, the clinical efficacy of antidepressant drugs[273)]. However, benzodiazepines might be a valuable adjunctive or complimentary therapy to the classical antidepressants, since symptoms such as psychic and somatic anxiety, agitation and insomnia are decreased (reviews in[280, 281)]).

The new triazolobenzodiazepine alprazolam also seems to possess antidepressant properties[282–285)]. It was most effective in depressed patients "with prominent anxiety symptoms"[286)]. In an open study in depressed patients refractory to previous treatment, alprazolam was effective in some endogenously depressed patients in a dose range of 4–7 mg/day, which is higher than the typical anxiolytic range of 1.5–3 mg/day[287)]. More controlled investigations are needed to define the antidepressant potential of alprazolam or other benzodiazepines.

## *7.3 Insomnia*

Benzodiazepines are of well-established efficacy in the short- or intermediate-term treatment of insomnia (in some cases up to 24 weeks). Intake of the drugs should be of limited duration, and daytime sedation as well as drug accumulation should be avoided. Thus, the short to moderate long acting benzodiazepines (see Table 2) should be favoured (reviews in[6, 272, 273)]). If patients complain about difficulty falling asleep, the absorption rate of the benzodiazepine is a critical determinant (see page 130). Very short-acting compounds, such as triazolam and midazolam will not cause residual effects[288, 289)] but might be less favourable than initially anticipated, since rebound insomnia has been reported to occur more frequently than with the moderate or long acting derivatives[290, 291]. However, in other studies with midazolam[292] or triazolam[293] no rebound phenomena have been observed. In general, it is evident that following withdrawal from prolonged treatment with a benzodiazepine in a certain percentage of individuals rebound insomnia will occur, which will last for several days or even some weeks. The time of the peak effect of sleep disturbances seems to depend on the rate of elimination, e.g. compounds with a short $t_{1/2}$ exhibit their maximal rebound within the first two nights and derivatives with longer $t_{1/2}$ after 3 to 5 days[294]. Therefore, it is mandatory that patients should be told that a period of disturbed sleep may be expected after stopping prolonged benzodiazepine intake. The development of tolerance (see also page 157) to the hypnotic effect within several days or weeks of treatment, represents an additional clinical problem discussed extensively in the literature[294].

## *7.4 Musculoskeletal Disorders*

The muscle relaxant properties of benzodiazepines (primarily diazepam) are an indication for the management of acute conditions such as trauma and tetanus[295], and for symptomatic relief of muscle spasms and spasticity (review in[273]).

## *7.5 Convulsions*

The main indication for the use of benzodiazepines in pediatric practice is for control of febrile convulsions. Diazepam (i.v., enemas) is the drug of first choice in reducing the severity of fits and preventing their recurrence. Intravenous diazepam or clonazepam have displaced phenytoin and phenobarbital as first line drugs in the management of status epilepticus[296]. Some data suggest an effective adjunctive role for oral benzodiazepines (diazepam, clonazepam) in the long-term treatment of certain epileptic seizure disorders[297]. In children, especially clonazepam has been used, but therapy might be limited by side effects and the apparent development of tolerance[298].

## *7.6 Alcohol Withdrawal*

Benzodiazepines have been advocated in the management of alcohol withdrawal, because of their efficacy and their low incidence of cardiovascular and respiratory depression. Acute symptoms (excitement, seizures) can be best controlled by intravenous administration of diazepam. Following the acute episodes, decreasing doses over the next few days have been proposed[299]. However, it has to be realized that the overall treatment program also includes supportive care and follow-up[300]. It should be kept in mind that cross-tolerance and -sensitivity between alcohol and the benzodiazepines exists. In addition, the danger that dependence might be transferred from alcohol to these drugs must be considered.

### *7.7 Premedication and Anesthesia*

The combined anxiolytic, sedative/hypnotic and muscle-relaxant properties of the benzodiazepines are of particular value in preoperative medication before surgical or endoscopic procedures and as induction agents before initiating general anesthesia. In addition, the induced anterograde amnesia is advantageous. Clinically acceptable short induction times can be only achieved with diazepam, flunitrazepam and midazolam. Midazolam seems to be the most promising derivative, since consciousness can be similarly titrated as with thiopental, but without concurrent hazards of cardiovascular and respiratory depression (review in[56)]).

## 8 Acute Toxicity and Adverse Effects

Benzodiazepines can be considered as very safe drugs, even huge overdoses are not associated with fatalities. They do not cause severe respiratory or cardiovascular depression, but only stupor, sleep (lasting 24 to 48 h) and usually a mild fall in blood pressure. During therapy the most common side effect is excessive depression of the central nervous system; clinical manifestations include drowsiness, impairment of intellectual function, reduced motor coordination and impairment of memory and recall, which appear to be simply a more intense expression of their desired pharmacological effects (reviews in[271–273)]). Dose- and age-related drowsiness occurring frequently during the first week(s) of treatment seems to wane, suggesting that tolerance outweights drug accumulation[301)]. Extremely seldom paradoxical effects (e.g. excitement, agitation, increased aggression and hostility) have been reported. Such reactions tend to be idiosyncratic and the population at risk has still to be defined[302)]. Especially after intravenous administration, benzodiazepines can cause some respiratory depression and a slight fall in blood pressure[303, 304)]. Following the injection of benzodiazepines, venous sequelae have great clinical importance and the normally marketed formulation of diazepam is the most harmful preparation (incidence between 22 and 39%). Injection into small hand and arm veins is more deleterious than into large antecubital veins. In addition, the solvent is very critical and the lowest frequency (2%) was found if diazepam is dissolved in a fat emulsion. The new water-soluble midazolam seems to be the best choice in this respect[56)].

Other side effects include weight gain, skin rash, impairment of sexual function, menstrual irregularities and rarely blood abnormalities. Benzodiazepines should be avoided in the first trimester of pregnancy, since it was claimed that diazepam intake was associated with a higher incidence of cleft lip in the fetus. However, this has not been substantiated in large retrospective surveys[305)].

### *8.1 Benzodiazepine and Driving*

Impairment of performance and morning-after residual effects ("hangover") are well known and very common with many benzodiazepines. Their intensities are dependent on the dose and the rate of elimination of the active compounds. The adverse reactions are especially hazardous if individuals are working on machines or riding a vehicle. There is no question that benzodiazepines impair the actual driving performance[306–308] and users of minor tranquilizers have a risk 4.9 times that of non-users with regard to road accidents[309]. Alcohol is the most common cause of traffic accidents. Psychotropic drugs (e.g. benzodiazepines) will enhance the deleterious effect of alcohol, and this interaction is still effective the next morning[310]. In 425 blood samples from people killed in motor vehicle accidents, alcohol was present in 51% and drugs (most commonly diazepam) were found in about 10% of the cases[311]. Thus, physicians should warn their patients about an impairment of driving skills when taking benzodiazepines.

## 8.2 Tolerance and Dependence

Benzodiazepines are used world-wide in tremendous amounts[273, 312]. The majority of patients taking these drugs appear to benefit clinically from them. Few patients escalate their dosage during prolonged treatment. The continued use of diazepam for 22 weeks[277] or for periods from 1 month to 16 years[278] seemed to retain its efficacy. However, there is now experimental[313–315] and clinical[290, 291, 316] evidence that tolerance occurs, which cannot be explained by pharmacokinetic phenomena but probably occurs at the receptor site[317].

Until recently, only few cases of dependence have been reported and it was concluded that the risk is low, but nevertheless benzodiazepines can produce psychological and physical dependence if given in excessive doses over a prolonged period, particularly to patients with unstable personality[318]. In the last few years data are accumulating from well-designed, placebo-controlled studies, which suggest that dependence occurs more often than initially thought. This dependence is characterized by withdrawal symptoms on stopping prolonged treatment with therapeutic doses of different benzodiazepines. The withdrawal syndrome often includes perceptual disturbances, weight loss, anorexia, insomnia, dizziness, autonomic symptoms, postural hypotension, anxiety, tension, apprehension, nausea, vomiting, tremor, muscle weakness. Occasionally, hyperthermia, muscle twitches, convulsions and confusional psychoses may occur. The withdrawal syndrome is more likely if the benzodiazepine has been taken for more than 4 months and if the drug is stopped suddenly. Therefore, dosage should be gradually reduced, and the temporary prescription of propranolol or clonidine may attenuate withdrawal symptoms. In addition, supportive psychotherapy will help the patients over the worst of the withdrawal period, which usually lasts about 2 weeks. Onset and time of the peak withdrawal reactions depend on the rate of the elimination of the active compounds (reviews in[271–274, 319, 320]). In some studies, withdrawal from long term treatment with therapeutic doses of benzodiazepines was accompanied with some form of discernible withdrawal reactions in all patients[272, 321, 322].

Whereas duration of continual treatment with benzodiazepines seems to be the most important determinant for withdrawal symptoms, e.g. the incidence was 5% if patients were treated for less than 8 months but increased to 43% if treated for more than 8 months[277], gradual tapering of the dose cannot prevent withdrawal reactions. Even if the withdrawal was gradual, double-blind and placebo-controlled, 44.4% of 36 patients who completed a recent study experienced true withdrawal phenomena on reducing their drugs. Eight other patients (22%) had pseudo-withdrawal reactions at a time when their drug treatment was unchanged. Patients with passive-dependent traits had a significantly greater prevalence of withdrawal reactions[323].

Whether the incidence of dependence is in the range of only a few percent or between 20 and 50% remains a matter of dispute, since the different patients studied so far might not represent the normal population. It would be important to detect patients at risk and to develop more sensitive measures of drug abuse potential. However, it should be realized by the prescribing physicians, the consuming patients and the health authorities that benzodiazepines can produce dependence, both at high and low dosage, in a significant part of our population.

The molecular and neurochemical mechanism(s) of abstinence symptoms after withdrawal of benzodiazepines remains speculative. During long-term exposure to anxiolytics, brain GABA synapses show evidence of adaptive changes, which after abrupt cessation lead to an acute reduction in GABA function. These changes could be responsible for tolerance and withdrawal phenomena[324]. In case endogenous ligand(s) exist for the GABA-benzodiazepine receptor complex, the up-and down regulation of such compound(s) also might be involved.

# 9 Conclusions and perspectives

A large and increasing number of different benzodiazepines is available. A logical problem derives from such complex situation: Are there any reliable guidelines for the proper choice of a particular drug for the variety of indications and requirements?

As we have learnt from the pharmacological and clinical data, the numerous derivatives are very similar in their action and can be replaced by each other in almost all instances. Thus, no more than 2 to 3 benzodiazepines should suffice in medical practice; only for special clinical needs, e.g. in the treatment of epilepsy or in anesthesiology one additional compound might be helpful. Since benzodiazepines differ mainly in their pharmacokinetics and the factors influencing the disposition of these drugs, kinetic properties (e.g. short, moderate, long acting derivatives; influence of age, liver function and comedication) also should be considered when selecting the "right" drug. Obviously, a more short and a more long acting derivative, dependent on the symptoms of the patient, should comprise the two or maximally three compounds needed. A very helpful criterion for the decision process of the prescribing physician is the available and well-documented information of the drug candidates.

The future will reveal whether more specific derivatives or compounds with different profiles of action can be developed and whether these turn out as valuable additions to the already existing large number of substances.

# 10 References


1. Sternbach, L. H.: J. Med. Chem. *22,* 1 (1979)
2. Childress, S. J., Gluckmann, M. I.: J. Pharm. Sci. *55,* 577 (1964)
3. Gerecke, M.: Brit. J. Clin. Pharmacol. *16,* 11 S (1983)
4. Römer, D., Büscher, H. H., Hill, R. C., Maurer, R., Petcher, T. J., Zeugner, H., Benson, W., Finner, E., Milkowski, W., Thies, P. W.: Nature *298,* 759 (1982)
5. Dundee, J. W.: Brit. J. Anaesth. *55,* 261 (1983)
6. Klotz, U., Kangas, L., Kanto, J.: Prog. Pharmacol. *3* (3), G. Fischer Verlag, Stuttgart, New York 1980
7. Greenblatt, D. J., Divoll, M., Abernethy, D. R., Ochs, H. R., Shader, R. I.: Clin. Pharmacokin. *8,* 233 (1983)
8. Schütz, H.: Benzodiazepines, Springer-Verlag, Berlin, Heidelberg 1982
9. Randall, L. O., Schallek, W., Heise, G. A., Keith, E. F., Bagdon, R. E.: J. Pharmacol. Exp. Ther. *129,* 163 (1960)
10. Harris, T. H.: J. Amer. Med. Assoc. *172,* 1162 (1960)
11. Tobin, J. M., Lewis, N. D. C.: J. Amer. Med. Assoc. *174,* 1242 (1960)
12. Kaim, S. C., Rosenstein, I. N.: Dis. Nerv. Syst. *21* (Suppl.), 46 (1960)
13. Randall, L. O., Kappell, B.: Pharmacological activity of some benzodiazepines and their metabolites, in: The benzodiazepines by Garattini, S., Mussini, E., Randall, L. O. (eds.), Raven Press, New York 1973
14. Garattini, S., Mussini, E., Randall, L. O. (eds.): The Benzodiazepines, Raven Press, New York 1973
15. Greenblatt, D. J., Shader, R. I.: Benzodiazepines in clinical practice, Raven Press, New York 1974
16. Costa, E. (ed.): The Benzodiazepines: From molecular biology to clinical practice, Raven Press, New York 1983
17. Usdin, E., Skolnick, P., Tallman Jr., J. F., Greenblatt, D., Paul, S. M. (eds.): Pharmacology of benzodiazepines, Macmillan Press, London 1982
18. Greenblatt, D. J., Shader, R. I., Abernethy, D. R.: N. Engl. J. Med. *309,* 354 (1983)
19. Bellantuono, C., Reggi, V., Tognoni, G., Garattini, S.: Drugs *19,* 195 (1980)
20. Klotz, U.: Deutsch. Ärztebl. *78,* 2227 (1981)
21. Klotz, U.: Krankenhauspharm. *3,* 63 (1982)
22. Lader, M., Petursson, H.: Drugs *25,* 514 (1983)
23. Costa, E.: Arzneim.-Forsch. *30 (I),* 858 (1980)
24. Möhler, H., Okada, T.: Science *198,* 849 (1977)
25. Squires, R. F., Braestrup, C.: Nature *266,* 732 (1977)
26. Hunkeler, W., Möhler, H., Pieri, L., Polc, P., Bonetti, E. P., Cumin, R., Schaffner, R., Haefely, W.: Nature *290,* 514 (1981)
27. Möhler, H., Richards, J. G.: Nature *294,* 763 (1981)
28. Braestrup, C., Nielsen, M., Olsen, C. E.: Proc. Natl. Acad. Sci. *77,* 2288 (1980)
29. Oakley, N. R., Jones, B. J.: Eur. J. Pharmacol. *68,* 381 (1980)
30. Tenen, S. S., Hirsch, J. D.: Nature *280,* 609 (1980)
31. Nutt, D. J., Cowen, P. J., Little, H. J.: Nature *295,* 436 (1982)
32. Polc, P., Bonetti, E. P., Schaffner, R., Haefely, W.: Naunyn-Schmiedeberg's Arch. Pharmacol. *321,* 260 (1982)
33. Ehlert, F. J., Roeske, W. R., Gee, K. W., Yamamura, H. I.: Biochem. Pharmacol. *32,* 2375 (1983)
34. Braestrup, C., Schmiechen, R., Nielsen, M., Petersen, E. N.: Benzodiazepine receptor ligands, receptor occupancy, pharmacological effect and GABA receptor coupling,in: Pharmacology of benzodiazepines by Usdin, E., Skolnick, P., Tallman Jr., J. F., Greenblatt, D., Paul, S. M. (eds.), Macmillan Press, London 1982
35. Braestrup, C., Nielsen, M.: Arzneim.-Forsch. *30 (I),* 852 (1980)
36. Tallman, J. F., Paul, S. M., Skolnick, P., Gallagher, D. W.: Science *207,* 274 (1980)
37. Study, R. E., Barker, J. L.: J. Amer. Med. Assoc. *247,* 2147 (1982)

38. Guidotti, A., Toffano, G., Costa, E.: Nature *275,* 553 (1978)
39. Dorow, R., Horowski, R., Paschelke, G., Amin, M., Braestrup, C.: Lancet II, 98 (1983)
40. Prado de Carvalho, L., Grecksch, G., Chapouthier, G., Rossier, J.: Nature *301,* 64 (1983)
41. Mendelson, W. B., Cain, M., Cook, J. M., Paul, S. M., Skolnick, P.: Science *219,* 414 (1983)
42. Olsen, R. W.: Ann. Rev. Pharmacol. Toxicol. *22,* 245 (1982)
43. Costa, E., Guidotti, A.: Ann. Rev. Pharmacol. Toxicol. *19,* 531 (1979)
44. Darragh, A., Lambe, R., O'Boyle, C., Kenny, M., Brick, I.: Psychopharmacol. *80,* 192 (1983)
45. Darragh, A., Lambe, R., Kenny, M., Brick, I.: Eur. J. Clin. Pharmacol. *24,* 569 (1983)
46. Gaillard, J.-M., Blois, R.: Brit. J. Clin. Pharmacol. *15,* 529 (1983)
47. Darragh, A., Scully, M., Lambe, R., Brick, I., O'Boyle, C., Downie, W. W.: Lancet II, 8 (1981)
48. Darragh, A., Lambe, R., Kenny, M., Brick, I., Taaffe, W., O'Boyle, C.: Brit. J. Clin. Pharmacol. *14,* 677 (1982)
49 Larson, G. F., Hurlbert, B. J., Wingard, D. W.: Anesth. Analg. *56,* 348 (1977)
50. Bidwai, A. V., Stanley, T. H., Roger, C., Riet, E. K.: Anesthesiology *51,* 256 (1979)
51. Avant, G. R., Speeg Jr., K. V., Freeman, F. R., Schenker, S., Berman, M. L.: Ann. Intern. Med. *91,* 53 (1979)
52. Pandit, U. A., Kothary, S. P., Samra, S. K., Domino, E. F., Pandit, S. K.: Anesth. Analg. *62,* 679 (1983)
53. Garber, J. G., Omnisky, A. J., Orkin, F. K., Quinn, P.: Anesth. Analg. *59,* 58 (1980)
54. Stirt, J. A.: Anesth. Analg. *60,* 767 (1981)
55. Arvidsson, S. B., Ekström-Jodal, B., Martinell, S. A. G., Niemand, D.: Lancet II, 1467 (1982)
56. Kanto, J., Klotz, U.: Ac. Anaesth. Scand. *26,* 554 (1982)
57. Sjövall, Kanto, J., Himberg, J. J., Hovi-Viander, M., Salo, M.: Eur. J. Clin. Pharmacol. *25,* 247 (1983)
58. Greenblatt, D. J., Koch-Weser, J.: N. Engl. J. Med. *295,* 542 (1976)
59. Gamble, J. A. S., Dundee, J. W., Assaf, R. A. E.: Anesth. *30,* 164 (1975)
60. Greenblatt, D. J., Shader, R. I., Koch-Weser, J., Franke, J.: N. Engl. J. Med. *291,* 1116 (1974)
61. Perry, P. J., Wilding, D. C., Fowler, R. C., Hepler, C. D., Caputo, J. F.: Clin. Pharmacol. Ther. *23,* 535 (1978)
62. Morgan, D. D., Robinson, J. D., Mendenhall, C. L.: Eur. J. Clin. Pharmacol. *19,* 279 (1981)
63. Korttila, K., Linnoila, M.: Brit. J. Anaesth. *47,* 857 (1975)
64. Greenblatt, D. J., Divoll, M., Harmatz, J. S., Shader, R. I.: J. Pharm. Sci. *71,* 248 (1982)
65. Moolenaar, F., Bakker, S., Visser, J., Huizinga, T.: Int. J. Pharmaceutics *5,* 127 (1980)
66. Klotz, U., Reimann, I. W.: Eur. J. Clin. Pharmacol. *26,* 223 (1984)
67. Arendt, R. M., Greenblatt, D. J., de Jong, R. H., Bonin, J. D., Abernethy, D. R., Ehrenberg, B. L., Giles, H. G., Seller, E. M , Shader, R. I.: J. Pharmacol. Exp. Ther. *227,* 98 (1983)
68. Ogura, C., Nakazawa, K., Majima, K.: Psychopharmacol. *68,* 61 (1980)
69. Oswald, I.: Brit. Med. J. *1,* 1167 (1979)
70. Carskadon, M. A., Seidel, W. F., Greenblatt, D. J., Dement, W. C.: Sleep *5,* 361 (1982)
71. Eberts, F. S., Philopoulos, Y., Reineke, L. M., Vliek, R. W.: Pharmacologist *22,* 279 (1980)
72. Greenblatt, D. J., Divoll, M., Abernethy, D. R., Moschitto, I. J., Smith, R. B., Shader, R. I.: Arch. Gen. Psych. *40,* 287 (1983)
73. Klotz, U., Ludwig, L., Ziegler, G.: II. World Conference on Clin. Pharmacol. Abstr. 530, Washington/D.C. 1983
74. Crevoisier, Ch., Heizmann, P., Wendt, G., Dubach, U. C.: II. World Conference on Clin. Pharmacol. Abstr. 45, Washington/D.C. 1983
75. Kaplan, S. A., Jack, M. L., Weinfeld, R. E., Glover, W., Weissman, L., Cotler, S.: J. Pharmacokin. Biopharm. *1,* 1 (1976)
76. Bechtel, W. D.: Brit. J. Clin. Pharmacol. *16,* 279 S (1983)
77. Greenblatt, D. J., Shader, R. I., McLeod, M., Seller, E. M.: Clin. Pharmacokin. *3,* 381 (1978)
78. Giudicelli, J. F., Berdeaux, A., Idrissi, N., Richer, C.: Brit. J. Clin. Pharmacol. *5,* 65 (1978)
79. Brogden, R. M., Heel, R. C., Speight, T. M., Avery, G. S.: Drugs *20,* 161 (1980)
80. Bertler, A., Lindgren, S., Magnusson, J.-O., Malmgren, H.: Psychopharmacol. *80,* 236 (1983)

81. Ochs, H. R., Steinhaus, E., Locniskar, A., Knüchel, M., Greenblatt, D. J.: Klin. Wschr. *60,* 411 (1982)
82. Klotz, U.: Krankenhausarzt *55,* 94 (1982)
83. Arendt, R., Ochs, H. R., Greenblatt, D. J.: Arzneim.-Forsch. *32,* 453 (1982)
84. Allen, M. C., Greenblatt, D. J., Arnold, J. D.: Psychopharmacol. *66,* 267 (1979)
85. Kangas, L., Kanto, J., Pakkanen, A.: Int. J. Clin. Pharmacol. Ther. Toxicol. *20,* 585 (1982)
86. Mahan, W. A., Inaba, T., Stone, R. M.: Clin. Pharmacol. Ther. *22,* 228 (1977)
87. Greenblatt, D. J., Locniskar, A., Shader, R. I.: Psychopharmacol. *80,* 178 (1983)
88. Eberts, F. S., Philopoulos, Y., Reinike, L. M., Vliek, R. W., Metzler, C. M.: Pharmacologist *19,* 165 (1977)
89. Ameer, B., Greenblatt, D. J.: Drugs *21,* 161 (1981)
90. Greenblatt, D. J., Divoll, M., Harmatz, J. S., Shader, R. I.: J. Pharm. Sci. *71,* 248 (1982)
91. Humpel, M., Jlli, V. , Milius, W., Wendt, H., Kurowski, M.: Eur. J. Drug Metab. Pharmacokin. *4,* 237 (1979)
92. Humpel, M., Nieuweboer, B., Milius, W., Hanke, H., Wendt, H.: Clin. Pharmacol. Ther. *28,* 673 (1980)
93. Allonen, H., Ziegler, G., Klotz, U.: Clin. Pharmacol. Ther. *30,* 653 (1981)
94. Klotz, U., Ziegler, G.: Clin. Pharmacol. Ther. *32,* 107 (1982)
95. Kangas, L., Breimer, D. D.: Clin. Pharmacokin. *6,* 346 (1981)
96. Yamazaki, Y., Iwai, T., Ninomiya, T., Kawahara, Y.: Ann. Rep. Sankyo Res. Lab. *32,* 104 (1980)
97. Pacifici, G. M., Placidi, G. F., Fornaro, P., Gomeni, R.: Eur. J. Clin. Pharmacol. *22,* 225 (1982)
98. Allen, M. D., Greenblatt, D. J., Harmatz, J. S., Shader, R. I.: J. Clin. Pharmacol. *19,* 445 (1979)
99. Smith, M. T., Evans, L. E. J., Eadie, M. J., Tyrer, J. H.: Eur. J. Clin. Pharmacol. *16,* 141 (1979)
100. Schwarz, H. J.: Brit. J. Clin. Pharmacol. *8,* 23S (1979)
101. Fucella, L. M., Bolcioni, G., Tamassia, V., Ferrario, L., Tognoni, G.: Eur. J. Clin. Pharmacol. *12,* 383 (1977)
102. Divoll, M., Greenblatt, D. J., Harmatz, J. S., Shader, R. I.: J. Pharm. Sci. *70,* 1104 (1981)
103. Heel, R. C., Brogden, R. N., Speight, T. M., Avery, G. S.: Drugs *21,* 321 (1981)
104. Eberts, F. S., Philopoulos, Reineke, L. M., Vliek, R. W.: Clin. Pharmacol. Ther. *29,* 81 (1981)
105. Pakes, G. E., Brogden, R. N., Heel, R. C., Speight, T. M., Avery, G. S.: Drugs *22,* 81 (1981)
106. Jochemsen, R., van Boxtel, C. J., Hermans, J., Breimer, D. D.: Clin. Pharmacol. Ther. *34,* 42 (1983)
107. Tallone, G., Ghirardi, P., Bianchi, M. C., Ravaccia, F., Bruni, G., Loreti, P.: Arzneim.-Forsch. *30,* 1021 (1980)
108. Elie, R., Caille, G., Levasseur, F. A., Gareau, J.: J. Clin. Pharmacol. *23,* 32 (1983)
109. Kales, A., Bixter, E. O., Soldatos, C. R., Vela-Bueno, A., Jacoby, J., Kales, J. D.: Clin. Pharmacol. *32,* 781 (1982)
110. Chasseud, L. F., Taylor, T., Brodie, R. R.: J. Pharm. Pharmacol. *32,* 652 (1980)
111. Welling, P. G.: J. Pharmacokinet. Biopharm. *5,* 291 (1977)
112. Melander, A.: Clin. Pharmacokin. *3,* 337 (1978)
113. Greenblatt, D. J., Allen, M. D., MacLaughlin, D. S., Harmatz, J. S., Shader, R. I.: Clin. Pharmacol. Ther. *24,* 600 (1978)
114. Divoll, M., Greenblatt, D. J., Ciraulo, D. A., Puri, S. K., Ho, I., Shader, R. I.: J. Clin. Pharmacol. *22,* 69 (1982)
115. Cenraud, B., Guyot, M., Levy, R. H., Brachet-Liermain, A., Morselli, P. L., Moreland, T. A., Loiseau, P.: Brit. J. Clin. Pharmacol. *16,* 728 (1983)
116. Melander, A., Danielson, K., Vessman, J., Wahlin, E.: Acta Pharmacol. Toxicol. *40,* 548 (1977)
117. Jusko, W. J., Gretch, M.: Drug Metab. Rev. *5,* 43 (1976)
118. Müller, W. E.: Klin. Wschr. *55,* 105 (1977)
119. Moschitto, L. J., Greenblatt, D. J.: J. Pharm. Pharmacol. *35,* 179 (1983)

120. Klotz, U.: Plasma protein binding of benzodiazepines, in: Progress in drug protein binding by Rietbrock, N., Woodcock, B. G. (eds.), Vieweg Verlag Braunschweig/Wiesbaden 1981
121. Abel, J. G., Sellers, E. M., Naranjo, C. A., Shaw, J., Kadar, D., Romach, M. K.: Clin. Pharmacol. Ther. *26,* 247 (1979)
122. Johnson, R. F., Schenker, S., Roberts, R. K., Desmond, P. V., Wilkinson, G. R.: J. Pharm. Sci. *68,* 1320 (1979)
123. Klotz, U., Antonin, K. H., Bieck, P. R.: Brit. J. Clin. Pharmacol. *4,* 85 (1977)
124. Korttila, K., Kangas, L.: Acta Pharmacol. Toxicol. *40,* 241 (1977)
125. Desmond, P. V., Roberts, R. K., Wood, A. J. J., Dunn, G. D., Wilkinson, G. R., Schenker, S.: Brit. J. Clin. Pharmacol. *9,* 171 (1980)
126. Naranjo, C. A., Sellers, E. M., Giles, H. G., Abel, J. G.: Brit. J. Clin. Pharmacol. *9,* 265 (1980)
127. Klotz, U., Reimann, I. W.: Clin. Pharmacokin. (in press; 1984)
128. Shader, R. I., Greenblatt, D. J., Harmatz, J. S., Franke, K., Koch-Weser, J.: J. Clin. Pharmacol. *17,* 709 (1977)
129. Roberts, R. K., Wilkinson, G. R., Branch, R. A., Schenker, S.: Gastroenterol. *75,* 479 (1978)
130. Klotz, U., Avant, G. R., Hoyumpa, A., Schenker, S., Wilkinson, G. R.: J. Clin. Invest. *55,* 347 (1975)
131. Ochs, H. R., Greenblatt, D. J., Divoll, M., Abernethy, D. R., Feyerabend, H., Dengler, H. J.: Pharmacol. *23,* 24 (1981)
132. Klotz, U., Müller-Seydlitz, P.: Brit. J. Clin. Pharmacol. *7,* 119 (1979)
133. Allen, M. C., Greenblatt, D. J., Harmatz, J. S., Shader, R. I.: Clin. Pharmacol. Ther. *28,* 196 (1980)
134. Shader, R. I., Greenblatt, D. J., Ciraulo, D. A., Divoll, M., Harmatz, J. S., Georgotas, A.: Psychopharmacol. *75,* 193 (1981)
135. Greenblatt, D. J., Divoll, M., Harmatz, J. S., MacLaughlin, D. S., Shader, R. I.: Clin. Pharmacol. Ther. *30,* 475 (1981)
136. Collier, P. S., Kawar, P., Gamble, J. A. S., Dundee, J. W.: Brit. J. Clin. Pharmacol. *13,* 602 P (1982)
137. Greenblatt, D. J., Divoll, M., Puri, S. K., Ho, I., Zinny, M. A., Shader, R. I.: Brit. J. Clin. Pharmacol. *12,* 631 (1981)
138. Greenblatt, D. J., Divoll, M., Abernethy, D. R., Moschitto, L. J., Smith, R. B., Shader, R. I.: Brit. J. Clin. Pharmacol. *15,* 303 (1983)
139. Moschitto, L. J., Greenblatt, D. J., Divoll, M., Abernethy, D. R., Smith, R. B., Shader, R. I.: Clin. Pharmacol. Ther. *29,* 267 (1981)
140. Shull, H. J., Wilkinson, G. R., Johnson, R., Schenker, S.: Ann. Int. Med. *84,* 420 (1976)
141. Greenblatt, D. J., Divoll, M., Harmatz, J. S., Shader, R. I.: J. Pharmacol. Exp. Ther. *215,* 86 (1980)
142. Kraus, J. W., Desmond, P. V., Marshall, J. P., Johnson, R. F., Schenker, S., Wilkinson, G. R.: Clin. Pharmacol. Ther. *24,* 411 (1978)
143. Greenblatt, D. J., Allen, M. C., Locniskar, A., Harmatz, J. S., Shader, R. I.: Clin. Pharmacol. Ther. *26,* 103 (1979)
144. Divoll, M., Greenblatt, D. J., Harmatz, J. S., Shader, R. I.: J. Pharm. Sci. *70,* 1104 (1981)
145. Castleden, C. M., George, C. F., Marcer, D., Hallett, C.: Brit. Med. J. *I,* 10 (1977)
146. Kangas, L., Iisalo, E., Kanto, J., Lehtinen, V., Pynnönen, S., Ruikka, I., Salminen, J., Sillanpää, M., Syvälahti, E.: Eur. J. Clin. Pharmacol. *15,* 163 (1979)
147. Jochemsen, R., van Beusekom, B. R., Spoelstra, P., Janssens, A. R., Breimer, D. D.: Brit. J. Clin. Pharmacol. *15,* 295 (1983)
148. Kanto, J., Kangas, L., Aaltonen, L., Hilke, H.: Internat. J. Clin. Pharmacol. Ther. Toxicol. *19,* 400 (1981)
149. Klotz, U.: Pharmacokinetics and pharmacodynamics in the elderly, in: Liver and Aging by K. Kitani (ed.), Elsevier Biomed. Press, Amsterdam 1982
150. Guidicelli, J. F., Tillement, J. P.: Clin. Pharmacokin. *2,* 157 (1977)
151. Greenblatt, D. J., Shader, R. I., Franke, K., MacLaughlin, D. S., Ransil, B. J., Koch-Weser, J.: Clin. Pharmacol. Ther. *22,* 893 (1977)
152. MacLeod, S. M., Giles, H. G., Bengert, B., Liu, F. F., Sellers, E. M.: J. Clin. Pharmacol. *19,* 15 (1979)

153. Divoll, M., Greenblatt, D. J., Ochs, H. R., Shader, R. I.: Anesth. Analg. *62,* 1 (1983)
154. Greenblatt, D. J., Abernethy, D. R., Locniskar, A., Harmatz, J. S., Limjuco, R. A., Shader, R. I.: Anesthesiol. *55,* 176 (1981)
155. Smith, R. B., Divoll, M., Gillespie, W. R., Greenblatt, D. J.: J. Clin. Psychopharmacol. *3,* 172 (1983)
156. Abernethy, D. R., Greenblatt, D. J., Divoll, M., Smith, R. B., Shader, R. I.: Clin. Pharmacol. Ther. *33,* 247 1983)
157. Abernethy, D. R., Greenblatt, D. J., Divoll, M., Harmatz, J. S., Shader, R. I.: J. Pharmacol. Exp. Ther. *217,* 681 (1981)
158. Abernethy, D. R., Greenblatt, D. J., Divoll, M., Shader, R. I.: J. Clin. Pharmacol. *23,* 369 (1983)
159. Abernethy, D. R., Greenblatt, D. J., Divoll, M., Shader, R. I.: J. Laborat. Clin. Med. *101,* 873 (1983)
160. Roberts, R. K., Wilkinson, G. R., Branch, R. A., Schenker, S.: Gastroenterol. *75,* 479 (1978)
161. Klotz, U., Antonin, K. H., Brügel, H., Bieck, P. R.: Clin. Pharmacol. Ther. *21,* 430 (1977)
162. Sellers, E. M., Greenblatt, D. J., Giles, H. G., Naranjo, C. A., Kaplan, H., MacLeod, S. M.: Clin. Pharmacol. Ther. *26,* 240 (1979)
163. Sellers, E. M., Sandor, P., Giles, H. G., Khouw, V., Greenblatt, D. J.: Brit. J. Clin. Pharmacol. *15,* 125 (1983)
164. Jochemsen, R., Nandi, K. L., Corless, D., Wesselman, J. G. J., Breimer, D. D.: Brit. J. Clin. Pharmacol. *16,* 299 S (1983)
165. Jochemsen, R., Joeres, R. P., Wesselman, J. G. J., Richter, E., Breimer, D. D.: Brit. J. Clin. Pharmacol. *16,* 315 S (1983)
166. Oder-Cederlöf, I., Vessman, J., Alván, G., Sjöqvist, F.: Acta Pharmacol. Toxicol. *40* (Suppl. I), 52 (1977)
167. Busch, U., Molzahn, M., Bozler, G., Koss, F. W.: Arzneim.-Forsch. *31,* 1507 (1981)
168. Kangas, L., Kanto, J., Forsstrom, J., Iisalo, E.: Clin. Nephrol. *5,* 114 (1976)
169. Verbeeck, R., Tjandramaga, T. B., Verberckmoes, R., de Schepper, P. J.: Brit. J. Clin. Pharmacol. *3,* 1033 (1976)
170. Verbeeck, R. K., Tjandramaga, T. B., de Schepper, P. J., Verberckmoes, R.: Brit. J. Clin. Pharmacol. *12,* 749 (1981)
171. Evers, J., Renner, E., Bechtel, W. D.: Brit. J. Clin. Pharmacol. *16,* 309 S (1983)
172. Ochs, H. R., Greenblatt, D. J., Kaschell, H. J., Klehr, U., Divoll, M., Abernethy, D. R.: Brit. J. Clin. Pharmacol. *12,* 829 (1981)
173. Eichelbaum, M.: Clin. Pharmacokin. *1,* 339 (1976)
174. Shenfield, G. M.: Clin. Pharmacokin. *6,* 275 (1981)
175. Scott, A. K., Khir, A. S. M., Bewsher, P. D., Hawksworth, G. M.: Brit. J. Clin. Pharmacol. *17,* 49 (1984)
176. Martyn, J. A. J., Greenblatt, D. J., Quinby, W. C.: Anesth. Analg. *62,* 293 (1983)
177. Kalow, W.: Clin. Pharmacokin. *7,* 373 (1982)
178. Ghoneim, M. M., Korttila, K., Chiang, C.-K., Jacobs, L., Schoenwald, R. D., Mewaldt, S. P., Kayaba, K.-O.: Clin. Pharmacol. Ther. *29,* 749 (1981)
179. Jusko, W. J.: Drug Metab. Rev. *9,* 221 (1979)
180. Norman, T. R., Fulton, A., Burrows, G. D., Maguire, K. P.: Eur. J. Clin. Pharmacol. *21,* 229 (1981)
181. Ochs, H. R., Greenblatt, D. J., Otten, H.: Klin. Wschr. *59,* 899 (1981)
182. Desmond, P. V., Roberts, R. K., Wilkinson, G. R., Schenker, S.: N. Engl. J. Med. *300,* 199 (1979)
183. Krauer, B., Krauer, F.: Clin. Pharmacokin. *2,* 167 (1977)
184. Perucca, E., Ruprah, M., Richens, A.: Proc. Brit. Pharmacol. Soc. 1st–3rd April, 276 P 1981
185. Lee, J. N., Chen, S. S., Richens, A., Menabawey, M., Chard, T.: Brit. J. Clin. Pharmacol. *14,* 551 (1982)
186. Moore, R. G., McBride, W. G.: Eur. J. Clin. Pharmacol. *13,* 275 (1978)
187. Rey, E., d'Athis, P., Giraux, P., de Lauture, D., Turquais, J. M., Chavinie, J., Olive, G.: Eur. J. Clin. Pharmacol. *15,* 175 (1979)
188. Bakke, O. M., Haram, K.: Clin. Pharmacokin. *7,* 353 (1982)

189. Rey, E., Giraux, P., d'Athis, P., Turquais, J. M., Chavinie, J., Olive, G.: Eur. J. Clin. Pharmacol. *15,* 181 (1979)
190. Kangas, L., Erkkola, R., Kanto, J., Eronen, M.: Eur. J. Clin. Pharmacol. *17,* 301 (1980)
191. Kangas, L., Kanto, J., Erkkola, R.: Eur. J. Clin. Pharmacol. *12,* 355 (1977)
192. Erkkola, R., Kanto, J.: Lancet I, 1235 (1972)
193. Humpel, M., Stoppelli, I., Milia, S., Raner, E.: Eur. J. Clin. Pharmacol. *21,* 421 (1982)
194. Morselli, P. L., Franco-Morselli, R., Bossi, L.: Clin. Pharmacokin. *5,* 485 (1980)
195. Tomson, G., Lunell, N. O., Sundwall, A., Rane, A.: Clin. Pharmacol. Ther. *25,* 74 (1979)
196. Abruzzo, C. W., Macosieb, T., Weinfeld, R., Rider, J. A., Kaplan, S. A.: J. Pharmacokin. Biopharm. *5,* 377 (1977)
197. Shader, R. I., Georgotas, A., Greenblatt, D. J., Harmatz, J. S., Allen, M. D.: Clin. Pharmacol. Ther. *24,* 308 (1978)
198. Chun, A. H., Carrigan, P. J., Hoffman, D. J., Kershner, M. S., Stuart, J. D.: Clin. Pharmacol. Ther. *22,* 329 (1977)
199. Shader, R. I., Ciraulo, D. A., Greenblatt, D. J., Harmatz, J. S.: Clin. Pharmacol. Ther. *31,* 180 (1982)
200. Greenblatt, D. J., Shader, R. I., Harmatz, J. S., Franke, K., Koch-Weser, J.: Clin. Pharmacol. Ther. *19,* 234 (1976)
201. Tsutsumi, E., Inaba, T., Mahon, W. A., Kalow, W.: Biochem. Pharmacol. *24,* 1361 (1975)
202. Dhillon, S., Richens, A.: Brit. J. Clin. Pharmacol. *12,* 591 (1981)
203. Routledge, P. A., Stargel, W. W., Kitchell, B. B., Barchowsky, A., Shand, D. G.: Brit. J. Clin. Pharmacol. *11,* 245 (1981)
204. Roberts, R. K., Desmond, P. V., Wilkinson, G. R., Schenker, S.: Clin. Pharmacol. Ther. *25,* 826 (1979)
205. Desmond, P. V., Patwardhan, R. V., Schenker, S., Hoyumpa, A. M.: Eur. J. Clin. Pharmacol. *18,* 275 (1980)
206. Giacomini, K. M., Swezey, S. E., Giacomini, J. C., Blaschke, T. F.: Life Sci. *27,* 771 (1980)
207. Brown, J. E., Kitchell, B. B., Bjornsson, T. D., Shand, D. G.: Clin. Pharmacol. Ther. *30,* 636 (1981)
208. Khoo, K.-C., Mendels, J., Rothbart, M., Garland, W. A., Colburn, W. A., Min, B. H., Lucek, R., Carbone, J. J., Boxenbaum, H. G., Kaplan, S. A.: Clin. Pharmacol. Ther. *28,* 368 (1980)
209. Sjö, O., Hvidberg, E. F., Naestoft, J., Lund, M.: Eur. J. Clin. Pharmacol. *8,* 249 (1975)
210. Lai, A. A., Levy, R. H., Cutler, R. E.: Clin. Pharmacol. Ther. *24,* 316 (1978)
211. Wilensky, A. J., Levey, R. H., Troupin, A. S., Moretti-Ojemann, L., Friel, P.: Clin. Pharmacol. Ther. *24,* 22 (1978)
212. Dhillon, S., Richens, A.: Brit. J. Clin. Pharmacol. *12,* 841 (1981)
213. Ochs, H. R., Greenblatt, D. J., Roberts, G.-M., Dengler, H. J.: Clin. Pharmacol. Ther. *29,* 671 (1981)
214. Patwardhan, R. V., Mitchell, M. C., Johnson, R. F., Schenker, S.: Hepatology *3,* 248 (1983)
215. Giles, H. G., Sellers, E. M., Naranjo, C. A., Frecker, R. C., Greenblatt, D. J.: Eur. J. Clin. Pharmacol. *20,* 207 (1981)
216. Abernethy, D. R., Greenblatt, D. J., Divoll, M., Arendt, R., Ochs, H. R., Shader, R. I.: N. Engl. J. Med. *306,* 791 (1982)
217. Jochemsen, R., van der Graaff, M., Boeijinga, J. K., Breimer, D. D.: Brit. J. Clin. Pharmacol. *13,* 319 (1982)
218. Ochs, H. R., Greenblatt, D. J., Knüchel, M.: Brit. J. Clin. Pharmacol. *16,* 743 (1983)
219. Mac Leod, S. M., Sellers, E. M., Giles, H. G., Billings, B. J., Martin, P. R., Greenblatt, D. J., Marsham, J. A.: Clin. Pharmacol. Ther. *24,* 583 (1978)
220. Sellers, E. M., Giles, H. G., Greenblatt, D. J., Naranjo, C. A.: Arzneim.-Forsch. *30,* 882 (1980)
221. Klotz, U., Anttila, V.-J., Reimann, I.: Lancet II, 699 (1979)
222. Klotz, U., Reimann, I.: N. Engl. J. Med. *302,* 1012 (1980)
223. Klotz, U., Reimann, I.: Eur. J. Clin. Pharmacol. *18,* 517 (1980)
224. Desmond, P. V., Patwardhan, R. V., Schenker, S., Speeg, Jr., K. F.: Ann. Int. Med. *93,* 226 (1980)

225. Patwardhan, R. V., Johnson, R. F., Sinclair, A. P., Schenker, S., Speeg, Jr., K. V.: Gastroenterol. *81,* 547 (1981)
226. Patwardhan, R. V., Yarborough, G. W., Desmond, P. V., Johnson, R. F., Schenker, S., Speeg, Jr., K. V.: Gastroenterol. *79,* 912 (1980)
227. Klotz, U., Reimann, I.: Clin. Pharmacol. Ther. *30,* 513 (1981)
228. Abernethy, D. R., Greenblatt, D. J., Divoll, M., Moschitto, L. J., Harmatz, J. S., Shader, R. I.: Psychopharmacol. *80,* 275 (1983)
229. Grigoleit, H.-G., Hajdu, P., Hundt, H. K. L., Koeppen, D., Malerczyk, V., Meyer, B. H., Müller, F. O., Witte, P. U.: Eur. J. Clin. Pharmacol. *25,* 139 (1983)
230. Ochs, H. R., Greenblatt, D. J., Gugler, R., Müntefering, G., Locniskar, A., Abernethy, D. R.: Clin. Pharmacol. Ther. *34,* 227 (1983)
231. Greenblatt, D. J., Abernethy, D. R., Divoll, M., Locniskar, J. S., Harmatz, J. S., Shader, R. I.: J. Pharm. Sci. *73,* 399 (1984)
232. Klotz, U., Reimann, I.: Klin. Wochenschr. *61,* 625 (1983)
233. Klotz, U., Reimann, I. W.: Pharm. Res. *2,* 59 (1984)
234. Hayes, S. L., Pablo, G., Radomski, T., Palmer, R. F.: N. Engl. J. Med. *296,* 186 (1977)
235. Laisi, U., Linnoila, M., Seppälä, T., Himberg, J. J., Mattila, M. J.: Eur. J. Clin. Pharmacol. *16,* 263 (1979)
236. Mac Leod, S. M., Giles, H. G., Patzalek, G., Thiessen, J. J., Sellers, E. M.: Eur. J. Clin. Pharmacol. *11,* 345 (1977)
237. Divoll, M., Greenblatt, D. J.: Pharmacol. *22,* 263 (1981)
238. Wills, R. J., Crouthamel, W. G., Iber, F. L., Perkal, M. B.: J. Clin. Pharmacol. *22,* 557 (1982)
239. Linnoila, M., Otterström, S., Anttila, M.: Ann. Clin. Res. *6,* 4 (1974)
240. Whiting, B., Lawrence, J. R., Skellern, G. G., Meier, J.: Brit. J. Clin. Pharmacol. *7,* 95 (1979)
241. Sellers, E. M., Naranjo, C. A., Giles, H. G., Frecker, R. C., Beeching, M.: Clin. Pharmacol. Ther. *28,* 638 (1980)
242. Taeuber, K., Badian, M., Brettel, H. F., Royen, T., Rupp, W., Sittig, W., Uihlein, M.: Brit. J. Clin. Pharmacol. *7,* 91 S (1979)
243. Davis, W. C., Ticku, M. K.: Molec. Pharmacol. *20,* 287 (1981)
244. Allen, C., Gough, K. R.: Brit. Med. J. *286,* 1181 (1983)
245. Nielsson, E., Himberg, J. J.: Acta Anaesthesiol. Scand. *26,* 9 (1982)
246. Grote, B., Doenicke, A., Kugler, J., Laub, M., Ott, H., Fichte, K., Suttmann, H., Zwisler, P.: Anaesthesist *30,* 627 (1981)
247. Mattila, M. J., Palva, E., Savolainen, K.: Med. Biol. *60,* 121 (1982)
248. Henauer, S. A., Hollister, L. E., Gillespie, H. K., Moore, F.: Eur. J. Clin. Pharmacol. *25,* 743 (1983)
249. Klotz, U., Ziegler, G., Reimann, I. W.: Eur. J. Clin. Pharmacol. (in press; 1984)
250. Braestrup, C., Nielsen, M., Olsen, C. F.: Proc. Natl. Acad. Sci. *77,* 2288 (1980)
251. Braestrup, C., Nielsen, M.: Lancet II, 1030 (1982)
252. Dasberg, H. H., van der Kleijn, E., Guelen, P. J. R., von Prag, H. M.: Clin. Pharmacol. Ther. *15,* 473 (1974)
253. Robin, A., Curry, S. H., Whelpton, R.: Psychol. Med. *4,* 388 (1974)
254. Curry, S. H.: Clin. Pharmacol. Ther. *16,* 192 (1974)
255. Gottschalk, L. A., Cohn, J. B.: Psychopharmacol. Bull. *14,* 39 (1978)
256. Bowden, C. L., Fisher, J. G.: J. Clin. Psychopharmacol. *2,* 110 (1982)
257. Fink, M., Irwin, P., Weinfield, R. E., Schwartz, M. A., Conney, A. H.: Clin. Pharmacol. Ther. *20,* 184 (1976)
258. Bittencourt, P. R. M., Wade, P., Smith, A. T., Richens, A.: Brit. J. Clin. Pharmacol. *12,* 523 (1981)
259. Smith, R. C., Dekirmenjian, H., Davis, J., Casper, R., Gosenfeld, L., Tsai, C.: Blood levels, mood and MHPG responses to diazepam in man, in: Pharmacokinetics of psychoactive drugs: blood levels and clinical response by L. A. Gottschalks, S. Merlis (eds.), Spectrum Publ. Inc., New York 1976
260. Bond, A. J., Hailey, D. M., Lader, M. H.: Brit. J. Clin. Pharmacol. *4,* 51 (1977)
261. Bond, A. J., Lader, M. H.: Psychopharmacol. *32,* 223 (1973)
262. Marks, I. M., Viswanathan, R., Lipsedge, M. S.: Brit. J. Psych. *121,* 493 (1972)

263. Garattini, S., Marcucci, F., Morselli, P. L., Mussini, E.: The significance of measuring blood levels of benzodiazepines, in: Biological effects of drugs in relation to their plasma concentrations by D. S. Davies, B. N. C. Prichard (eds.), Macmillan, London 1973
264. Tansella, M., Siciliani, O., Burti, L., Schiavon, M., Zimmermann-Tansella, C., Gerna, M., Tognoni, G., Morselli, P. L.: Psychopharmacol. *41,* 81 (1975)
265. Tansella, M., Zimmermann-Tansella, C., Ferrario, L., Preziati, L., Tognoni, G., Lader, M.: Pharmakopsych. Neuropsychopharmakol. *11,* 68 (1978)
266. Gottschalk, L. A., Kaplan, S. A.: Compr. Psych. *13,* 519 (1972)
267. Lin, K.-M., Friedel, R. O.: Am. J. Psych. *136,* 1 (1979)
268. Hvidberg, E. F., Dram, M.: Clin. Pharmacokin. *1,* 161 (1976)
269. Kangas, L., Kanto, J., Syvälahti, E.: Acta Pharmacol. Toxicol. *41,* 65 (1977)
270. Amrein, R.: Zur Pharmakokinetik und zum Metabolismus von Flunitrazepam, in: Rohypnol (Flunitrazepam): Pharmakologische Grundlagen-Klinische Anwendung von F. W. Ahnefeld, H. Bergmann, C. Burri, W. Dick, M. Halmágyi, G. Hossli, E. Rügheimer (eds.), Springer-Verlag, Berlin 1978
271. Greenblatt, D. J., Shader, R. I., Abernethy, D. R.: N. Engl. J. Med. *309,* 410 (1983)
272. Lader, M., Petursson, H.: Drugs *25,* 514 (1983)
273. Bellantuono, C., Reggi, V., Tognoni, G., Garattini, S.: Drugs *19,* 195 (1980)
274. Rickels, K.: Benzodiazepine in the treatment of anxiety, in: Pharmacology of benzodiazepines by E. Usdin, P. Skolnick, J. F. Tallman Jr., D. Greenblatt, S. M. Paul (eds.), Macmillan Press, London 1982
275. Rickels, K.: Psychopharmacol. *58,* 1 (1978)
276. Downing, R. W., Rickels, K.: Psychopharm. Bull. *18,* 37 (1982)
277. Rickels, K., Case, W. G., Downing, R. W., Winokur, A.: J. Amer. Med. Assoc. *250,* 767 (1983)
278. Hollister, L. E., Conley, F. K., Britt, R. H., Suer, L.: J. Amer. Med. Assoc. *246,* 1568 (1981)
279. Fabre, L. F., McLendon, D. M., Stephens, A. G.: J. Int. Med. Res. *9,* 191 (1981)
280. Schatzberg, A. F., Cole, J. O.: Arch. Gen. Psychiat. *35,* 1359 (1978)
281. Schatzberg, A. F., Cole, J. O.: Brit. J. Clin. Pharmacol. *11,* 17 S (1981)
282. Fabre, L. F.: Curr. Ther. Res. *19,* 661 (1976)
283. Fabre, L. F., McLendon, D. M.: Curr. Ther. Res. *27,* 474 (1980)
284. Rickels, K., Cohen, D., Csanalosi, I., Harris, H., Koepke, H., Werblowski, J.: Curr. Ther. Res. *32,* 157 (1982)
285. Dawson, G. W., Jue, S. G., Brogden, R. N.: Drugs *27,* 132 (1984)
286. Feighner, J. P.: Benzodiazepines as antidepressants: A triazolo-benzodiazepine used to treat depression, in: Modern problems in Pharmacopsychiatry by H. Lehmann (ed.), S. Karger Verlag, Basel 1981
287. Schatzberg, A. F., Altesman, R. I., Cole, J. O.: An update of the use of benzodiazepines in depressed patients, in: Pharmacology of benzodiazepines by E. Usdin, P. Skolnick, J. F. Tallman Jr., D. Greenblatt, S. M. Paul (eds.), Macmillan Press, London 1982
288. Borbély, A. A., Loepfe, M., Mattmann, P., Tobler, I.: Arzneimittelforsch. *33,* 1500 (1983)
289. Ziegler, G., Ludwig, L., Klotz, U.: Brit. J. Clin. Pharmacol. *16,* 81 S (1983)
290. Kales, A., Soldatos, C. R., Bixler, E. O., Kales, J. D.: Pharmacol. *26,* 121 (1983)
291. Kales, A., Soldatos, C. R., Bixler, E. O., Goff, P. J., Vela-Bueno, A.: Pharmacol. *26,* 138 (1983)
292. Vogel, G. W., Vogel, F.: Brit. J. Clin. Pharmacol. *16,* 103 S (1983)
293. Nicholson, A. N.: Brit. J. Clin. Pharmacol. *9,* 223 (1980)
294. Oswald, I., French, C., Adam, K., Gilham, J.: Brit. Med. J. *284,* 860 (1982)
295. Dasta, J. F., Brier, K. L., Kidwell, G. A., Schonfeld, S. A., Couri, D.: South. Med. J. *74,* 278 (1981)
296. Delgado-Escueta, A. V., Wasterlain, C., Treiman, D. M., Porter, R. J.: N. Engl. J. Med. *306,* 1337 (1982)
297. Delgado-Escueta, A. V., Treiman, D. M., Walsh, G. O.: N. Engl. J. Med. *308,* 1576 (1983)
298. Browne, T. R.: N. Engl. J. Med. *299,* 812 (1978)
299. Sellers, E. M., Kalant, H.: N. Engl. J. Med. *294,* 757 (1976)
300. Shaw, J. M., Kolesar, G. S., Sellers, E. M., Kaplan, H. L., Sandor, P.: J. Clin. Psychopharmacol. *1,* 382 (1981)

301. Ochs, H. R., Greenblatt, D. J., Eckardt, B., Harmatz, J. S., Shader, R. I.: Clin. Pharmacol. Ther. *33,* 471 (1983)
302. Hall, R. C. W., Zisook, S.: Brit. J. Clin. Pharmacol. *11,* 99 S (1981)
303. Rao, S., Sherbaniuk, R. W., Prasad, K., Lee, S. J. K., Sproule, B. J.: Clin. Pharmacol. Ther. *14,* 182 (1973)
304. Morel, D., Forster, A., Gardez, J.-P., Suter, P. M., Gemperle, M.: Arzneimittelforsch. *31,* 2264 (1981)
305. Hartz, S. C., Heinonen, O. P., Shapiro, S., Siskind, V., Slone, D.: N. Engl. J. Med. *292,* 726 (1975)
306. de Gier, J. J., Hart, B. J., Nelemans, F. A., Bergman, H.: Psychopharmacol. *73,* 340 (1981)
307. O'Hanlon, J. F., Haak, T. W., Blaauw, G. J., Riemersma, J. B. J.: Science *217,* 79 (1982)
308. Betts, T. A., Birtle, J.: Brit. Med. J. *285,* 852 (1982)
309. Skegg, D. C. G., Richards, S. M., Doll, R.: Brit. Med. J. *1,* 917 (1979)
310. Seppala, T., Linnoila, M., Mattila, M. J.: Drugs *17,* 389 (1979)
311. Vine, J., Watson, T. R.: Med. J. Aust. *1,* 612 (1983)
312. Tyrer, P.: Brit. J. Psychiat. *137,* 576 (1980)
313. File, S. E.: Psychopharmacol. *77,* 284 (1982)
314. File, S. E.: Eur. J. Clin. Pharmacol. *81,* 637 (1982)
315. Rosenberg, H. C., Smith, S., Chiu, T. H.: Life Sci. *32,* 279 (1983)
316. Aranko, K., Mattila, M. J., Seppälä, T.: Brit. J. Clin. Pharmacol. *15,* 545 (1983)
317. Greenblatt, D. J., Shader, R. I.: Drug Metabol. Rev. *8,* 12 (1978)
318. Marks, J.: The benzodiazepines: Use, overuse, misuse, abuse. MTP Press, London 1972
319. Owen, R. T., Tyrer, P.: Drugs *25,* 385 (1983)
320. Böning, J., Schrappe, O.: Dtsch. Ärztebl. *81,* 211 and 279 (1984)
321. Petursson, H., Lader, M. H.: Brit. Med. J. *283,* 643 (1981)
322. Lader, M., Petursson, H.: Neuropharmacol. *22,* 527 (1983)
323. Tyrer, P., Owen, R., Dawling, S.: Lancet I, 1402 (1983)
324. Cowen, P. J., Nutt, D. J.: Lancet II, 360 (1982)

# Interferon Gamma

**Holger Kirchner**

Institute of Virus Research, German Cancer Research Center Heidelberg, DKFZ, Im Neuenheimer Feld 280, D-6900 Heidelberg, Federal Republic of Germany

In recent years, there has been significant progress in interferon research. Among the subtypes of interferon, interferon gamma (IFN γ) has been the least well studied for many years but only recently, significant progress has been made also in this area. IFN γ is a prototype of two important classes of substances; of the lymphokines, and of the interferons. The genomic structure of IFN γ is now known, and it has been the first lymphokine defined by the DNA sequence. In the following, we wish to summarize the recent literature on IFN γ but for a better understanding of the subject we need to start with two short chapters on the interferons and lymphokines.



Progress in Clinical Biochemistry
and Medicine, Vol. 1

## General Introduction

In recent years, there has been significant progress in interferon research and this has been covered in a number of reviews[1–3]. Among the subtypes of interferon, interferon gamma (IFN γ) has been the least well studied for many years but only recently, significant progress has been made also in this area. The progress of IFN γ has been competently reviewed several times by L. Epstein[4–6]. These outstanding reviews have served to update the specialists in the interferon field whereas it is the purpose of this article to give a more general account of the state of affairs in the research on IFN γ.

IFN γ is a prototype of two important classes of substances; of the lymphokines, and of the interferons. The genomic structure of IFN γ is now known, and it has been the first lymphokine defined by the DNA sequence. In the following, we wish to summarize the recent literature on IFN γ but for a better understanding of the subject we need to start with two short chapters on the interferons and lymphokines.

## Interferons

Interferon as an antiviral molecule has been discovered in 1957 by Isaacs and Lindenmann[7] during their studies on viral interference, a phenomenon which has been known before. Isaacs and Lindenmann, however, have discovered that some forms of viral interference were caused by a protein for which they coined the term "interferon". As it turned out, this molecule was biologically highly active. Since only small amounts were produced, advances in purification and characterization of the molecule(s) have been slow. However, very rapid progress has been made during the last few years.

As it appears, cells of every species make their own individual interferons, different from the interferons of other species. Best characterized are human interferons. To date, three major classes of human interferons have been identified (in parentheses the old names are given):

IFN α (leucocyte interferon)
IFN β (fibroblast interferon)
IFN γ (immune interferon)

IFN α in the classical scheme has been produced in cultures of white blood cells, and production was induced by viruses. Human IFN α is now known to represent a whole group of proteins that are the products of a multi-gene family and show about 70% homology on the level of their DNA. The human IFN α gene family contains at least 20 distinct members. Some of these genes are non-allelic variants, others are allelic variants. Sometimes, it can be clearly established which one is the case: for example, a genome segment may contain two genes in tandem, clearly showing they are non-allelic pairs. Extensive differences in 3′ non-coding regions also suggest that two genes are most likely non-allelic. In other cases, two DNA segments each containing one IFN α gene may be identical over their entire length, 20 Kb or more of DNA, except for a few nucleotide changes; the two genes are then most likely allelic. In many cases, a decision is not easy because recently duplicated DNA segments containing an IFN α gene, while in fact being non-allelic, may be difficult to distinguish from an allelic pair until a sufficiently long DNA segment is examined and a linkage map established.

The IFNs α share about 40% homology with IFN β which appears to exist in a single molecular form. However, from their analyses of the mRNAs, Seghal et al.[8] reported heterogeneity of human IFN β mRNA species. IFN β is in vitro produced by fibroblasts stimulated by viruses or by synthetic inducers such as polyinosinic-polycytidylic acid (Poly I : Poly C). Human IFN γ also appears to exist in a single molecular form and – as will be detailed below – is a product of a special subclass of lymphocytes – the (thymic-dependent) T lymphocytes.

The genome of human IFN γ shows no homology with the genomes of the α or β IFNs and, in addition the principal difference in the genomic structure is that the IFN γ gene possesses several introns whereas the genes of the other interferon subtypes are devoid of introns[9].

The genes of all major types of human interferon have now been cloned and expressed in bacteria. It was possible to produce large quantities and to provide purified material for clinical tests. The first clinical tests have recombinant bacterial interferon shown to be tolerated by patients[10].

Interferons, although initially discovered and studied as antiviral compounds, during the last years exhibited a plethora of additional effects on cells. These sometimes have been called "non-antiviral" or "cellular" effects. They are too numerous to be covered by this short introduction, the reader is referred to the article of Gresser[11]. One example of the "non-antiviral" effects of interferons is the antiproliferative effect on cells[12, 13]. Very interesting are the effects of interferons on the cellular components of the immune system[14] for example the enhanced expression of immunologically relevant cell membrane constituents, and the activation of macrophages and Natural Killer cells (NK cells). These effects are a central issue in the research on IFN γ and will be discussed below.

Interferons have been demonstrated to possess anti-tumor effects in experimental animals[15] and also in patients with certain forms of neoplastic diseases[16]. The reasons basis for these anti-tumor effects are at present poorly understood and perhaps a combination of different mechanisms cause the in vivo anti-tumor effect. It may be an anti-proliferative effect on tumor cells themselves in combination with the activation of defense cells (macrophages, NK cells) capable of destroying tumor cells. The anti-tumor effects of interferons certainly merit many additional investigations. Only limited data are available in regard to the anti-tumor effects of IFN γ because sufficiently purified preparations only very recently have become available.

Interferons in addition to displaying an antiviral effect in vitro, also have antiviral effects in vivo, again both in experimental models of animals and in patients with selected viral diseases. Similarly to the tumor situation, the mechanisms of the in vivo antiviral effects are quite unclear and probably represent a combination of different effects.

One has to make it clear that the in vitro antiviral effect of interferon in reality is an anti-cellular effect. Unlike neutralizing antibodies, interferons do not act on viruses themselves but render cells incapable of virus replication. Cells have to be pretreated with interferon in order to be protected and cellular RNA synthesis is a prerequisite of the antiviral effect. The basis of the in vitro antiviral effect of interferon is still incompletely understood and may differ between virus systems. Important data, however, have been collected during recent years indicating that viral protein translation may be one major target of the interferon effect. At least three enzymes are induced in interferon-treated cells, including a phosphodiesterase, a 2′,5′-oligoadenylate synthetase, and a protein kinase[2,3]. The latter two are dependent of dsRNA and constitute two distinct pathways for the inhibition of mRNA translation in cell extracts. We will further discuss these aspects in regard to the antiviral activity of IFN $\gamma$.

Interferon inducers is the term most commonly used for substances capable of inducing interferon in vitro or in vivo. Some years ago, these substances have appeared to hold great promise[17] but subsequently have faced many problems. For example, usually they are quite toxic. Since, in addition, there has been much progress in preparing interferons themselves for clinical use, there is currently not much emphasis on preclinical or clinical studies of interferon inducers. However, the situation in regard to the clinical use of interferon is quite complex: there are many different subtypes of interferon and basically nothing is known about dosages, injection, routes, pharmacokinetics etc. Thus, it may take years till these issues will be settled and the induction of the bodies own interferon system still represents a reasonable therapeutic alternative. In fact, the latter approach may be superior to exogenous interferon therapy.

The best characterized mode of interferon induction is by viruses, although viruses as interferon inducers may not be useful clinically. Besides viruses, bacteria, and more or less defined bacterial products are also capable of interferon induction. Further inducers represent quite an array of different compounds, both of high and of low molecular weight, which have been reviewed[17].

Many interferon inducers work both in vivo and in vitro, but there are unexplained situations where substances that are highly active under in vivo conditions, function poorly, if at all, in vitro – and vice-versa. There are also unexplained differences between species. The situation in regard to IFN $\gamma$ is a special one: exclusively substances that activate T cells are capable of inducing IFN $\gamma$. Most methods to induce IFN $\gamma$ have been established in vitro, and much further work will be required to develop protocols of in vivo induction of IFN $\gamma$.

## General Introduction on Lymphokines

As referred to above, IFN $\gamma$ by definition, not only is a representative of the interferons, but also a lymphokine. In fact, it was the first lymphokine the molecular structure of which has been identified. Lymphokines are defined as substances produced by lymphocytes upon activation. Usually, one also includes products of monocytes/macrophages and sometimes uses the term monokines to indicate this fact. The other part of the definition of these molecules is that they themselves act on the cellular components of the immune system for example on T lymphocytes or macrophages. Thus, lymphokines are immunoregulatory molecules. Some of these are fairly well defined such as interleukin II (IL-II). Others are defined poorly, they represent "factors" that occur in the supernatants of activated lymphocytes and have biologic activities in certain indicator systems. Undoubtedly, even these factors will be better defined in the future. At present it is not clear if there are numerous different lymphokines or if there are only a few with a multitude of immunoregulatory effects.

This reviewer certainly does not intend to cover the field of lymphokines which has been competently done[18, 19]. In the following we only wish to give a few examples:

Interleukin I (IL-I) is a product of macrophages which is defined and measured by its capacity to cause proliferation of and IL-2 production by T lymphocytes[20].

IL-II (which is also called T cell growth factor; TCGF) is a product of T lymphocytes which is essential for growing T cells in long-term culture[21]. It is an exciting concept that for continuous growth T lymphocytes are capable of producing their own growth supporting molecules. Recently, a cDNA coding for human IL-2 has been cloned. The DNA sequence codes for a polypeptide which consists of 153 amino acids including a putative signal sequence[22].

Another example for the group of lymphokines is the socalled macrophage activating factor (MAF) which also is a product of T lymphocytes and is capable of activating macrophages (for example to become cytotoxic for tumor cells)[23]. With MAF a problem becomes evident that is often encountered in lymphokine research. Besides MAF, interferons (including IFN γ) are capable of macrophage activation and often it is not clear if in a crude preparation MAF or interferon cause activation of the macrophages. In fact, at present there is some controversy if indeed a MAF molecule can be clearly distinguished from interferon or not. Lymphokines are usually collected from the supernatant fluid of activated lymphocytes and there is always concomitant production of a number of different lymphokines (and factors) and of course of IFN γ.

Yet another lymphokine is lymphotoxin[24] which also at times has been considered to be identical with IFN γ. Recently, however, Yip et al.[25] have presented evidence for the presence of a potent lymphotoxin as a contaminant in crude preparations of IFN γ. Rubin and his colleagues[26] have obtained divergent findings and they state that the anticellular activity of human IFN γ preparations is mediated by the interferon itself.

## Brief History of IFN γ

IFN γ was discovered by Wheelock in 1965 who first described an antiviral activity in the supernatant of human lymphocytes stimulated with the mitogen PHA (Phytohemagglutinin derived from Phaseolus vulgaris)[27]. The antiviral protein produced shared many of the then known properties of interferon. Wheelock already had drawn attention to one property of IFN γ which is utilized to distinguish it from other interferon, i.e. its lability at pH 2. Subsequently, there have been two additional reports of other stimuli known to activate human lymphocytes being capable of inducing the production of IFN γ[28, 29]. The old term "immune interferon" was coined by R. Falcoff et al.[30] who, in addition, demonstrated that anti-lymphocyte globulin (which is mitogenic, too) is capable of inducing the production of IFN γ in lymphocyte cultures. Many years later, a monoclonal anti-T cell antibody was described to induce interferon in human lymphocyte cultures[31]. Also quite late the possible detection of IFN γ in the supernatant of the human mixed lymphocyte culture (MLC) was documented by several groups[32–34].

In the murine system interferon production in the allogeneic MLC has been described early[35] but only years later, this situation was analyzed in detail by two groups[36, 37]. Induction of interferon by a specific antigen has been analyzed in murine systems, both in vivo and in vitro[38, 39]. Stobo et al.[40] have first shown that IFN γ production in murine spleen cell cultures induced by PHA is a T-cell dependent phenomenon. In this regard, an interesting set of data has been presented by Wietzerbin and her collaborators[41] who have found spleen cells of homozygous nude mice possessing no thymus and thus, lacking mature T lymphocytes, capable of producing IFN γ when stimulated by PHA.

Lately, new tools have become available in cellular immunology, particularly monoclonal antibodies to characterize T cell subsets, both in the human and the murine system. Equally important was the development of T cell lines with defined specificities that can be grown in tissue culture for prolonged periods of time. Following the initial reports of Marcucci et al.[42] and Nathan et al.[43] several groups showed that such lines can be stimulated by specific antigens or by mitogens to produce IFN γ[44–46]. These data will be discussed in detail below. It is now quite clear that T cells without the need for other cell types can produce high titers of IFN γ. Generally, it has been difficult to detect the production of IFN γ in vivo, except for one protocol when mice were pretreated with BCG and rechallenged with PPD[39].

Characterization of the biological effects of IFN γ originally has met difficulties since pure preparations of IFN γ have not been available. However, recently the genomic structure of human IFN γ has been identified and the gene has been cloned and expressed in bacteria[9]. In the near future, IFN γ will be available in a highly purified form for further exploration of its biological effects and its potential clinical value.

## Detection and Standardization of IFN γ

In the future conceivably radioisotope assays for interferon will be developed. At present, interferons are measured by biological assays, all of which are based on the antiviral effects of interferon. The principle of all these is similar: a cell culture is pretreated with the test fluids (or by interferon standards) and subsequently, is infected by a suitable virus, for example Vesicular stomatitis virus (VSV). After an appropriate period of time (i.e. the time which is required for virus replication in untreated controls) either the virus yield may be measured (which is laborious when many samples are to be tested) or the cytopathic effect evaluated. In any variation of the test the samples are tested in several dilutions and the endpoint (i.e. the interferon titers) can be estimated by the dilution of the test fluid that does no longer inhibit viral replication.

One *International Unit* (IU) of interferon is defined as the amount that inhibits viral replication by 50%. Since these values differ between cell lines and depend on tissue culture conditions, it is necessary to have International Standards for interferon measurements. At the time of this writing, International Standards for IFN γ (of any animal species) were not available. Thus, comparing data with respect to interferon gamma titers as reported in papers from different laboratories is of limited value, since they largely depend on the sensitivity of the cell line used for the assay. In our laboratory, internal standards are used, for example a IFN γ induced by PHA in mouse spleen cell cultures (which has a fairly low titer) when estimating the potential of new inducers or different types of cell cultures used for interferon production.

IFN γ has been known for long to differ in certain physicochemical properties from the other interferons. For example, IFN γ looses activity after treatment at pH 2. This was evident from the very first studies since it was standard method to treat interferons at pH 2 in order to destroy the virus used for induction, and not to carry the virus into the interferon assay system. Additional characteristics of human IFN γ include it's heat-lability, the sensitivity to SDS treatment and the lack of cross-protection observed when bovine cells are tested. These and additional properties will be discussed below.

Antisera are available now against IFN γ of several species, which are of great help for determination whether an interferon produced by white blood cells is indeed IFN γ or a representative of the group of the IFNs α. A monoclonal antibody against human IFN γ was reported first by Hochkeppel and de Ley[47].

Recently, an acid-labile *IFN α* subtype was described, which is produced by immune human lymphocytes upon addition of *Influenza virus*[48].

## Inducers of IFN γ

Above we have mentioned a few situations in which IFN γ is produced. In the following, we will elaborate further on the inducers of IFN γ.

## In vitro Induction of IFN γ

Almost all systems known to produce IFN γ, represent in vitro cultures of leucocytes. All stimulatory mechanisms in common represent mechanisms that are known otherwise to result in lymphocyte activation as measured by the induction of DNA synthesis and mitosis. This does, however, not necessarily indicate that DNA synthesis is a prerequisite of interferon induction. In fact, the production of IFN $\gamma$ has been found to occur in cultures in which DNA synthesis was blocked.

## Nonspecific Mitogens

These substances are known to polyclonally activate lymphocytes. In contrast to antigen stimulation which causes the proliferation of individual cell clones, mitogens activate a number of clones simultaneously. The prototypes of these mitogens are PHA (phytohemagglutinin from *Phaseolus vulgaris*) and Con A (concanavalin A from *Concanavalia ensiformis*), both established T cell mitogens. There are numerous similar lectins some of which have been analyzed comparatively in regard to the induction of IFN $\gamma$ by Yip and his coworkers[49]. Our experience was similar to these authors, i.e. in principle no mitogen is superior to PHA in its interferon inducing capacity. However, Rönneblom et al.[50] have reported *Lens culinaris* agglutinin to be a mitogen particularly well suited to induce IFN $\gamma$ in human leucocytes. In Con A-activated leucocyte cultures IFN $\gamma$ production occurs within 24 h after addition of the mitogen[51]. Con A (and PHA) were found to activate T cells (with the consequence of IFN $\gamma$ production) in various types of leucocytes of various species. We will, however, restrict our discussion to data in human and murine systems. The sequence of events that occur when resting lymphocytes (that are cells in the Go phase of the cell cycle) are treated by a mitogen such as PHA or Con A has been thouroughly studied over the past 20 years and competently reviewed[42]. It is obvious that numerous biochemical events are triggered in these cells, culminating in DNA synthesis and mitosis. RNA and protein synthesis are also induced, and IFN $\gamma$ is only one out of a number of proteins to be produced.

## Phorbolester

Phorbolester (diterpene esters), for example PMA (phorbol myristate acetate), also abbreviated TPA (12-O-tetradecanoylphorbol-13-acetate) are cocarcinogenic compounds that have met great interest because of this property and because of a wide spectrum of important effects on cells. TPA was found to be a T cell mitogen in human leucocyte cultures[53] and to induce small quantities of interferon when tested in cultures of normal human lymphocytes. It is not mitogenic for normal mouse lymphocytes[54]. However, in certain murine tumor cells high titers of IFN $\gamma$ were induced by TPA[55]. Furthermore, useful protocols of IFN $\gamma$ induction have been developed with TPA used in combination with conventional mitogens, such as Con A or PHA. TPA also has been found to increase the production of IFN $\gamma$ in the human allogeneic mixed lymphocyte culture[56].

## Mitogens of Bacterial Origin

A number of bacterial-derived mitogens were used to induce interferon gamma. Some research groups have, for example, preferred *Staphyloccus enterotoxin A* or *Staphyloccus enterotoxin B* over the conventional mitogens to induce IFN $\gamma$ in human and murine leucocyte cultures[57, 58]. A drawback of these experiments is, that this material is toxic and difficult to remove from the interferons produced. Saito et al.[59] have induced IFN $\gamma$ in mouse spleen cells by OK-432, a preparation of *Streptococcus pyogens*.

Recently, a potent mitogen was found in the supernatant of cultures of *Mycoplasma arthritidis*, which stimulates DNA synthesis in human and murine T lymphocytes. Most interestingly, in the murine system, this lymphoproliferative response is under the genetic control of the I region of the major histocompatibility locus[60]. It is not known if in the human system a similar dependency exists. Evidence has been obtained that besides stimulating DNA synthesis the mycoplasma mitogen induces IFN $\gamma$[61].

A problem common to all mitogens derived from microorganisms is that sometimes it is difficult to decide if they truly represent polyclonal activators or perhaps antigens to which the majority of the blood donors have been sensitized. Also, often one encounters the problem that crude preparations derived from microorganisms are capable of alternate induction of IFN $\alpha$ and of IFN $\gamma$ in human peripheral blood mononuclear leucocyte cultures[62].

## Induction of IFN $\gamma$ by Specific Antigens

Human leucocyte cultures of immune donors have been shown by Green et al.[28] to elaborate IFN $\gamma$ when exposed to specific antigens such as tetanus toxoid or PPD. In the mouse, it has been shown that IFN $\gamma$ is produced in vivo when animals sensitized with BCG are restimulated by PPD. This interferon was detected in the serum[39] but subsequently Sonnenfeld et al.[63] have reported that spleen cells taken out from such treated animals and cultivated in vitro also produce IFN $\gamma$. In another study, T cells were obtained from the peritoneal cavity, a cell population which is known to be highly reactive in test systems of cell-mediated immunity[38]. Peritoneal exudate T cells from mice immune to *Listeria monocytogenes* were also stimulated in vitro with *Listeria* antigen causing the production of high titers of IFN $\gamma$[64, 65].

In quite a number of studies, cellular immunity against viruses has been measured by the production of IFN $\gamma$ in human leucocyte cultures. However, these experiments are open to misinterpretations because viruses, such as *Herpes simplex virus* (HSV) also cause the production of alpha interferon in leucocyte cultures. Green et al.[66] in a careful study, have shown that both IFN $\alpha$ and IFN $\gamma$ are produced in human leucocyte cultures upon addition of HSV. Similar data have been reported by Kelsey et al.[67] who studied the production of IFN $\alpha$ and IFN $\gamma$ by spleen cells from cytomegalovirus-infected mice.

## Antisera as Inducers of IFN $\gamma$

Falcoff et al.[30] reported anti-lymphocyte serum or -globulin known to be mitogenic to induce the production of IFN $\gamma$ in human lymphocyte cultures. More recently, it was found that a monoclonal anti-T cell antibody directed against the majority of human T cells (anti OKT-3) is mitogenic and capable of induction of IFN $\gamma$[31]. Several other monoclonal anti-T cell antibodies do not share these properties including the antibodies reacting with the helper cell and suppressor cell subset. At present the mechanisms by which anti-OKT-3 activates lymphocytes are not known, but one might expect to find a clue as to nature of the receptors involved in lymphocyte activation and IFN $\gamma$ induction from further studies of the interaction between anti-OKT 3 and T cells.

## Induction of IFN $\gamma$ by Alloantigens

When lymphocytes of two unrelated donors (or of two inbred animal strains) are cocultivated in vitro, a sequence of events occurs collectively known as the mixed lymphocyte culture (MLC) – (for a review see reference 68). The study of these events has attracted immunologists for long since they are considered to represent an in vitro model of allograft rejection. Best studied among the facettes of these phenomena are lymphoproliferation and the development of specific cytotoxic T lymphocytes (CTL). Despite an early report of Gifford et al.[35] there were no detailed early studies of the production of IFN $\gamma$ in the murine MLC. Later on this issue has been taken up by two groups of investigators[36, 37] and subsequently in a collaboration between the two[69]. The following points have emerged from these studies:

1. The production of IFN $\gamma$ in the MLC is a function of T-lymphocytes
2. This function is independent of proliferation and thus, treatment of one of the reacting populations with mitomycin C, as commonly done to establish a one-way MLC, is not effective.
3. Differences both in the K- or D end of the H-2 locus as well as differences in the I region are capable of eliciting the production of IFN $\gamma$.
4. Even differences in minor histocompatibility antigens alone are causing the production of IFN $\gamma$[70].

Three groups of investigators independently studied IFN $\gamma$ production in the human MLC[32–34]. This phenomenon was considerably more difficult to study than the murine MLC since only low interferon titers were observed after several days of culture. Like in the mouse experiments, it appeared that there was no possibility to do a one-way MLC, in other words, interferon production occured even when both reacting cell populations were irradiated before culture[34]. Further experiments demonstrated that IFN $\gamma$ was produced in sibling and unrelated individual combinations that expressed HLA-A, HLA-B or HLA-DR region incompatibility.

A set of papers have been published by Ito and his coworkers[71–73]. In these experiments mice were immunized with allogeneic cells in vivo and subsequently, spleen cells were cocultured in vitro with the sensitizing antigen. In these studies, the author also succeeded in enumeration of IFN γ producing cells induced by allogeneic stimulation and proved that they were indeed T lymphocytes.

There is yet a debate, if the production of IFN γ simply represents a side effect of alloantigen recognition or if it does indeed have an immunoregulatory role in the reaction in which it is produced. The data of Farrar et al.[74] which will be discussed below have suggested that IFN γ does play a role in the activation of CTL.

## Production of IFN γ in the Syngeneic MLC

The term "syngeneic MLC" (S-MLC) describes a phenomenon which attracted the attention of immunologists, as a model of autoimmunity (for a review see reference 75). In the S-MLC, T cells are cocultured with non-T of the same donor (or inbred mouse strain) after enrichment of these subpopulations by appropriate in vitro techniques. A marked lymphoproliferative response is measured under these conditions, both in the human and the murine S-MLC. In recent studies the production of IFN γ could be detected in S-MLC of a variety of strains of mice[76,77]. In studies of the human S-MLC we have failed to detect the production of interferon[33]. Argov et al.[78] however, have cocultured human T cells in vitro with autologous B cells or monocytes and have detected the production of IFN γ in the supernatant.

## Mixed Lymphocyte Tumor Cell Interaction

Yet another experimental situation in which the production of interferon has been observed are cocultures between lymphocytes and tumor cells, the so-called "mixed lymphocyte-tumor cell interaction" (MLTI). In this situation, interferon production occured[79–82] while the subtype of the interferon produced remained uncertain. In the studies of Trinchieri et al.[79] it appeared that a mixture of different interferon subtypes was produced whereas the studies of Timonen et al.[80] and Peter et al.[81] suggested the interferon represented IFN α.

All these studies suffer from a serious drawback, since in the majority of MLTI-experiments mycoplasma contamination of the tumor cells had to be held responsible for interferon induction[83]. Subsequently, we have found mycoplasmas themselves are capable of interferon induction in leucocytes[84], an observation which was confirmative to previous work of others[85]. In our studies, we have observed the production of IFN α/β in mouse spleen cell cultures when treated by *Mycoplasma arginini*. Kumar and his colleagues[86] subsequently have reported the production of IFN γ in cocultures of mouse spleen cells and mycoplasma-infected tumor cells. Tamida et al.[87] also reported the production of IFN γ in mixed cultures of mouse spleen cells and tumor cells although the authors did not specify if mycoplasmata or perhaps endogenous retrovirus were responsible for interferon induction.

The above-mentioned findings were reminiscent of the work of Cole et al.[60] who found that *M. arthritidis* produced a T cell mitogen for mouse spleen cells. In recent studies, this mycoplasma species was found to cause the production of IFN $\gamma$ in mouse spleen cell cultures[61].

We conclude from these data that mycoplasmas are both capable of the induction of IFN $\gamma$ and of IFN $\alpha/\beta$ in murine spleen cells depending on the species of the microorganisms. This may explain why some investigators find the production of IFN $\gamma$ in the MLTI, whereas others have described the production of IFN $\alpha/\beta$.

Lately, however, we have confirmed the data of Timonen et al.[80] and of Peter et al.[81] who reported the cell line K562 induces the production of interferon in cultures of human leucocytes despite the fact that the tumor cells appear to be free of mycoplasma in repeated and extensive testing[88]. Since the line K562 is of myeloid origin, we have tested a number of similar cell lines and have obtained the same type of result for three of them. Our tentative conclusion is that differentiation antigens on myeloid tumor cell lines cause the production of IFN $\gamma$ when recognized by lymphocytes.

## Additional methods of In vitro Induction of IFN $\gamma$

Dianzani and his collaborators[89, 90] have shown treatment of human lymphocytes by galactose oxidase or by a calcium ionophore to be capable of inducing the production of IFN $\gamma$. Ennis and Meager[91] described the production of high titers of IFN $\gamma$ by antigenic stimulation of human lymphocytes with *Influenza virus*.

## Spontaneous Production of IFN $\gamma$

In classical interferon work, there has been a long-standing debate if spontaneous interferon, i.e. interferon produced without the deliberate addition of an inducer to the cell culture, may occur (of course one has to consider that the media itself contain an inducer). Now, the general conclusion is that interferons are produced only when the cells are induced. However, certain types of lymphoblastoid cell lines spontaneously produce low titers of interferon (they produce high titers when induced by a virus).

In our own experience, and as far as we have traced the literature, there is no indication that normal lymphocytes, when put in culture without a mitogen (or another type of stimuli), spontaneously produce IFN $\gamma$. This applies to the types of long-term T cell cultures, we and others have studied, and also to the type of double stimulation protocols that have been studied in our laboratory (see below). There has been one report of spontaneous IFN $\gamma$ production in a T cell tumor line[92], whereas other T cell tumors had to be induced, commonly by TPA[93].

## Role of IL-2 in the Production of IFN $\gamma$

There are two reports that IL-2 is capable of the induction of IFN $\gamma$[94, 95]. Kasahara et al.[94] described human IL-2 which was free of lecitin and IFN activity induced human peripheral T lymphocytes to produce IFN $\gamma$. Handa et al.[95] found mouse NK cell clones growing in IL-2 to elaborate IFN $\gamma$ into the culture fluid. Addition of Con A to cultures in IL-2-free media induced no IFN production.

## In Vivo Production of IFN $\gamma$

After IFN $\gamma$ had been discovered by Wheelock in vitro in the supernatant fluid of cultivated human lymphocytes[27] it was also detected in the serum of mice immunized by BCG and rechallenged by PPD[39]. This interferon shared the properties of the in vitro produced IFN $\gamma$ particularly the sensitivity to treatment at pH 2. Differences between different mouse strains in the production of this interferon have been observed[96].

There have been reports of the occurence of IFN $\gamma$ in the serum of patients with autoimmune diseases[97] but some doubt on the subtype of this interferon occured. Preble et al.[98] suggested the interferon represented an unusual subtype of IFN $\alpha$, since it was neutralized by a specific antiserum against IFN $\alpha$, but nervertheless it was acid-labile.

## Producer Cells of IFN $\gamma$

To this reviewer without doubt T cells are the producers of IFN $\gamma$ and they principally can do so without the need for other cell types. At present, there is no convincing evidence that any other lymphoid cell besides the T lymphocyte is capable of the production of IFN $\gamma$ but this opinion may have to be revised in the future.

There is no indication that macrophages by themselves are capable of IFN $\gamma$ production[99] but it was shown quite early that macrophages are providing help to normal resting T cells for optimal stimulation and interferon production[100]. There is no doubt that interferon production by normal resting T cells is enhanced by the addition of macrophages and small numbers of macrophages may even be essential. However, there is also no doubt that cloned T cell lines that are 100% pure T cells are capable of the production of high titers of IFN $\gamma$ in the absence of any additional cell type[101].

Recently, we have reviewed these latter aspects in detail[102] and we have concluded that, resting T cells probably differ from activated (continuously cycling) T cells in that the former require the help of macrophages for lymphocyte activation and IFN $\gamma$ production whereas the latter don't. We have also previously demonstrated continuously growing T cell to produce much higher titers of IFN $\gamma$ when they are stimulated by Con A, and that higher mitogen doses are required for this induction as compared to interferon induction in resting spleen cells[42].

Rapid progress in the field of cellular immunology has been made during recent years, and one of the newly established findings are subsets of T lymphocytes that can be clearly defined by cell surface markers and by their functions (e.g. helper T cells, suppressor T cells, cytotoxic T cells). There are conflicting data in the literature as to which T cell subsets are the producer of IFN $\gamma$. This issue is quite controversial[103–105] and the reader is referred to a recent review[102] for a detailed discussion of individual data from different laboratories.

In our opinion, the data from different series of experiments collectively suggest that in the human and in the murine system both the helper T cells and the cytotoxic T cells are capable of responding to mitogenic or antigenic stimulation with production of IFN $\gamma$. Marcucci et al.[106] and Conta et al.[107] in addition established several mouse T cell lines of different phenotype and clearly showed all of these were capable of the production of IFN $\gamma$. Guerne et al.[108] positively selected Lyt $2^+$ mouse T lymphocytes and described that IFN $\gamma$ is released in large amounts mainly, if not exclusively by Lyt-$2^+$ cells.

As stated above, at present, there is little evidence to support the possibility of cells other than T lymphocytes to produce IFN $\gamma$. One report[109] stated that human „large granular lymphocytes" – probably identical with NK cells, are producing IFN $\gamma$ when stimulated with *Influenza virus* but no data are available that such NK cells repond to conventional mitogens with interferon production. This issue is clouded, since recently human leucocytes when exposed to *Influenza virus* in vitro were shown to produce an unusual subtype of *IFN* $\alpha$ which is acid-labile[39]. A similar subtype of interferon has also been found in the blood of patients with certain autoimmune diseases[98]. Prior to these two reports, all IFN $\alpha$ – in contrast to IFN $\gamma$ – were generally accepted to be acid-stable.

Thymocytes (also belonging to the T cell lineage) were found to function poorly in many assays of cellular immunity and accordingly attempts to induce the production of interferon in cultures of thymocytes initially have been negative[40]. More recently, however, Marcucci et al.[110] showed that the corticosteroid-resistant subpopulation of mouse thymocytes was capable of the production of IFN $\gamma$ upon stimulation with PHA. Reem et al.[111] have reported human thymocytes, when grown in conditioned medium derived from B cell lines to produce substantial titers of IFN $\gamma$.

## In vivo and in vitro Regulation of IFN γ Production

There is little doubt that the IFN γ system such as other compartments of the body's defense system is in vivo regulated in a complex fashion. However, to date little is known about the mechanisms of the in vivo control of the production of IFN γ. In vivo experiments demonstrated pretreatments of mice with certain bacteria *(L. monocytogenes, C. parvum, B. pertussis)* to increase the production of IFN γ[112–115]. When animals are induced to produce interferon, a hyporeactivity status ensues during which they are resistant to induction[116]. A hyporeactivity factor is also produced after induction of IFN γ in mice sensitized with BCG[117].

There have been a number of reports on the control of production of IFN γ by in vitro regulation[118]. In vitro, the addition of TPA[49], or of different preparations of thymic hormones[119, 120] was described to result in increased titers of IFN γ. Otherwise, little is known about "superinduction" or "priming" for the production of IFN γ. Their relevance for the production of IFN α and IFN β has been summarized[3]. Northoff et al.[121] have shown *in vitro* preincubation of human lymphocytes leads to a greater production of IFN γ, similarly as observed for IL-2.

There is only a limited number of studies on the effect of drugs on in vitro production of IFN γ[122, 123].

## Large-Scale Production of IFN γ

For years, it has not been possible to produce significant amounts of human IFN γ purified to a degree to be used for clinical testing. The reason for these failures was that the titers, that could be induced in the starting material, were exceedingly low. Later on, the gene of human IFN γ was cloned and expressed in bacteria and there is no doubt the material produced by the bacteria will be available for clinical tests in the near future.

At the time of this writing, it is not established if the amino acid sequence of the natural gamma interferon is indeed identical with the one of the bacterial product. However, it is worth to recall that several natural IFNs α have been found to differ from the clonal material by lacking the 10 COOH-terminal amino acids suggested by the DNA sequence[124]. It is important to realize such differences may have important consequences regarding pharmacokinetics[125]. Thus, it may be still useful to improve the methods of producing „natural" IFN γ in lymphocyte cultures (such may be established from blood donations). For this approach too, significant progress has been made during recent years.

First, the data of Yip et al.[49] have shown the yield of IFN γ in human leucocyte cultures to be increased by adding certain diterpene esters, such as TPA, together with one of the conventional mitogens such as PHA or Con A. This type of protocol is now used routinely to prepare human IFN γ for purification and characterization of the molecule. There seem to be species differences because normal mouse lymphocytes do not produce increased titers of IFN γ when TPA is added together with a mitogen. Interestingly, however, murine tumor cells produced IFN γ upon addition of TPA (even without an additional mitogen)[55].

Another type of protocol for the induction of high titers of IFN γ was developed in the murine system by Marcucci et al. Actually, these authors developed two protocols[126, 127]. In one of these, mouse spleen cells are grown a few days in the presence of IL-2 and then restimulated with Con A. In the other protocol, also performed with mouse spleen cells, these are prestimulated with Con A for several days, then washed, and restimulated with Con A. Maybe, the principle of both assays is similar, since also in the latter protocol IL-2, is generated which may change the cells in a way so that they are capable of producing high titers of IFN γ. It has to be recalled that Kasahara et al.[94] and Handa et al.[95] have reported IL-2 to be capable of inducing the production of IFN γ.

With both protocols of Marcucci et al.[126, 127] certainly much higher titers of IFN γ can be induced as compared to freshly explanted leucocytes.

Finally, following the report of Marcucci et al.[42] it was repeatedly shown, that "continuous T cell lines" – CTL – are capable of producing IFN γ when stimulated appropriately. CTL are now used widely as an experimental tool for studies of cellular immunology. They are grown either in the presence of IL-2 or they are repeatedly restimulated by the specific antigen. They grow as long as the appropriate signal is provided, but otherwise they will die in culture. Such CTL are capable of interferon production upon addition of both specific antigens or nonspecific mitogens.

More recently, also certain T cell tumors were demonstrated to be capable of interferon production[128, 129]. In addition, hybridomas were established by hybridization of human lymphocytes with 6-thio-guanine-resistant mutant cells. The lymphocytes came from peripheral blood and were previously stimulated with Con A. The mutant cell line was derived by irradiation of a cloned human cutaneous T lymphoma line[130]. Up to 1330 units of interferon were produced spontaneously by the hybrids and, on induction with TPA, IFN γ synthesis was enhanced 8–16 fold.

## Properties of IFN γ

Initially, IFN γ was found to be distinct from the classical interferons (then called type 1 interferon, now called IFN α and IFN β) by the fact that IFN γ was labile at pH 2. Besides that it was heat-labile at 56 °C (1 h) and instable after treatment with SDS[49]. Lateron, IFN γ was found to be antigenically distinct from the other interferons and specific antisera were raised[131]. Another interesting feature of human IFN γ is, its strict species-specificity. There was no activity detectable on any of the available cell species sensitive to IFN α and IFN β[132].

Besides these findings, some years ago there were no major differences apparent between the biological effects of IFN γ and other interferons, but it has to be recalled that a that time it represented great troubles to produce significant amounts of IFN γ. Thus, IFN γ was an antiviral inhibitor broadly active against a multitude of viruses. The antiviral effect required preincubation of the target cells and depended on cellular RNA synthesis. Furthermore, IFN γ appeared – like other interferons – to be a glycoprotein with a molecular weight of about 40–70 000[133], that was de novo synthesized upon addition of the inducer. Thus, the elaboration of IFN γ could be inhibited by the addition of inhibitors of RNA or protein synthesis.

Earlier in this paper, we have inferred interferon gamma is synthesized whenever T cells are activated in a polyclonal or a clonally restricted manner, and that all these situations are known to result also in DNA synthesis of the lymphocytes. A number of studies have, however, shown lymphocyte DNA synthesis is *not* a requirement for the production of IFN γ.

Lately, important progress was made in the research on IFN γ. The DNA sequence of human IFN γ was reported, and the amino acid sequence has been deduced from the latter[9]. This sequence consists of 146 amino acids with a total molecular weight of approximately 17 000. The IFN γ gene contains three introns, a repetitive DNA element and is not highly polymorphic. The evidence of Gray and Goeddel[9] suggests that this was the only gene for IFN γ and the resolution of IFN γ into two components, as reported by Yip et al.[134], was probably the result of post-translational processing of the protein.

The recombinant IFN γ is an unglycosylated polypeptide carrying two potential N-glycosylation sites. Obviously, the molecular weight of the native glycosylated IFN γ is higher.

Trent et al.[135] localized the IFN γ gene to the long arm of chromosome 12 (q 24.1), whereas the human IFN α and IFN β genes were localized to the short arm of chromosome 9 (p 21 → pter).

## Purification of IFN γ

Several purification procedures for human IFN γ have been described. Langford et al.[136] reported a combination of chromatographic steps on Amicon Matrex Blue, controlled pore glass (CPG) and Ultragel; the final specific activity of the purified material was between $10^{5.3}$ and $10^{6.3}$ units/mg of protein. Wiranowska-Stewart et al.[137] reported the use of CPG and poly (U) Sepharose for the partial purification of IFN γ to specific activities of approximately $10^6$ units per mg of protein. Compared to these methods, the purification sequence of Yip et al.[138] achieved a similar degree of purity by a combination of CPG and Con A-Sepharose. Their third step was gel filtration in Bio-Gel P-200 resulting in about 10 fold purification. Thus, the final specific activity of Yip's preparation was about $10^7$ units/mg of protein. From these data the authors conclude the specific activity of pure IFN γ will likely be similar to that of other interferons, i.e. in the range of $10^8$–$10^9$ units/mg of protein.

De Ley et al.[132] have similar to the work of Langford et al.[136], used absorption to CPG and elution by ethylene glycol. Upon gel filtration the biologic activity of their Con A-induced interferon was eluted together with a protein moiety of an apparent molecular weight of 45 000.

## Receptors for IFN γ

Interferons bind to specific receptors on the cell surface and elicit a variety of cellular responses. Direct evidence for specific interferon receptors was presented by Aguet[139)] who showed biologically active $^{125}$I-labelled mouse interferon binds to sensitive L 1210-S cells but not to interferon-resistant L 1210 R cells. Branca and Baglioni[140)] labelled cloned human IFN α with $^{125}$I for binding assays with human cells. They determined the specificity of different human IFNs α for the cellular sites which bind cloned IFN α and found that IFN γ does not compete for binding whereas all IFNs α and β do. Thus, their report shows IFN γ possessing a receptor, distinct from the common receptor for IFN α and β.

## Biologic Properties of IFN γ

Above, we have inferred that all major biologic effects attributed to IFN α and β also have been observed with preparations of IFN γ. Thorough investigations using pure preparations, however, were started only recently. Since IFN γ differs considerably from other interferons in its genomic structure and its amino acid sequence, and since it uses a special receptor, one may expect that future research will unravel differences between the biologic effects of IFN γ and other interferons.

## Antiviral Effects of Interferon

Interferons exhibit antiviral effects, both in vitro and in vivo. This is part of their definition. The assay system and the calculation of International Units is based on their in vitro antiviral effect. The basis of the in vivo antiviral effect at present is incompletely understood because, besides the protective effects of interferon on the targets of viral replication, effects involving the cells of the antiviral defense system appear to play a role.

The antiviral effects of the IFN α and β have been extensively studied over the past years. The antiviral effects of interferons are in reality cellular effects. Unlike neutralizing antibodies acting on virions themselves, interferons act on the cells and change their biochemistry in a way that virus replication is no longer possible. There are quite a few indications of enzymes that are induced in interferon-treated cells critically involved in the development of the antiviral state. But also different cellular mechanisms appear to be responsible for the protection observed when the replication of different viruses is studied. At the time of this writing, there is no indication of a principal difference between the antiviral effect of IFN γ and that of other interferons. However, it needs to be stressed again that the IFN γ used in previous studies was quite impure.

Tan et al.[141] have mouse-human cell hybrids shown to respond to the antiviral effect of exogenous human interferon only if human chromosome 21 was present, and they would lose this response if chromosome 21 was lost from the hybrid. Several genes are known to be carried by chromosome 21, including the antiviral gene AVC. Epstein and Epstein[142] demonstrated the antiviral expression of IFN γ was also mediated by the same locus.

## Induction of Polypeptides

Hovanessian et al.[143] used two-dimensional polyacrylamide gel electrophoresis to show that mouse IFN β and IFN γ both induce the synthesis of polypeptides having molecular weights of 60 K, 67 K, 88 K and 120 K. By using one-dimensional polyacrylamid gels, Rubin and Gupta[144] demonstrated polypeptides of 56 K, 67 K, 88 K and 120 K are induced in human fibroblasts by human IFN α and IFN γ. An 80 K polypeptide, induced by IFN α, could not be detected after induction by IFN γ. Weil et al.[145] have found that IFN γ induces significant synthesis of at least 12 polypeptides, six of which were not induced by IFN α or IFN β.

## Induction of Enzymes in Interferon Treated Cells

Most of the alterations in cells caused by interferon treatment depend on RNA and protein synthesis and interferons were shown to enhance the level of various messenger RNAs and proteins. Among the enzymes induced by interferon is $2'5'(A)_n$-synthetase. Upon its activation by dsRNA, this enzyme generates 2′5′ linked oligoadenylates from ATP. These in turn activate RNase L, a latent endoribonuclease (that cleaves single-stranded RNAs e.g. mRNA and ribosomal RNA at preferred sites). Another enzyme induced by interferon is a protein kinase that, inhibits protein synthesis if activated by dsRNA. This kinase phosphorylates and thereby impairs the activity of a peptide chain initiation factor.

These enzymes were discovered in experiments with extracts from IFN-treated cells. It is hypothesized that the role, the dsRNA plays in virus infected cells, is exerted by partially ds intermediated or side products of viral RNA synthesis. It is in line with the assumed function of the above-mentioned enzymes in IFN action that viral RNA and protein accumulation are among the processes of virus replication being impaired in cells treated with interferons.

Several laboratories have studied the induction of enzymes in cells after treatment with IFN γ[146–148].

## Antiproliferative Effects of Interferon

Interferons have an antiproliferative effect on cells in vitro and presumably this effect is at least partially responsible for the in vivo antitumor effects of the interferons (see below). Whereas tumor cells or transformed cells appear to be particularly sensitive to the action of interferon, there is no doubt that the growth of tissue cultures of "normal" cells may also be inhibited by interferon. As viruses vary in their sensitivity to interferon, so do cells. Several studies have shown tumor cells to vary in sensitivity to the antiproliferative effects of interferon[149, 150]. The basis of the antiproliferative effect of the interferons is incompletely understood. The reader is referred to the paper by Brouty-Boyé[13].

There have been several recent reports that (relatively impure) preparations of IFN $\gamma$ share this antiproliferative effect. Rubin and Gupta[144] demonstrated that preparations of IFN $\gamma$ had a reversible cytostatic effect on normal human fibroblasts at 10 units/ml. However, transformed cells such as HeLa exhibited extensive cell death, indicating IFN $\gamma$ may have a cytocidal effect on certain tumor cells. These data are noteworthy since when working with the other interferons, it was the general experience that interferons were antiproliferative (cytostatic) but not cytotoxic. However, the presence of a lymphotoxin in the IFN $\gamma$ preparation might have contributed to the effect. This has been suggested by Yip et al.[25] but was refuted by Rubin et al.[26]. The studies of Blalock et al.[151] demonstrated that on the basis of a unit of antiviral activity purified human immmune interferon had about 20 and 100 time more anticellular activity than purified fibroblast or leucocyte interferon, respectively.

## Immunoregulatory Effects of Interferon

The effects of interferons on the cellular components of the immune system have been noted quite a few years ago. They include a significant number of divergent effects and both stimulatory and inhibitory effects were observed. These were found to be shared by a variety of interferon subtypes. Before dwelling on these divergent effects, it is perhaps worthy to reiterate that all situations in which IFN $\gamma$ is produced involve specific or nonspecific activation of the T lymphocytes. As lined out above, these situations include antigenic restimulation, stimulation by alloantigens and polyclonal activation by mitogens. Based upon the well-known effects of interferon on immune reactivities a major role was postulated for IFN $\gamma$ as an immunoregulatory molecule in the reactions in which it is produced, for example in cytotoxicity against allogeneic cells (see below). However, such a role is still far from being proven. Possibly, some of the effects caused by exogenous addition of interferon, such as the activation of NK cells, are also activated by interferon in the reactions in which it is produced. Thus, IFN $\gamma$ may have the function of being a nonspecific amplifier molecule of the immunologically specific defense.

Furthermore, this function may not be restricted to specific antigenically driven reactions since certain mitogens occur in pathogens such as the mycoplasma mitogen described by Cole et al.[60], or staphylococcal toxins, and thus, IFN $\gamma$ may also be produced during primary defense mechanisms.

In the following, we wish to briefly review the effects of interferon on the immune system.

## Enhancement of the Expression of Cell Surface Components

One of the most interesting effects of interferon is represented by its effect on cell membranes. This includes the enhanced expression of a number of immunologically relevant membrane components such as the Fc receptors[152] and most notably, of the histocompatibility antigens[153, 154]. It was demonstrated by molecular hybridization that the latter effect was preceeded by an increase in the amount of HLA mRNA in the cell. The results of Revel's laboratory[155, 156] have shown IFN $\gamma$ to share the ability to induce (2′–5′) oligo (A) synthetase and HLA proteins but there is a dissociation between the two effects. HLA synthesis is stimulated at concentrations much lower than those needed for the synthetase induction. As referred to above, there are several cellular proteins that are induced differently by IFN $\alpha/\beta$ and IFN $\gamma$.

## Effects of Interferon on in Vivo Delayed Type Hypersensitivity

In extensive and careful studies, De Maeyer and associates[157] documented the effects of pure IFN $\alpha/\beta$ on in vivo delayed type hypersensitivity (DTH) reactions in mice. They presented evidence for exogenous IFN $\alpha/\beta$ to influence the afferent and efferent pathways of DTH in the mouse. The effect of interferon on the afferent pathways (sensitization) is dependent on the timing of administration; when given 24 h before sensitization, interferon acts inhibitory, whereas when given a few hours after the antigen, interferon actually enhances sensitization. We are not aware of similar studies testing the effects of pure IFN $\gamma$ on in vivo delayed type hypersensitivity.

## In Vitro Immunosuppression by Interferon

One of the first "non-antiviral" effects of interferon reported was the antiproliferative effect on PHA-induced lymphocyte activation[158]. Thorley-Lawson[159] has reported that lymphocytes of newborns were relatively resistant to this effect of interferon. We are not aware of a study in which such an effect has been tested with IFN γ. However, it has to be realized that mitogen-stimulated lymphocytes themselves produce IFN γ. Perhaps this IFN γ plays a regulatory role by inhibiting excessive proliferation. If this were true, one would expect an increase in lymphoproliferation after the addition of anti-IFN γ.

It has been observed that the dose response curves for mitogen-induced lymphoproliferation show an inhibition at higher doses. Furthermore, the optimal doses for induction of IFN γ usually are higher than those required for inducing lymphoproliferation. Thus, the inhibition of lymphoproliferation at high mitogen doses may be caused by the endogenous production of IFN γ.

Immunosuppression by interferon has also been observed in the in vitro antibody response to various antigens e.g. SRBC. Several authors reported that preparations of IFN γ were potent inhibitors of these responses[160–162]. However, one has to realize that the IFN γ used at this time contained other lymphokines besides interferon.

Another effect of interferon was observed in vitro that, is the reversion of antigen- and mitogen induced leucocyte migration inhibition. Szigeti et al.[163] suggested that interferon acted in this system both by affecting directly the migration of leucocytes and that it also blocks the elaboration of leucocyte migration inhibition factor by lymphocytes. We are not aware of a study in which IFN γ was investigated in this test system.

## Stimulatory Effects of Interferon on the Cellular Components of the Immune System

One of the first examples of a stimulatory effect of interferon on immunocompetent cells was the enhancing effect on specific T cell cytotoxicity[164]. However, at that time the NK cells were not yet known. It is also worthy to recall that augmentation of nonspecific cytotoxicity by interferon had been shown even before[165]. The latter report was indeed pioneering the field well ahead of the important later discovery of activation of NK cells by interferon[166]. At the time of this writing, it remains unclear to the reviewer whether the stimulatory effect on the cytotoxic reaction observed by Lindahl et al.[164], was indeed caused by augmentation of T cell killing or perhaps by recruiting of additional non-specific activity of NK cells.

More recently, Farrar et al.[74] proposed that IFN γ may be involved in the sequence of events in the MLC leading to the activation of specific T killer cells. They postulated the induction of CTL involves a cell-factor interaction in which IL-1 (macrophage-derived) stimulates T cells to produce IL-2, which in run stimulates other T cells to produce IFN γ and become cytotoxic.

Klein et al.[167] reported spleen cells from nude mice less than 2 mo of age required an exogenous source of IL-2 for the activation of cytotoxic effector cells and for the production of IFN γ. The response of spleen cells from nude mice older than 9 mo did not require the addition of exogenous IL-2. Thus, it appeared that cytotoxic T cell functions and the release of IFN $\gamma$ were closely associated immunologic events.

## Activation of Macrophages by Interferon

One of the first non-antiviral effects of interferon reported was the stimulation of the phagocytic capacity of macrophages[168]. Later on, interferon treatment was found to also cause a general activation of macrophages[169], similar to another lymphokine termed "macrophage activating factor" (MAF).

The term macrophage activation is used a bit loosely for a number of phenomena including an increase in the phagocytic and in the secretory functions of macrophages. Macrophage activation most notably includes the induction of a cytotostatic and/or cytotoxic function of macrophages for tumor cells.

Meltzer et al.[170] reported a T cell line producing two factors, activating macrophages for cytotoxicity. One was indistinguishable from IFN $\gamma$, the other was clearly distinct (no antiviral activity, pH 2 stable and unaffected by anti-IFN $\gamma$). However, Kelso et al.[171] had reported that out of 72 T cell clones 68 produced detectable quantities of MAF. Production of IFN $\gamma$ and MAF was not dissociated for any of the clones tested.

The studies of Russell et al.[172] demonstrated the activation of macrophages to be a two-step event. The first of these ("priming") does not cause the expression of cytolytic activity, instead it primes macrophages to respond to a signal that then triggers the onset of killing. The studies of Pace et al.[173] have shown recombinant mouse IFN $\gamma$ to induce the priming step in macrophage activation for tumor cell killing.

## Activation of NK Cells by Interferon

The enhancement of immunologically non-specific cellular cytotoxicity by interferon has been demonstrated before the discovery of the NK cells[165]. NK cell activity could be increased by a number of compounds including immunomodulators and viruses[174]. The effect common to these compounds was the induction of interferon and it is now very well documented that interferon itself causes the activation of NK cells. The studies of Senik et al.[175] have shown that IFN $\gamma$ stimulates mouse NK cells in a manner comparable to IFN $\alpha/\beta$.

These are a number of additional interesting aspects of interferon-mediated activation of NK cells.

**1.** Obviously, interferon activates both macrophages and NK cells (probably also K cells). Thus, care has to be taken that one uses properly defined cell populations in these experiments.

**2.** Although in most experimental instances the induction of interferon and the activation of NK cells are correlated, there are situations in which NK cell activation is observed *in the absence* of measurable titers of interferon[176]. It may be, that in these experiments interferon is produced locally but in concentrations too low to be detected in the fluid recovered for testing. There are, however, indications that certain viruses (or viral structures) are capable of NK cell activation without induction of interferon. Thus, perhaps there exists an additional pathway of NK cell activation which is independent of interferon.

**3.** Conceivably, there are other lymphokines capable of activating NK cells besides interferon. There have been several reports about IL-2 being capable of sustaining the growth of NK cells[177]. No indication is found, however, that IL-2 *activates* NK cells. Interestingly, Handa et al.[95] could show that IL-2 is capable of inducing the production of interferon gamma in cultured NK cells (without addition of a conventional mitogen such as PHA or Con A).

**4.** It has to be recalled that besides activating NK cells interferon has yet another important effect in the cytotoxic reaction. Treatment of the target cells by interferon *protects* these against the lytic activity of NK cells[178, 179]. This effect was documented with different types of interferon, including IFN γ.

## Production of IFN γ as a Measure of Cell-Mediated Immunity

During recent years, methods have been developed to measure cell-mediated immunity by in vitro tests. These methods are yet too complicated for the clinical laboratory but it is obvious that measuring the cell-mediated immune status may turn out to be at least as important as the determination of antibody titers. There are three reaction principles commonly used to measure cell-mediated immunity including lymphoproliferation, lymphocytotoxicity, and lymphokine production. We have pointed out above that IFN γ is produced whenever T lymphocytes are activated by specific antigens including viral antigens. Thus, determination of IFN γ was suggested to be of use for measuring specific antiviral cell-mediated immunity. Examples are determinations during recurrence and interval in patients suffering from recurrent infection with *Herpes simplex virus.*

However, there are serious experimental problems because viruses, such as HSV, also induce the prodution of IFN α in leucocyte cultures. A careful analysis of this situation was presented by Green et al.[66] who found IFN α is produced early in leucocytes of both seropositive *and* seronegative donors upon addition of HSV. IFN γ is produced later, particularly in cultures of T lymphocytes (plus added macrophages). The production of IFN γ is restricted to cultures of seroimmune donors. When using *Listeria* organisms as antigens in cultures of human leucocytes, both IFN α and IFN γ were found to be produced similarly[62].

Therefore it appears that there is a delicate balance in leucocyte cultures in regard to the production of interferon subtypes depending on the cellular composition and the origin of the leucocytes. The type of antigenic preparation used for induction will probably also turn out to be important.

It requires purified T cells for the production of IFN $\gamma$. However, these need to be reconstituted by small numbers of macrophages[180]. At present, it is not quite clear what the contribution of T cells and macrophages are in the production of IFN $\alpha$. Finally, it still has to be considered that there are acid-labile subtypes of IFN $\alpha$[48] that previously might have been misinterpreted using the sensitivity at pH 2 as the main criterion for an interferon to be classified as IFN $\gamma$. Since only IFN $\gamma$ production reflects specific immunity, whereas the production of IFN $\alpha$ is immunologically non-specific, we do not advocate the assay of IFN $\gamma$ production as measure of cell-mediated immunity.

## Anti-Tumor Effects of Interferon

Interferon have been shown to exhibit antiproliferative effects on tumor cells in vitro (see above). In addition, early animal experiments by Gresser et al.[15] documented the in vivo antitumor effects of interferons. The mechanisms of this in vivo effect are incompletely understood but at least four mechanisms may contribute:

1. The antiproliferative effect on tumor cells themselves[13] (see above).
2. The enhancement of the expression of "tumor antigens" which makes the tumor cells more vulnerable to the attack by immunocytes. However, this has been investigated only for melanoma cells and controverial results have been reported[181, 182].
3. Interferons were demonstrated to activate at least two components of the cellular defense system, i.e. the macrophages and the NK cells, both of which are capable of destroying tumor cells (see above).
4. Finally, data from a number of laboratories have suggested that prolonged treatment of tumor cells by interferon in vitro causes a reversion of the malignant phenotype as for example reflected by decreased growth in soft agar[183]. Similar changes conceivably occur in vivo, leading to a decreased capacity of tumor cells for invasion and metastasis. However, this issue may be more complex since in vitro experiments of a different type have shown that IFN treatment caused *an enhancement* of the invasive capacity of Ewing sarcoma cells[184].

## Anti-Tumor Effects of IFN $\gamma$

Data on antitumor effects of IFN $\gamma$ are scant and this work has been performed before pure IFN $\gamma$ was available. However, Crane et al.[185] have shown that in mice inhibition of tumor development was achied with an amount of IFN $\gamma$ a hundred times less than was required with a preparation of IFN $\alpha/\beta$. Fleischmann et al.[186] have described preparations of mouse IFN $\gamma$ significantly enhancing or potentiating the anti-tumor effect of mouse virus-induced interferon, when the interferons were used in combined therapy. Data from the same laboratory[187] have similarly shown that mixed preparations of IFN α/β and IFN γ interacted with cells synergistically to cause the developement of much higher level of antiviral activity than was expected on the basis of their activities.

## Interferons in Clinical Cancer Research

Highly purified interferons from genetic engineering are now available and are being tried in clinical tests, but information on these is still limited. Earlier clinical tests were performed with conventional preparations of IFN α usually prepared by Dr. Kari Cantell from Helsinki (Finland). More recently, IFNs α prepared from lymphoblastoid cell lines have also become available for clinical tests. The most recent update of clinical data was reported by Quesada and the reader is referred to this competent overview[16)]. Without doubt interferon treatment may cause objective responses of tumor growth in vivo in patients with different forms of malignancies albeit the responses are irregular and unpredictable, and limited to a small percentage of patients in any given treatment series. Furthermore, interferon treatment at this time is in no instance superior to established treatment schedules. However, there is hope that by careful analysis of the in vivo effects of interferons (dosage problem, subtype problem, pharmacokinetics) in the not too distant future some interferons may be useful therapeutic agents for some tumors.

## Side Effects of Interferon Therapy

This side effects of interferon therapy have also been carefully reviewed by Quesada[16)]. Administration of interferon is generally well tolerated by the majority of the patients. Acute side effects consist mainly of fever, chills, malaise and headaches, generally described by the patients as influenza-like symptoms. Asthenia is the most frequent subacute symptom in patients receiving interferon for longer than seven to ten days. Cantell et al.[188)] and Rohatiner et al.[189)] independently reported electroencephalographic abnormalities in patients receiving doses above 100 MU of interferon. All symptoms are completely reversible within 7 to 15 days of discontinuation of interferon. The most consistent laboratory finding is depression of the peripheral blood counts. Interferon-induced leukopenia occurs within 12 to 24 h following administration and a proportional decrease in neutrophils and lymphocytes is observed. No direct toxicity to the bone marrow has been observed clinically.

## Clinical Use of IFN γ

Only recently, IFN γ was made available for clinical studies because of the difficulty in purifying a stable compound. Clinical studies were performed at the MD Anderson Hospital at Houston[190]. In this study, intramuscular or intravenous bolus injection of up to 54 MU of conventionally prepared IFN γ produced little clinical toxicity. Antitumor activity was demonstrated in one patient with metastatic renal carcinoma out of a group of 18 patients. Most probably, in the near future the first clinical studies of recombinant human IFN γ will be performed.

## Role of Interferon in the Pathogenesis of Disease

Since the interferon system appears to be a carefully controlled system, it is not unreasonable to expect that perturbations of this system may also be involved in the pathogenesis of diseases. The studies of Gresser et al. in animal models[191] have documented progressive glomerulonephritis in mice treated at birth with interferon. At the time of this writing, there is little evidence to indicate a dysregulation of the interferon system may be involved in the pathogenesis of human diseases.

Most investigators agree that in the serum of healthy persons interferon cannot be detected. However, Hooks et al.[97] have detected IFN γ in the serum of patients with autoimmune disease, including systemic lupus erythematosus and Sjögren syndrome. Similarly, Fujii et al.[192] have detected IFN γ in the circulation of patients with Behcets disease. Cultures of peripheral lymphocytes of these patiens produced IFN γ in vitro. Preble et al.[98] have reported the presence of an unusual acid-labile IFN α in systemic lupus erythematodes. This interferon was shown to be IFN α by neutralization with a specific antiserum, affinity column chromatography and antiviral activity on bovine cells. A similar type of human IFN α was observed in the serum of homosexual men with Kaposi's sarcoma and lymphadenopathy[193].

The latter studies suggest that hyperproduction of interferon may be involved in the development of autoimmune disease. There are only few reports of defects in the IFN γ system. Bryson et al.[194] have reported a deficiency of IFN γ production in leucocytes of normal newborns.

## Conclusions

1. IFN γ is produced whenever T lymphocytes are activated, including immunologically specific activation by antigens or alloantigens and nonspecific activation by polyclonal activators such as the plant mitogens PHA and Con A. Mitogens do not only occur in plants but also in pathogenetic microorganisms

such as *mycoplasmata*. There is yet little evidence for cells other than T cells to produce IFN γ albeit new data in the future may prove the contrary.

2. IFN γ is a representative of two groups of biologically important molecules, the lymphokines and the interferons. So far IFN γ is the only lymphokine having an antiviral effect. However, the antiviral effect may have been a coincidental discovery and the more important role of IFN γ may be in immunoregulation.
3. The same may be true for other interferons as well, since besides their antiviral effect they have a multitude of ("non-antiviral", "cellular") effects.
4. The classical interferons are the IFNs α and IFN β. The former are the products of a multigene family and there are up to 20 subtypes that are distinct in their genomic structure. Only one gene is known for IFN β, but based on studies of the mRNA there may be also different subtypes of IFN β. IFNs α and IFN β show homology and it has been suggested that the genes are derived from a common ancestor.
5. Human IFN γ has little or no homology with the above-mentioned interferons, and the gene has a different structure, i.e. it contains introns (which are lacking in the other interferon genes).
6. Also the cellular receptors for human and murine IFN γ are different from the common receptors for IFN α and β.
7. Based on items 4 and 5 one expects to find considerable differences between the biological effects of IFN γ and those of the classical interferons, but such information ist still scant.
8. Thus, basically all the effects that have been reported for the classical interferons have also been observed with preparations of IFN γ but it needs to be stressed that relatively impure preparations of IFN γ have been used in previous work and pure preparations have become available only recently.
9. It was shown in several laboratories that besides those polypeptides observed in cells treated by classical interferons, certain proteins are induced uniquely by IFN γ.
10. IFN γ clearly has an antiviral effect in vitro, appearing to be quite similar in its mechanisms to the antiviral effects of other interferons. Thus, IFN γ was demonstrated to induce the same enzyme systems as previously described for IFN α and IFN β. There are practically no data on in vivo antiviral effects of IFN γ.
11. IFN γ, like other interferons enhances the expression of histocompatibility antigens on the cell surface which again points to the important role of IFN γ in immunoregulation.
12. IFN γ activates NK cells and macrophages for cytotoxicity. In fact, IFN γ may be identical with MAF, although the latter point is still controversial.
13. Earlier studies with relatively impure preparations suggested that IFN γ has a greater immunosuppresive potential than other interferon subtypes.
14. The same has been claimed for the in vitro antiproliferative and the in vivo antitumor effect. It was also reported that combinations of IFN α/β and IFN γ have an anti-tumor effect greater than additive. However, the impurities in these early preparations of IFN γ have to be stressed again, and there is work from at least one group that the preparations of IFN γ contained potent lymphotoxins as well.

15. Recently, the gene for human IFN γ has been cloned and expressed in bacteria, and sufficiently pure material is now available. Besides this, better methods for conventional production of IFN γ in lymphocyte cultures have been developed so that one might also be able to define the structure of natural IFN γ.
16. It can be foreseen that with pure IFN γ exciting data in biologic test-systems will be obtained and that also the first clinical tests will be performed. The side effects of IFN γ will have to be determined and whether they are perhaps less severe than those of the other interferons.
17. Despite the great interest in these future clinical tests, it may turn out to be even more exciting to further unravel the biologic role of IFN γ, both in the body's defense systems and in the pathogenesis of disease. Manipulation of the interferon (and/or lymphokine) systems may also lead to therapeutic implications.

## Acknowledgement

The dedicated editorial assistance of M. Kulka is gratefully acknowledged.

## References

1. Stewart, W. E.: The interferon system, Wien – New York, Springer 1979
2. Lengyel, P.: Ann. Rev. Biochem. *51,* 251 (1982)
3. Seghal, P., Pfeffer, L. M., Tamm, I.: Interferon and Its Inducers, in: Chemotherapy of Viral Infections (eds.) Came, P. E., Caliguiri, L. A., p. 205 Springer 1982
4. Epstein, L. B.: Fed. Proc. *40,* 56 (1981)
5. Epstein, L. B.: Interferon-Gamma: Is it really different from the other interferons? Interferon 3 (ed.) Gresser, I., p. 13 1981
6. Epstein, L. B., Weil, J., Lucas, D. O., Cox, D. R., Epstein, C. J.: The biology and properties of interferon-gamma: An overview, studies of the production by T lymphocyte subsets, and analysis of peptide synthesis and antiviral effect in trisomy 21 and diploid human fibroblasts, in: The Biology of the Interferon System (eds.) De Maeyer, E., Galasso, G., Schellekens, H., p. 247, Elsevier/North Holland Biomedical Press 1981
7. Isaacs, A., Lindenmann, J.: Proc. Royal Soc. London *147,* 258 (1957)
8. Sehgal, P. B., Sagar, A. D., Braude, I. A., Smith, D.: Heterogeneity of Human α and β Interferon mRNA Species, in: The Biology of the Interferon System (eds.) De Maeyer, E., Galasso, G., and Schellekens, H., p. 48, Elsevier/North Holland, Biomedical Press 1981
9. Gray, P. W., Goeddel, D. V.: Nature *298,* 859 (1982)
10. Quesada, J. R., Gutterman, J. U.: J. Nat. Cancer Inst. *70,* 1041 (1983)
11. Gresser, I.: Cell Immunol. *34,* 406 (1977)
12. Paucker, K., Cantell, K., Henle, W.: Virology *21,* 22 (1962)
13. Brouty-Boyé, D.: Lymphokine Reports (Pick, E., ed.) *1,* 99 (1980).
14. Sonnenfeld, G.: Lymphokine Reports (Pick, E., ed.) *1,* 113 (1980)
15. Gresser, U., Bouraldi, C., Levy, J. P., Fontaine-Brouty-Boyé, D., Thomas, M. T.: Proc. Natl. Acad. Sci. USA *63,* 51 (1969)
16. Queseda, J. R.: The Cancer Bull. *35,* 30 (1983)

17. Stringfellow, D. A.: Interferon and Interferon Inducers. Series Modern. Pharmacology vo. 17, New York and Basel, Marcel Dekker 1980
18. Waksman, B. H.: Overview: Biology of the Lymphokines, in: The Biology of the Lymphokines (eds.) Cohen, S., Pick, E., Oppenheim, J. J., p. 585, New York, Academic Press 1979
19. De Weck, A., Kristensen, F., Landy, M. (eds.): Biochemical Chracterization of Lymphokines. London, New York, Academic Press 1980
20. Lachman, L. B.: Fed. Proc. *42,* 2639 (1983)
21. Gillis, S.: J. Clin. Immunol. *3,* 1 (1983)
22. Taniguchi, T., Matsui, H., Fujita, T., Takaoka, C., Kashima, N., Yoshimoto R., Hamuro, J.: Nature *302,* 305 (1983)
23. Lohman-Matthes, M.-L., Ziegler, F. G., Fischer, H.: Eur. J. Immunol. *3,* 56 (1973)
24. Ruddle, N. H., Waksman, B. H.: J. Exp. Med. *128,* 1267 (1968)
25. Yip, Y. K., Anderson, P., Stone-Wolff, D. S., Barrowclough, B. S., Urban, C., Vilcek, J.: Structure-function studies with human interferon-gamma, in: Interferons, UCLA Symposia on Molecular and Cellular Biology (eds.) Merigan, T. C., Friedman, R. M., New York, Academic Press 1982
26. Rubin, B. Y., Bartal, A. H., Anderson, S. L., Millet, S. K., Hirshaut, Y., Feit, C.: J. Immunol. *130,* 1019 (1983)
27. Wheelock, E. F.: Science *149,* 310 (1965)
28. Green, J. A., Cooperband, S. R., Kibrick, S.: Science *164,* 1415 (1969)
29. Friedman, R. M., Cooper, L. H.: Proc. Soc. Exp. Biol. Med. *125,* 901 (1967)
30. Falcoff, R.: J. Gen. Virol. *16,* 251 (1972)
31. Wussow v., P., Platsoucas, C. D., Wiranowska-Stewart M., Stewart, W. E.: J. Immunol. *127,* 1197 (1981)
32. Perussia, B., Mangoni, L., Engers, H., Trinchieri, G.: J. Immunol. *125,* 1589 (1980)
33. Manger, B., Zawatzky, R., Kirchner, H., Kalden, J. R.: Transplant. *32,* 149 (1981)
34. Andreotti, P. E., Cresswell, P.: Human Immunol. *3,* 109 (1981)
35. Gifford, G. E., Tibor, A., Peavy, D. L.: Infect. and Immun. *3,* 164 (1971)
36. Landolfo, S., Marcucci, F., Giovarelli, M., Viano, I., Forni, G.: Immunogenetics *9,* 245 (1979)
37. Kirchner, H., Zawatzky, R., Schirrmacher, V.: Eur. J. Immunol. *9,* 97 (1979)
38. Milstone, L. M., Waksman, B. H.: J. Immunol. *111,* 1914 (1973)
39. Youngner, J. S., Salvin, S. B.: J. Immunol. *111,* 1914 (1973)
40. Stobo, J., Green, I., Jackson, L., Baron, S.: J. Immunol. *112,* 1589 (1974)
41. Wietzerbin, J., Stefanos, R., Falcoff, M., Lucero, L., Catinot, L., Falcoff, E.: Infect. and Immun. *21,* 966 (1978)
42. Marcucci, F., Waller, M., Kirchner, H., Krammer, P.: Nature *291,* 79 (1981)
43. Nathan, I., Groopman, J. E., Quan, S. G., Bersch, N., Goide, D. W.: Nature *292,* 842 (1981)
44. Morris, A. G., Lin, Y.-L., Askonas, B. A.: Nature *295,* 150 (1982)
45. McKimm-Breschkin, J. L., Mottram, P. L., Thomas, W. R., Miller, J. F. A. P.: J. Exp. Med. *155,* 1204 (1982)
46. Klein, J. R., Raulet, D. H., Pasternack, M. S., Bevan, M. J.: J. Exp. Med. *155,* 1198 (1982)
47. Hochkeppel, H. K., de Ley, M.: Nature *296,* 258 (1982)
48. Balkwill, F. R., Griff, D. B., Band, H. A., Beverley, P. C. L.: J. Exp. Med. *157,* 1059 (1983)
49. Yip, Y. K., Pang, R. H. L., Oppenheim, J. D., Nachbar, M. S., Henrikson, D., Zerebeckyz-Eckhard, I., Vilcek, J.: Infect. Immun. *34,* 131 (1981)
50. Rönnblom, L., Funa, K., Ersson, B., Alm, G. V.: Scand, J. Immunol. *16,* 327 (1982)
51. Ng., W. S., Ng., M. H., Inoue, M., Tan, Y. H.: Clin. Exp. Immunol. *44,* 594 (1981)
52. Wedner, H. J., Parker, C. W.: Prog. Allergy *20,* 195 (1976)
53. Touraine, J.-L., Hadden, J. W., Touraine, F., Hadden, E. M., Estensen, R., Good, R. A.: J. Exp. Med. *145,* 460 (1977)
54. Abb, J., Bayliss, G. J., Deinhardt, F.: J. Immunol. *122,* 1639 (1979)
55. Landolfo, S., Arnold, B., Suzan, M.: J. Immunol. *128,* 2807 (1982)
56. Andreotti, P. E.: J. Immunol. *129,* 91 (1982)
57. Osborne, L. C., Georgiades, J. A., Johnson, H. M.: Infect. Immun. *23,* 80 (1979)

58. Von Wussow, P., Chen, Y.-S., Wiranowska-Stewart, M., Stewart, W. E. II.: J. Interferon Res. *2* (1982)
59. Saito, M., Ebina, T., Koi, M., Yamaguchi, T., Kwade, Y., Ishida, N.: Cell. Immunol. *68,* 187 (1982)
60. Cole, B. C., Sullivan, G. J., Daynes, R. A., Sayed, I. A., Ward, J. R.: J. Immunol. *128,* 2013 (1982)
61. Cole, B. C.: Yale J. Biol. Med. (in press)
62. Nakane, A., Minagawa, T.: J. Immunol. *126,* 2139 (1981)
63. Sonnenfeld, G., Mandel, A. D. Merigan, T. C.: Immunology *36,* 883 (1979)
64. Scott, J. W., Finke, J. H., Hsu, L., Proffit, M. R.: J. Interferon Res. *2,* 111 (1982)
65. Kaufmann, S. H. E., Hahn, H., Berger, R., Kirchner, H.: Eur. J. Immunol. *13,* 265 (1983)
66. Green, J. A., Tze-Jou, Y., Overall, J. C., Jr.: J. Immunol. *127,* 1192 (1981)
67. Kelsey, D. K., Overall, J. C., Glasgow, L. A.: Infect. Immun. *36,* 651 (1982)
68. Bach, F. H., Bach, M. L., Sondel, P. M. Nature *259,* 273 (1976)
69. Landolfo, S., Marcucci, F., Schirrmacher, V., Kirchner, H.: J. Interferon Res. *1,* 339 (1981)
70. Schirrmacher, V., Landolfo, S., Zawatzky, R., Kirchner, H.: Invasion and Metastasis *1,* 175 (1981)
71. Ito, Y., Aoki, H., Kimura, Y., Takano, M., Maeno, K., Shimokata, K.: Infect. Immun. *28,* 542 (1980)
72. Ito, Y., Aoki, H., Kimura, Y., Shimokata, K., Maeno, K.: Infect. Immun. *31,* 879 (1981)
73. Ito, Y., Shimokata, K., Maeno, K.: Infect. Immun. *37,* 427 (1982)
74. Farrar, W. L., Johnson, H. M., Farrar, J. J.: J. Immunol. *126,* 1120 (1981)
75. Battisto, J. R., Ponzio, N. M.: Prog. Allergy *28,* 160 (1981)
76. Lattime, E. C., Macphail, S., von Wussow, P., Stutman, O.: Behring Inst. Mitt. *72,* 47 (1983)
77. Berger, R., Landolfo, S., Balzi, E., Kirchner, H.: Immunobiology *165,* 225 (1983)
78. Argov, S., Cantell, K., Klein, E., Klein, G.: Cell. Immunol. *76,* 196 (1983)
79. Trinchieri, G., Santoli, D., Dee, R. R., Knowles, B. B.: J. Exp. Med. 1299 (1977)
80. Timonen, T., Saksela, E., Virtanen, I., Cantell, K.: Eur. J. Immunol. *10,* 422 (1980)
81. Peter, H. H., Dalügge, H., Zwatzky, R., Euler, S., Leibold, W., Kirchner, H.: Eur. J. Immunol. *10,* 547 (1980)
82. Olstad, R., Degré, M., Seljelid, R.: Scand. J. Immunol. *13,* 605 (1981)
83. Beck, J., Engler, H., Kirchner, H.: J. Immunol. Methods *38,* 63 (1980)
84. Beck, J., Brunner, H., Marcucci, F., Kirchner, H.: J. Interferon Res. *2,* 31 (1982)
85. Cole, B. C., Overall, J. C., Lombardi, P. S., Glasgow, L. A.: Infect. Immun. *14,* 88 (1976)
86. Kumar, V., Lust, J., Gifaldi, A., Bennett, M., Sonnenfeld, G.: Immunobiology *165,* 445 (1983)
87. Tomida, M., Yamamoto, Y., Hozumi, M., Kawade, Y.: J. of Interferon Res. *2,* 271 (1982)
88. Berger, R., Knapp, W., Kirchner, H.: Int. Journal Cancer, in press
89. Dianzani, F., Monahan, T. M., Georgiades, J., Alperin, J. B.: Infect Immun. *29,* 561 (1980)
90. Dianzani, F., Monahan, T. M., Santiani, M.: Infect. Immun. *36,* 915 (1982)
91. Ennis, F. A., Meager, A.: J. Exp. Med. *154,* 1279 (1981)
92. Morris, A. G., Morser, J., Meager, A.: Infect. Immun. *35,* 533 (1982)
93. Benjamin, W. R., Steeg, P. S., Farrar, J. J.: Proc. Natl. Acad. Sci. USA *79,* 5379 (1982)
94. Kasahara, T., Hooks, J. J., Dougherty, S. F., Oppenheim J. J.: *130,* 1784 (1983)
95. Handa, K., Suzuki, R., Matsui, H., Shimizu, Y., Kumagai, K.: J. Immunol. *130,* 988 (1983)
96. Tyring, S. K., Lefkowitz, S. S.: Proc. Soc. Exp. Biol. Med. *164,* 519 (1980)
97. Hooks, J. J., Moutsopoulos, J. M., Geis, S. A., Stahl, M. S., Decker, J. L., Notkins, A. L.: New Engl. J. Med. *301,* 5 (1981)
98. Preble, O. T., Black, R. J., Friedman, R. M.: Science *216,* 429 (1982)
99. Epstein, L. B., Gupta, S.: J. Clin. Immunol. *1,* 186 (1981)
100. Epstein, L. B., Cline, M. J., Merigan, T. C.: J. Clin. Invest. *50,* 744 (1971)
101. Matsuyama, M., Sugamura, K., Kawade, Y., Hinuma, Y.: J. Immunol. *129,* 450 (1982)
102. Kirchner, H., Marcucci, F.: The producer cells of interferon in leucocyte cultures, in press
103. Chang, T.-W., Testa, D., Kung, P. C., Perry, L., Dreskin, H. J., Goldstein, G.: J. Immunol. *128,* 585 (1982)
104. O'Malley, J. A., Nussbaum-Blumenson, A., Sheedy, D., Grossmayer, B. J., Ozer, H.: J. Immunol. *128,* 2522 (1982)

105. Abb, J., Abb, H., Deinhardt, F.: Med. Microbiol. Immunol. *3* (1983)
106. Marcucci, F., Klein, B., Altevogt, P., Landolfo, S., Kirchner, H., J. Immunol., in press
107. Conta, B. S., Powell, M. B., Ruddle, N. H.: J. Immunol. *130,* 2231 (1983)
108. Guerne, P.-A., Piguet, P.-F., Vassalli, P.: J. Immunol. *130,* 2225 (1983)
109. Djeu, J. Y., Stocks, N., Zoon, K., Stanton, G. J., Timonen, T., Herberman, R. B.: J. Exp. Med. *156,* 1222 (1982)
110. Marcucci, F., Kirchner, H., Resch, K.: J. Interferon Res. *1,* 87 (1980)
111. Reem, G. H., Cook, L. A., Henriksen, D. M., Vilcek, J.: Infect. Immun. *37,* 216 (1982)
112. Havell, E. A., Spitalny, G. L., Patel, P. J.: J. Exp. Med. *156,* 112 (1982)
113. Okamura, H., Kawaguchi, K., Shoji, K., Kawade, Y.: Infect. Immun. *38,* 440 (1982)
114. Digel, W., Marcucci, F., Kirchner, H.: J. Interferon Res. *3,* 65 (1983)
115. Kirchner, H., Weyland, A., Storch, E.: J. Interferon Res. *3,* 351 (1983)
116. Youngner, S. J., Stinebring, W. R.: Nature *208,* 456 (1965)
117. Minagawa, T., Ho, M.: Infect. Immun. *22,* 371 (1978)
118. Johnson, H. M.: Antiviral Res. *1,* 36 (1981)
119. Shoham, J., Eshel, I., Aboud, M., Salzberg, S.: J. Immunol. *125,* 54 (1980)
120. Svedersky, L. P., Hui, A., May, L., McKay, P., Stebbing, N.: Eur. J. Immunol. *12,* 244 (1982)
121. Northoff, H., Hooks, J. J., Jordan, G., Oppenheim, J. J.: Behring Inst. Mitt. *67,* 90 (1980)
122. Abb, J., Abb, H., Deinhardt, F.: Immunopharmacol. *4,* 303 (1982)
123. Kalman, V. K., Klimpel, G. R.: Cell. Immunol. *78,* 122 (1983)
124. Levy, W. P., Rubinstein, M., Shively, J., Del Valle, U., Lai, C.-Y., Moschera, J., Brink, L., Gerber, L., Stein, S., Pestka, S.: Proc. Natl. Acad. Sci. USA *78,* 6186 (1981)
125. Sarkar, F. H.: Antiviral Res. *2,* 103 (1982)
126. Marcucci, F., Nowak, M., Krammer, P., Kirchner, H.: J. Gen. Virol. *60,* 195 (1982)
127. Marcucci, F., Klein, B., Kirchner, H., Zawatzky, R.: Eur. J. Immunol. *12,* 787 (1982)
128. Le, J., Prensky, W., Henriksen, D., Vilcek, J.: Cell. Immunol. *72,* 157 (1982)
129. Hooks, J. J., Haynes, B. F., Detrick-Hooks, B., Diehl, L. F., Gerrard, T. L., Fauci, A. S.: Blood *59,* 198 (1982)
130. Le, J., Vilcek, J., Saxinger, C., Prensky, W.: Proc. Natl. Acad. Sci. USA *79,* 7857 (1982)
131. Stefanos, S., Catinot, L., Wietzerbin, J., Falcoff, E.: J. gen. Virol. *50,* 225 (1980)
132. De Ley, M., van Damme, J., Claeys, H., Weening, H., Heine, J. W., Billiau, A., Vermylen, C., de Somer, P.: Eur. J. Immunol. *10,* 877 (1980)
133. Yip, Y. K., Barrowclough, B. S., Urban, C., Vilcek, J.: Science *215,* 411 (1982)
134. Yip, Y. K., Barrowclough, B. S., Urban, C., Vilcek, J.: Proc. Natl. Acad. Sci. USA *79,* 1820 (1982)
135. Trent, J. M., Olson, S., Lawn, R. M.: Proc. Natl. Acad. Sci. USA *79,* 7809 (1982)
136. Langford, M. P., Georgiades, J. A., Stanton, G. J., Dianzani, F., Johnson, H. M.: Infect. Immun. *26,* 36 (1979)
137. Wiranowska-Stewart, M., Lin, L. S., Braude, I. A., Stewart, W. E. II: Mol. Immunol. *17,* 625 (1980)
138. Yip, Y. K., Pang, H. L., Urban, C., Vilcek, J.: Proc. Natl. Acad. Sci. USA *78,* 1601 (1981)
139. Aguet, M., Belardelli, F., Blanchard, B., Marcucci, F., Gresser, I.: Virology *117,* 541 (1982)
140. Branca, A. A., Baglioni, C.: Nature *294,* ,768 (1981)
141. Tan, Y., Schneider, H., Tischfield, J., Epstein, C. J., Ruddle, F. H.: Science *185,* 132 (1974)
142. Epstein, L. B., Epstein, C. J.: J. Infect. Dis. *133,* 56 (1976)
143. Hovanessian, A. G., Meurs, E., Aujean, O., Vaquero, C., Stefanos, S., Falcoff, F.: Virology *104,* 195 (1980)
144. Rubin, B. Y., Gupta, S. L.: Proc. Natl. Acad. Sci. USA *77,* 5928 (1980)
145. Weil, J., Epstein, J., Epstein, L. B.: Nature *301,* 437 (1983)
146. Hovanessian, A. G., La Bonnardiere, C., Falcoff, E.: J. Interferon Res. *1,* 125 (1980)
147. Ohtsuki, D., Torres, B. A., Johnson, H. M.: Biochem. Biophys. Res. Commun. *104,* 422 (1982)
148. Tomita, Y., Cantell, K., Kuwata, T.: Int. J. Cancer *30,* 161 (1982)
149. Adams, A., Strander, H., Cantell, K., J. Gen. Virol. *28,* 207 (1975)
150. Hilfenhaus, J., Damm, H. Karges, H. E., Manthey, K. F.: Arch. Virol. *51,* 87 (1976)
151. Blalock, J. E., Georgiades, J. A., Langford, M. P., Johnson, H. M.: Cell. Immunol. *49,* 390 (1980)

152. Fridman, W. H., Gresser, I., Bandu, M. T., Aguet, M., Néauport-Sautes, C.: J. Immunol. *124,* 2436 (1980)
153. Lindahl, P., Leary, P., Gresser, I.: Eur. J. Immunol. *4,* 779 (1974)
154. Heron, O., Hokland, M., Berg, K.: Proc. Natl. Acad. Sci. USA *75,* 6215 (1978)
155. Fellous, M., Nir, U., Wallach, D., Merlin, G., Rubinstein, M., Revel, M.: Proc. Natl. Acad. Sci. USA *79,* 3082 (1982)
156. Wallach, D., Fellous, M., Revel, M.: Nature *299,* 833 (1982)
157. De Maeyer, E., De Maeyer-Guignard, J.: J. Immunol. *130,* 2392 (1983)
158. Lindahl-Magnusson, P., Leary, P., Gresser, I.: Nature *237,* 121 (1972)
159. Thorley-Lawson, D. A.: J. Immunol. *126,* 829 (1981)
160. Sonnenfeld, G., Mandel, A. D., Merigan, T. C.: Cell. Immunol. *34,* 193 (1977)
161. Virelizier, J. L., Chan, E. L., Allison, A. C.: Clin. exp. Immunol. *30,* 299 (1977)
162. Lucero, M. A., Wietzerbin, J., Stefanos, S., Billardon, C., Falcoff, E., Fridman, W. H.: Cell. Immunol. *54,* 58 (1980)
163. Szigeti, R., Masucci, M. G., Masucci, G., Klein, E., Klein, G.: Nature *288,* 594 (1980)
164. Lindahl, P., Leary, P., Gresser, I.: Proc. Natl. Acad. Sci. *69,* 721 (1972)
165. Svet-Moldavsky, G. J., Chernyakhovskaya, I. Y.: Nature *215*, 1299 (1967)
166. Gidlund, M. A., Örn, H., Wigzell, A., Senik, A., Gresser, I.: Nature *273,* 759 (1978)
167. Klein, J. R., Bevan, M. J.: J. Immunol. *130,* 1780 (1983)
168. Huang, K. Y., Donahoe, R. M., Gordon, F. B., Dressler,H. R.: Infect. Immun. *4,* 481 (1971)
169. Schultz, R. M., Papamatheakis, J. D., Chirigos, M. A.: Science *197,* 674 (1977)
170. Meltzer, M. S., Benjamin, W. R., Farrar, J. J.: J. Immunol. *129,* 2802 (1982)
171. Kelso, A., Glaserbrook, A. L., Kanagawa, O., Brunner, K. T.: J. Immunol. *129,* 550 (1982)
172. Russell, S. W., Doe, W. F., McIntosh, A. T.: J. Exp. Med. *146,* 1511 (1977)
173. Pace, J. L., Russell, S. W., Torres, B. A., Johnson, H. M., Gray, P. W.: J. Immunol. *130,* 2011 (1983)
174. Oehler, J. R., Lindsay, L. R., Nunn, M. E., Holden, H. T., Herberman, R. T.: Int. J. Cancer *21,* 210 (1978)
175. Senik, A., Stefanos, S., Kolb, J. P., Lucero, M., Falcoff, E.: Ann. Immunol. (Inst. Pasteur) *131 c,* 349 (1980)
176. Kirchner, H., Engler, H., Schröder, C. H., Zawatzky, R., Storch, E.: J. Gen. Virol. *64,* 437 (1983)
177. Dennert, G.: Nature *287,* 47 (1980)
178. Welsh, R. M., Kärre, K., Hansson, M., Kunkel, L. A., Kiessling, R. W.: J. Immunol. *126,* 219 (1981)
179. Trinchieri, G., Granato, D., Perussia, B.: J. Immunol. *126,* 335 (1981)
180. Arbeit, R. D., Leary, P. L., Levin, M. J.: Infect. Immunol. *35,* 383 (1982)
181. Imai, K., Ng, A.-K., Glassy, M. C., Ferrone, S.: J. Immunol. *127,* 505 (1981)
182. Liao, S.-K., Kwong, P. C., Khosravi, M., Dent, P. B.: J. Natl. Cancer Inst. *68,* 19 (1982)
183. Hicks, J., Morris, A. G., Burke, D. C.: J. Cell. Sci. *49,* 225 (1981)
184. Siegal, G. P., Thorgeirsson, U. P., Russi, R. G., Wallace, D. M., Liotta, L. A., Berger, S. L.: Proc. Natl. Acad. Sci. USA *79,* 4064 (1982)
185. Crane, Jr. J. L., Glasgow, L. A., Kern, E. R., Youngner, J. S.: J. Natl. Cancer Inst. *61,* 871 (1978)
186. Fleischmann, Jr. W. R., Kleyn, K. M., Baron, S.: J. Natl. Cancer Inst. *65,* 936 (1980)
187. Fleischmann, W. R., Georgiades, J. A., Osborne, L. C., Johnson, H. M.: Infect. Immun. *26,* 248 (1979)
188. Cantell, K., Mattson, K., Miiranen, A., Kauppinne, H.-L., Livanainen, M., Bergstrom, L., Farkkila, M., Holsti, L. R.: presented at the Third Annual International Congress for Interferon Research, Miami, 1–3 Nov. 1982
189. Rohatiner, A. Z., Prior, P. F., Burtin, A. C., Smith, A. T., Balkwill, F. R., Lister, T. A.: Br. J. Cancer *47,* 419 (1983)
190. Gutterman, J. U., Rios, A., Quesada, J. R., Rosenblum, M.: presented at The Third Annual International Congress for Interferon Research, Miami, 1–3 Nov. 1982
191. Gresser, I., Maury, C., Tovey, M., Morel-Maroger, L., Pontillon, F.: Nature *263,* 420 (1976)
192. Fujii, N., Minagawa, T., Nakane, A., Kato, F., Ohno, S.: J. Immunol. *130,* 1683 (1983)

193. DeStefano, E., Friedman, R. M., Friedman-Kien, A. E., Goedert, J. J., Henriksen, D., Preble, O. T., Sonnabend, J. A., Vilcek, J.: J. Infect. Dis. *146,* 451 (1982)
194. Bryson, Y. J., Winter, H. S., Gard, S. E., Fischer, T. J., Stiehm, R.: Cell. Immunol. *55,* 191 (1980)

11976606R00124

Made in the USA
San Bernardino, CA
08 December 2018